Variation and Evolution in Plants and Microorganisms

TOWARD A NEW SYNTHESIS 50 YEARS AFTER STEBBINS

Francisco J. Ayala, Walter M. Fitch, and Michael T. Clegg,
Editors

NATIONAL ACADEMY PRESS
Washington, D.C.

NATIONAL ACADEMY PRESS • 2101 Constitution Avenue, N.W. • Washington, D.C. 20418

This volume is based on the National Academy of Sciences' Colloquium on the Variation and Evolution in Plants and Microorganisms: Toward a New Synthesis 50 Years after Stebbins. The articles appearing in these pages were contributed by speakers at the colloquium and have not been independently reviewed. Any opinions, findings, conclusions, or recommendations expressed in this volume are those of the authors and do not necessarily reflect the views of the National Academy of Sciences.

Library of Congress Cataloging-in-Publication Data

Variation and evolution in plants and microorganisms : toward a new synthesis 50 years after Stebbins / Francisco J. Ayala, Walter M. Fitch, and Michael T. Clegg, editors.
p. cm.
Includes bibliographical references and index.
ISBN 0-309-07075-9 (hardcover) — ISBN 0-309-07099-6 (pbk.)
1. Plants—Evolution—Congresses. 2. Plants—Variation—Congresses. I. Ayala, Francisco José, 1934- II. Fitch, Walter M., 1929- III. Clegg, Michael T., 1941- IV. Colloquium on the Variation and Evolution in Plants and Microorganisms: Toward a New Synthesis 50 Years After Stebbins (2000 : Beckman Center of the National Academies)
QK980 .V37 2000
581.3'8--dc21

00-010861

Additional copies of this report are available from National Academy Press, 2101 Constitution Avenue, N.W., Lockbox 285, Washington, D.C. 20055; (800) 624-6242 or (202) 334-3313 (in the Washington metropolitan area); Internet, http://www.nap.edu

Printed in the United States of America

THE NATIONAL ACADEMIES

National Academy of Sciences
National Academy of Engineering
Institute of Medicine
National Research Council

The **National Academy of Sciences** is a private, nonprofit, self-perpetuating society of distinguished scholars engaged in scientific and engineering research, dedicated to the furtherance of science and technology and to their use for the general welfare. Upon the authority of the charter granted to it by the Congress in 1863, the Academy has a mandate that requires it to advise the federal government on scientific and technical matters. Dr. Bruce M. Alberts is president of the National Academy of Sciences.

The **National Academy of Engineering** was established in 1964, under the charter of the National Academy of Sciences, as a parallel organization of outstanding engineers. It is autonomous in its administration and in the selection of its members, sharing with the National Academy of Sciences the responsibility for advising the federal government. The National Academy of Engineering also sponsors engineering programs aimed at meeting national needs, encourages education and research, and recognizes the superior achievements of engineers. Dr. William A. Wulf is president of the National Academy of Engineering.

The **Institute of Medicine** was established in 1970 by the National Academy of Sciences to secure the services of eminent members of appropriate professions in the examination of policy matters pertaining to the health of the public. The Institute acts under the responsibility given to the National Academy of Sciences by its congressional charter to be an adviser to the federal government and, upon its own initiative, to identify issues of medical care, research, and education. Dr. Kenneth I. Shine is president of the Institute of Medicine.

The **National Research Council** was organized by the National Academy of Sciences in 1916 to associate the broad community of science and technology with the Academy's purposes of furthering knowledge and advising the federal government. Functioning in accordance with general policies determined by the Academy, the Council has become the principal operating agency of both the National Academy of Sciences and the National Academy of Engineering in providing services to the government, the public, and the scientific and engineering communities. The Council is administered jointly by both Academies and the Institute of Medicine. Dr. Bruce M. Alberts and Dr. William A. Wulf are chairman and vice chairman, respectively, of the National Research Council.

Preface

"The present book is intended as a progress report on [the] synthetic approach to evolution as it applies to the plant kingdom" (Stebbins, 1950, p. *ix*). With this simple statement, G. Ledyard Stebbins formulated the objectives of *Variation and Evolution in Plants* (Stebbins, 1950), published in 1950, the last of a quartet of classics that, in the second quarter of the twentieth century, set forth what became known as the "synthetic theory of evolution" or "the modern synthesis." The other books are Theodosius Dobzhansky's *Genetics and the Origin of Species* (1937), Ernst Mayr's *Systematics and the Origin of Species* (1942) and George Gaylord Simpson's *Tempo and Mode in Evolution* (1944). The pervading conceit of these books is the molding of Darwin's evolution by natural selection within the framework of rapidly advancing genetic knowledge. Stebbins said it simply: "In brief, evolution is here visualized as primarily the resultant of the interaction of environmental variation and the genetic variability recurring in the evolving population" (Stebbins, 1950, p. *xi*).

Variation and Evolution in Plants distinctively extends the scope of the other books to the world of plants, as explicitly set in the book's title. Dobzhansky's perspective had been that of the geneticist and he set the tone for the others, Mayr's that of the zoologist and systematist, and Simpson's that of the paleobiologist. All four books were outcomes of the famed Jesup Lectures at Columbia University. Plants, with their unique genetic, physiological and evolutionary features, had all but been left completely out of the synthesis until that point. In 1941, the eminent botanist Edgar Anderson, had been invited to write botany's analogue to Mayr's *Systematics and the Origin of the Species* and to publish it jointly with Mayr's book. Anderson did not fulfill the task, and Stebbins was thereafter invited to deliver the Jesup Lectures in 1947. *Variation and Evolution in Plants* is the outgrowth of the Lectures.

The mathematical underpinnings of the modern synthesis were set between 1918 and 1931 by R. A. Fisher (1930) and J.B.S. Haldane (1932) in Britain, and Sewall Wright (1931) in the United States. According to Darwin, evolutionary change occurs by natural selection of small individual differences appearing every generation within any species. Any change effected by selection is typically small but they amount to major change over time. Thomas Huxley and Francis Galton, among Darwin's most dedicated supporters, argued instead that evolution occurs by selection of discontinuous variations, or sports; evolution proceeds rapidly by discrete leaps. Natural selection operating only upon gradual differences among individuals, could hardly account, in Huxley's view, for the gaps between existing species evident in the paleontological record. According to Galton, evolution proceeds by "jerks," some of which imply considerable organic change, rather than as a smooth and uniform process.

In the latter part of the nineteenth century, the biometricians Karl Pearson and W.F.R. Weldon believed, like Darwin, in the primary importance of common individual differences. Like other geneticists, William Bateson argued, rather, for the primary importance of discontinuous variations. The controversy was acrimonious. The rediscovery of Mendelian inheritance in 1900 might have served as the common grounds to resolve the conflict. Instead, the dispute between biometricians and geneticists extended to continental Europe and to the United States. Bateson was the champion of the Mendelians, many of whom accepted the mutation theory proposed by De Vries (1900), and denied that natural selection played a major role in evolution. The biometricians for their part argued that Mendelian characters were sports of little significance for the evolutionary process. Fisher, Haldane, and Wright advanced theoretical models of evolutionary processes based on the natural selection of genetic changes (mutations) that are individually small, but are cumulatively of great consequence.

Theodosius Dobzhansky first and, then, Mayr, Simpson, and Stebbins (and, less notably, many others) completed the mathematicians' theoretical propositions with a wealth of biological knowledge and empirical support. Stebbins was particularly suited to bring in the evidence from plants. He was born in 1906 and had become interested in natural history from childhood. He started botanizing in his early teens while a student at Cate School in Santa Barbara (California). As an undergraduate at Harvard (1924–1928) he came under the influence of Merritt Lyndon Fernald (1873–1950), a charismatic teacher and distinguished botanist, whom Stebbins accompanied on field trips to study the New England flora. In 1928, Stebbins became a graduate student at Harvard and worked on the cytology, geographic variation, and seed development of *Antennaria*, a genus that bore several apomictic species that could be collected in nearby localities. The distinctive evolutionary role of vegetative reproduction in plants would remain a focus of interest to the end of his life.

The 17 papers that follow were presented at a colloquium sponsored by the National Academy of Sciences, "Variation and Evolution in Plants and Microorganisms. Towards a New Synthesis 50 Years after Stebbins." The colloquium, held at the Beckman Center of the National Academies, in Irvine, California, January 27–29, 2000, sought to celebrate the 50th anniversary of the publication of Stebbins' classic. Professor Stebbins, although frail for the last few years, intended to attend the colloquium. Alas, he became ill around Christmas time and died on January 19, 2000, a few days before the colloquium was held, just about two weeks after his 94th birthday, on January 6. The "Appreciation" (Chapter 1) was read by Peter Raven, after dinner on January 28th, at the time and place that had been reserved for Stebbins' own words. The 16 papers following the "Appreciation" are organized into five successive sections: Early Evolution and the Origin of Cells, Virus and Bacterial Models, Protoctist Models, Population Variation, Trends and Patterns in Plant Evolution.

We are grateful to the National Academy of Sciences for the generous grant that financed the colloquium, to the staff of the Arnold and Mabel Beckman Center for their skill and generous assistance, and to Mrs. Denise Chilcote, who performed the administrative functions of the colloquium with skill and dedication.

REFERENCES

De Vries, H. (1900) Sur les unités des caractères spécifiques et leur application á l'étude des hybrides. *Rev. Gen. Bot.* **12,** 257–271.

Dobzhansky, Th. (1937) *Genetics and the Origin of Species* (Columbia University Press, New York); 2nd Ed., 1941; 3rd Ed., 1951.

Fisher, R. A. (1930) *The Genetical Theory of Natural Selection* (Clarendon, Oxford).

Haldane, J. B. S. (1932) *The Causes of Evolution* (Harper, New York).

Mayr, E. (1942) *Systematics and the Origin of Species* (Columbia University Press, New York).

Simpson, G. G. (1944) *Tempo and Mode in Evolution* (Columbia University Press, New York).

Stebbins, G. L. (1950) *Variation and Evolution in Plants* (Columbia University Press, New York).

Wright, S. (1931) Evolution in mendelian populations. *Genetics* **16,** 97–159.

Contents

Part I

EARLY EVOLUTION AND THE ORIGIN OF CELLS

Darwin noticed the sudden appearance of several major animal groups in the oldest known fossiliferous rocks. "If [my] theory be true, it is indisputable that before the lowest Cambrian stratum was deposited . . . the world swarmed with living creatures," he wrote, noting that he has "no satisfactory answer" to the "question why we do not find rich fossiliferous deposits belonging to these assumed earliest periods" (Darwin, 1859, ch. 10). In "Solution to Darwin's Dilemma: Discovery of the Missing Precambrian Record of Life" (Chapter 2), J. William Schopf points out that, one century later, one decade after the publication of Stebbins' *Variation and Evolution in Plants* (Stebbins, 1950), the situation had not changed. The known history of life extended only to the beginning of the Cambrian Period, some 550 million years ago. This state of affairs would soon change, notably due to three papers published in *Science* in 1965 by E.S. Barghoorn and S.A. Tyler (1965), Preston Cloud (1965), and E.S. Barghoorn and J.W. Schopf (1965). Schopf tells of the predecessors who anticipated or made possible the work reported in the three papers, and of his own and others' contributions to current knowledge, which places the oldest fossils known, in the form of petrified cellular microbes, nearly 3,500 million years ago, seven times older than the Cambrian and reaching into the first quarter of the age of the Earth.

Lynn Margulis, M.F. Dolan, and R. Guerrero set their thesis right in the title of their contribution: "The Chimeric Eukaryote: Origin of the Nucleus from the Karyomastigont in Amitochondriate Protists" (Chapter 3). The karyomastigont is an organellar system composed at least of a nucleus with protein connectors to one (or more) kinetosome. The ancestral eukaryote cell was a chimera between a thermoacidophilic archaebacterium and a heterotrophic eubacterium, a "bacterial consortium" that evolved into a heterotrophic cell, lacking mitochondria at first. Cells with free nuclei evolved from karyomastigont ancestors at least five times, one of them becoming the mitochondriate aerobic ancestor of most eukaryotes. These authors aver that only two major categories of organisms exist: prokaryotes and eukaryotes. The Archaea, making a third category according to Carl Woese and others (Woese *et al.*, 1990), should be considered bacteria and classified with them.

The issue of shared genetic organelle origins is also a subject, if indirect, of the paper by Jeffrey D. Palmer and colleagues ("Dynamic Evolution of Plant Mitochondrial Genomes: Mobile Genes and Introns, and Highly Variable Mutation Rates," Chapter 4). The mitochondrial DNA (mtDNA) of flowering plants (angiosperms) can be more than 100 times larger than is typical of animals. Plant mitochondrial genomes evolve rapidly in size, both by growing and shrinking; within the cucumber family, for example, mtDNA varies more than six fold. Palmer and collaborators have investigated more than 200 angiosperm species and uncovered enormous pattern heterogeneities, some lineage specific. The authors reveal numerous losses of mt ribosomal protein genes (but only rarely of respiratory genes), virtually all in some lineages; yet, most ribosomal protein genes have been retained in other lineages. High rates of functional transfer of mt ribosomal protein genes to the nucleus account for many of the loses. The authors show that plant mt genomes can increase in size, acquiring DNA sequences by horizontal transfer. Their striking example is a group I intron in the mt *cox1* gene, an invasive mobile element that may have transferred between species more than 1,000 independent times during angiosperm evolution. It has been known for more than a decade that the rate of nucleotide substitution in angiosperm mtDNA is very low, 50–100 times lower than in vertebrate mtDNA. Palmer *et al.* have now discovered fast substitution rates in *Pelargonium* and *Plantago*, two distantly related angiosperms.

REFERENCES

Barghoorn, E. S. & Schopf, J. W. (1965) Microorganisms from the Late Precambrian of central Australia. *Science* 150, 337–339.

Barghoorn, E. S. & Tyler, S. A. (1965) Microorganisms from the Gunflint chert. *Science* 147, 563–577.

Cloud, P. (1965) Significance of the Gunflint (Precambrian) microflora. *Science* 148, 27–45.

Darwin, C. (1859) *On the Origin of Species by Means of Natural Selection* (Murray, London).

Stebbins, G. L. (1950) *Variation and Evolution in Plants* (Columbia University Press, New York).

Woese, C. R., Kandler, O., & Wheelis, M. L. (1990) Towards a natural system of organisms: Proposal for the domains Archaea, Bacteria, and Eucarya. *Proc. Natl. Acad. Sci.* USA 87, 4576–4579.

1

G. Ledyard Stebbins (1906–2000) — An Appreciation

PETER H. RAVEN

Ledyard Stebbins, born January 6, 1906, was deeply fond of nature all his life, starting with his experiences around Seal Harbor, Maine, at about four years old. When he was still young, his mother contracted tuberculosis, and the family moved west to try to find a more healthy climate for her—first to Pasadena, then to Colorado Springs. An important formative period of Ledyard's life was spent at Cate School, in Santa Barbara, where he studied for four years. During those years, he was, by his own account, shy and relatively unpopular; but he learned to ride horseback, explored the Santa Ynez Mountains, and fell under the influence of the botanist Ralph Hoffmann, who taught him much about the plants and natural history of that lovely area.

Enrolling in Harvard University in 1924, Ledyard at first had difficulty defining his major, but the summer between his freshman and sophomore years was spent investigating the plants around Bar Harbor, Maine, the family home, and brought him into contact with Edgar T. Wherry, professor of botany from the University of Pennsylvania and a specialist in ferns, who encouraged his botanical interests. When he enrolled for the fall semester of 1925 at Harvard, he had decided to pursue a botanical career. But during his time at Harvard, his love of classical music, which was to be an important element for the remainder of his life, was awakened and nourished, as he participated in music classes and

Missouri Botanical Garden, P.O. Box 266, St. Louis, MO 63166 choruses, and was encouraged by some powerful and encouraging faculty members and students. Continuing on in the Harvard Graduate School, Ledyard was caught in the cross-fire between those, like Merrit Lyndon Fernald, who took a classical view of botany and plant classification, and the more modern approaches of Karl Sax, who was applying cytogenetic principles to developing a deeper understanding of plants. Thanks to judicious efforts by Paul Mangelsdorf and others, his dissertation was finally approved; but it was a struggle; he graduated in 1931.

One of the key events in Ledyard's early career was his attending the International Botanical Congress at Cambridge, England, in 1930; there he met Edgar Anderson, who was to become a lifetime friend and colleague; Irene Manton; and C.D. Darlington, whose classical "Recent Advances of Cytology," was still in the future. These and other contacts greatly encouraged his interest in and enthusiasm for botany and botanists, which was to be sustained for the rest of his life.

After he obtained his Ph.D., Ledyard Stebbins spent the years 1931–1935 at Colgate University, which he described years later as unhappy years, but it is not clear why this was the case. With an associate, Professor Percy Sanders, he undertook the cytogenetic study of Paeonia, which was the first of a series of essentially biosystematic investigations of diverse plant groups that were to characterize the remainder of his research career. During this time, he discovered complex structural heterozygosity in the western North American species of the genus, an exciting find that was to fuel his enthusiasm for further cytogenetic investigations.

In 1935, Professor Ernest Brown Babcock of the University of California, Berkeley, offered Stebbins a research position in connection with his investigations of the genus Crepis, which he accepted with alacrity. Met at the train station by his fellow Harvard student Rimo Bacigalupi, he plunged into this project with enthusiasm. Also at Berkeley, he began his lifetime preoccupation with Democratic politics, working actively in the 1936 Roosevelt election, and from there onward. After four years on Professor Babcock's grant, Stebbins was appointed to the faculty at Berkeley, and began to teach a course in the principles of evolution, which helped him to generalize his thoughts and finally to his preparing the classical work, "Variation and Evolution in Plants," whose 50th anniversary we are celebrating in this symposium.

In his research efforts, Ledyard began investigating the American species of Crepis (most species of the genus are Eurasian, but there are some very interesting offshoots in North America), outlining the evolutionary features of this group as a pillar complex of polyploids, with the base chromosome number 2n = 22, but widespread polyploids, characteristically apomictic, linking their more narrowly-distributed diploid progenitors. He also began investigating grasses, first Bromus and then Triticeae, with the objective of developing perennial grasses that would provide forage on the dry rangelands of California, and which eventually led to his extensive studies of the genus Dactylis, which he pursued throughout its native range in western Eurasia and North Africa in the decades to follow. Never successful, this quest nonetheless led Stebbins to many interesting discoveries, and broadened the scope of his knowledge of the details of evolution in plants in such a way as to expand the coverage of and insights provided in his landmark book.

In the early 1940s, Stebbins began working actively with Carl Epling on the genetics of Linanthus parryae, an annual of the Mohave Desert in which the prevalence of white or blue flowers in individual populations was held at the time to have resulted from random drift. He also started an active association with Theodosius Dobzhansky, centering around Dobzhansky's efforts at Mather; he regarded Epling, Dobzhansky, and Edgar Anderson as his closest and most influential professional associates.

In 1947, Ledyard Stebbins spent three months at Columbia University in New York, delivering the Jesup Lectures; and these lectures, expanded and elaborated, became *Variation and Evolution in Plants,* the most important book on plant evolution of the 20th century. I first met him in 1950, on a Sierra Club outing, and he was as encouraging to me at the age of 14 as I could have imagined. It seemed to me later that his own rather unhappy and lonely childhood led him naturally to an appreciation for young people, a lifetime interest in connection with which he made significant contributions to the lives of many young scholars. I maintained a strong friendship with him for the remaining half-century of his life.

The first period of Ledyard Stebbins' botanical life extended from 1925, when his serious interest in plants was kindled at Harvard, to 1935, when he arrived at Berkeley; the second, highly productive period, from there to 1950, when "Variation and Evolution in Plants" was published. In that same year, he answered an invitation from the University to establish a department of genetics at the Davis campus, and entered the third period of his professional life. And my, how he loved Davis, its growth, its variety, and its accessibility to all. He was proud of his work at Davis, proud of the growing campus as it matured, pleased with his own contributions, and always contented living there. In 1971, after Dobzhansky's retirement from Rockefeller University, he was influential in recruiting both Dobzhansky and his associate Francisco Ayala, to Davis, where they made outstanding contributions. With retirement, he traveled widely, for example, teaching in Chile during the time of the 1973 coup, and visiting Australia, Africa, Europe, and other parts of the world in teaching, visiting with his colleagues, and, as always, enjoying students.

2

Solution to Darwin's Dilemma: Discovery of the Missing Precambrian Record of Life

J. WILLIAM SCHOPF

In 1859, in *On the Origin of Species,* Darwin broached what he regarded to be the most vexing problem facing his theory of evolution—the lack of a rich fossil record predating the rise of shelly invertebrates that marks the beginning of the Cambrian Period of geologic time (≈550 million years ago), an "inexplicable" absence that could be "truly urged as a valid argument" against his all embracing synthesis. For more than 100 years, the "missing Precambrian history of life" stood out as one of the greatest unsolved mysteries in natural science. But in recent decades, understanding of life's history has changed markedly as the documented fossil record has been extended seven-fold to some 3,500 million years ago, an age more than three-quarters that of the planet itself. This long-sought solution to Darwin's dilemma was set in motion by a small vanguard of workers who blazed the trail in the 1950s and 1960s, just as their course was charted by a few pioneering pathfinders of the previous century, a history of bold pronouncements, dashed dreams, search, and final discovery.

Department of Earth and Space Sciences, Institute of Geophysics and Planetary Physics (Center for the Study of Evolution and the Origin of Life), and Molecular Biology Institute, University of California, Los Angeles, CA 90095-1567

This paper was presented at the National Academy of Sciences colloquium "Variation and Evolution in Plants and Microorganisms: Toward a New Synthesis 50 Years After Stebbins," held January 27–29, 2000, at the Arnold and Mabel Beckman Center in Irvine, CA.

In 1950, when Ledyard Stebbins' *Variation and Evolution in Plants* first appeared, the known history of life—the familiar progression from spore-producing to seed-producing to flowering plants, from marine invertebrates to fish, amphibians, then reptiles, birds, and mammals—extended only to the beginning of the Cambrian Period of the Phanerozoic Eon, roughly 550 million years ago. Now, after a half-century of discoveries, life's history looks strikingly different—an immense early fossil record, unknown and assumed unknowable, has been uncovered to reveal an evolutionary progression dominated by microbes that stretches seven times farther into the geologic past than previously was known. This essay is an abbreviated history of how and by whom the known antiquity of life has been steadily extended, and of lessons learned in this still ongoing hunt for life's beginnings.

PIONEERING PATHFINDERS

Darwin's Dilemma

Like so many aspects of natural science, the beginnings of the search for life's earliest history date from the mid-1800s and the writings of Charles Darwin (1809–1882), who in *On the Origin of Species* first focused attention on the missing Precambrian fossil record and the problem it posed to his theory of evolution: "There is another . . . difficulty, which is much more serious. I allude to the manner in which species belonging to several of the main divisions of the animal kingdom suddenly appear in the lowest known [Cambrian-age] fossiliferous rocks . . . If the theory be true, it is indisputable that before the lowest Cambrian stratum was deposited, long periods elapsed . . . and that during these vast periods, the world swarmed with living creatures . . . [But] to the question why we do not find rich fossiliferous deposits belonging to these assumed earliest periods before the Cambrian system, I can give no satisfactory answer. The case at present must remain inexplicable; and may be truly urged as a valid argument against the views here entertained" (Darwin, 1859, Chapter X).

Darwin's dilemma begged for solution. And although this problem was to remain unsolved—the case "inexplicable"—for more than 100 years, the intervening century was not without bold pronouncements, dashed dreams, and more than little acid acrimony.

J. W. Dawson and the "Dawn Animal of Canada"

Among the first to take up the challenge of Darwin's theory and its most vexing problem, the missing early fossil record, was John William

Dawson (1820–1899), Principal of McGill University and a giant in the history of North American geology. Schooled chiefly in Edinburgh, Scotland, the son of strict Scottish Presbyterians, Dawson was a staunch Calvinist and devout antievolutionist (O'Brien, 1971).

In 1858, a year before publication of Darwin's opus, specimens of distinctively green- and white-layered limestone collected along the Ottawa River to the west of Montreal were brought to the attention of William E. Logan, Director of the Geological Survey of Canada. Because the samples were known to be ancient (from "Laurentian" strata, now dated at about 1,100 million years) and exhibited layering that Logan supposed too regular to be purely inorganic (Fig. 1), he displayed them as possible "pre-Cambrian fossils" at various scientific conferences, where they elicited spirited discussion but gained little acceptance as remnants of early life.

In 1864, however, Logan brought specimens to Dawson who not only confirmed their biologic origin but identified them as fossilized shells of giant foraminiferans, huge oversized versions of tiny calcareous protozoal tests. So convinced was Dawson of their biologic origin that a year later, in 1865, he formally named the putative fossils *Eozoon canadense*, the "dawn animal of Canada." Dawson's interpretation was questioned almost immediately (King and Rowney, 1866), the opening shot of a fractious debate that raged on until 1894 when specimens of *Eozoon* were found near Mt. Vesuvius and shown to be geologically young ejected blocks of limestone, their "fossil-like" appearance the result of inorganic alteration and veining by the green metamorphic mineral serpentine (O'Brien, 1970).

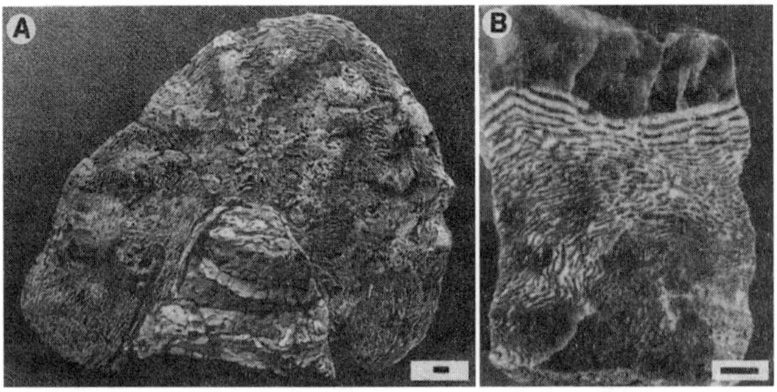

FIGURE 1. Eozoon canadense, the "dawn animal of Canada," as illustrated in Dawson's The Dawn of Life (Dawson, 1875), (A) and shown by the holotype specimen archived in the U.S. National Museum of Natural History, Washington, DC (B). (Bars = 1 cm.).

Yet despite the overwhelming evidence, Dawson continued to press his case for the rest of his life, spurred by his deeply held belief that discovery of his "dawn animal" had exposed the greatest missing link in the entire fossil record, a gap so enormous that it served to unmask the myth of evolution's claimed continuity and left Biblical creation as the only answer: "There is no link whatever in geological fact to connect Eozoon with the Mollusks, Radiates, or Crustaceans of the succeeding [rock record] . . . these stand before us as distinct creations. [A] gap . . . yawns in our imperfect geological record. Of actual facts [with which to fill this gap], therefore, we have none; and those evolutionists who have regarded the dawn-animal as an evidence in their favour, have been obliged to have recourse to supposition and assumption" (Dawson, 1875, p. 227). (In part, Dawson was right. In the fourth and all later editions of *The Origin*, Darwin cited the great age and primitive protozoal relations of *Eozoon* as consistent with his theory of evolution, just the sort of "supposition and assumption" that Dawson found so distressing.)

C. D. Walcott: Founder of Precambrian Paleobiology

Fortunately, Dawson's debacle would ultimately prove to be little more than a distracting detour on the path to progress, a redirection spurred initially by the prescient contributions of the American paleontologist Charles Doolittle Walcott (1850–1927).

Like Dawson before him, Walcott was enormously energetic and highly influential (Yochelson, 1967, 1997). He spent most of his adult life in Washington, DC, where he served as the CEO of powerful scientific organizations—first, as Director of the U.S. Geological Survey (1894–1907), then Secretary of the Smithsonian Institution (1907–1927) and President of the National Academy of Sciences (1917–1923). Surprisingly, however, Walcott had little formal education. As a youth in northern New York State he received but 10 years of schooling, first in public schools and, later, at Utica Academy (from which he did not graduate). He never attended college and had no formally earned advanced degrees (a deficiency more than made up for in later life when he was awarded honorary doctorates by a dozen academic institutions).

In 1878, as a 28-year-old apprentice to James Hall, Chief Geologist of the state of New York and acknowledged dean of American paleontology, Walcott was first introduced to stromatolites—wavy layered mound-shaped rock masses laid down by ancient communities of mat-building microbes—Cambrian-age structures near the town of Saratoga in eastern New York State. Named *Cryptozoon* (meaning "hidden life"), these cabbagelike structures (Fig. 2) would in later years form the basis of Walcott's side of a nasty argument known as the "*Cryptozoon* controversy."

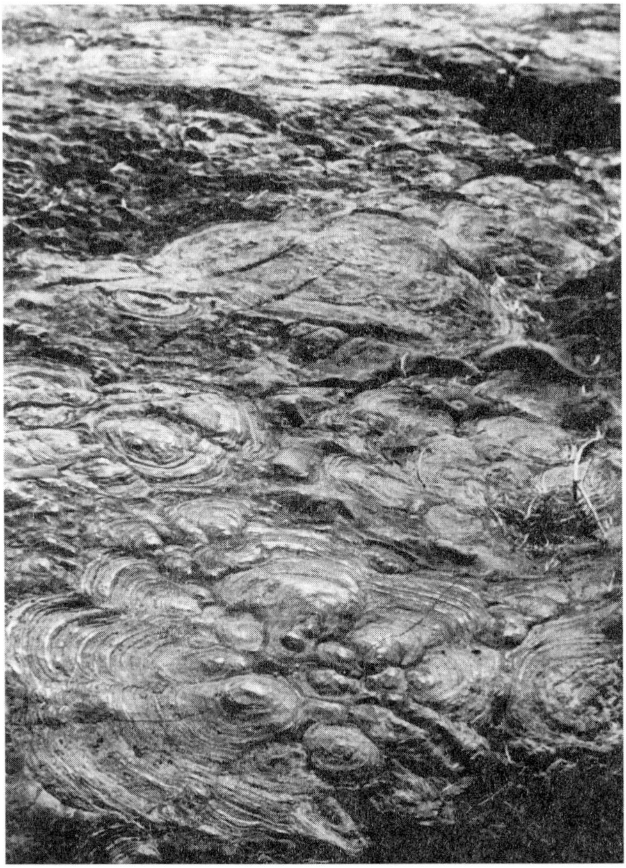

FIGURE 2. Cryptozoon reefs near Saratoga, NY. (Photo by E. S. Barghoorn, November, 1964.)

A year later, in July, 1879, Walcott was appointed to the newly formed U.S. Geological Survey. Over the next several field seasons, he and his comrades charted the geology of sizable segments of Arizona, Utah, and Nevada, including unexplored parts of the Grand Canyon, where in 1883 he first reported discovery of Precambrian specimens of *Cryptozoon* (Walcott, 1883). Other finds soon followed, with the most startling in 1899—small, millimeter-sized black coaly discs that Walcott named *Chuaria* and interpreted to be "the remains of . . . compressed conical shell[s]," possibly of primitive brachiopods (Walcott, 1899). Although *Chuaria* is now known to be a large single-celled alga, rather than a shelly invertebrate, Walcott's specimens were indeed authentic fossils, the first true cellularly preserved Precambrian organisms ever recorded.

After the turn of the century, Walcott moved his field work northward along the spine of the Rocky Mountains, focusing first in the Lewis Range of northwestern Montana, from which he reported diverse stromatolitelike structures (Walcott, 1906) and, later, chains of minute cell-like bodies he identified as fossil bacteria (Walcott, 1915). His studies in the Canadian Rockies, from 1907 to 1925, were even more rewarding, resulting in discovery of an amazingly well-preserved assemblage of Cambrian algae and marine invertebrates—the famous Burgess Shale Fauna that to this day remains among the finest and most complete samples of Cambrian life known to science (Walcott, 1911; Gould, 1989).

Walcott's contributions are legendary—he was the first discoverer in Precambrian rocks of *Cryptozoon* stromatolites, of cellularly preserved algal plankton (*Chuaria*), and of possible fossil bacteria, all capped by his pioneering investigations of the benchmark Burgess Shale fossils. The acknowledged founder of Precambrian paleobiology (Schopf, 1970), Walcott was first to show, nearly a century ago and contrary to accepted wisdom, that a substantial fossil record of Precambrian life actually exists.

A. C. Seward and the *Cryptozoon* Controversy

The rising tide in the development of the field brought on by Walcott's discoveries was not yet ready to give way to a flood. Precambrian fossils continued to be regarded as suspect, a view no doubt bolstered by Dawson's *Eozoon* debacle but justified almost as easily by the scrappy nature of the available evidence. Foremost among the critics was Albert Charles Seward (1863–1941), Professor of Botany at the University of Cambridge and the most widely known and influential paleobotanist of his generation. Because practically all claimed Precambrian fossils fell within the purview of paleobotany—whether supposed to be algal, like *Cryptozoon* stromatolites, or even bacterial—Seward's opinion had special impact.

In 1931, in *Plant Life Through the Ages*, the paleobotanical text used worldwide, Seward assessed the "algal" (that is, cyanobacterial) origin of *Cryptozoon* as follows: "The general belief among American geologists and several European authors in the organic origin of *Cryptozoon* is . . . not justified by the facts. [Cyanobacteria] or similar primitive algae may have flourished in Pre-Cambrian seas and inland lakes; but to regard these hypothetical plants as the creators of reefs of *Cryptozoon* and allied structures is to make a demand upon imagination inconsistent with Wordsworth's definition of that quality as 'reason in its most exalted mood'" (Seward, 1931, pp. 86–87).

Seward was even more categorical in his rejection of Walcott's report of fossil bacteria: "It is claimed that sections of a Pre-Cambrian limestone from Montana show minute bodies similar in form and size to cells and cell-chains of existing [bacteria]. . . . These and similar contributions . . . are by no means convincing. . . . We can hardly expect to find in Pre-Cambrian rocks any actual proof of the existence of bacteria . . . " (Seward, 1931, p. 92).

Seward's 1931 assessment of the science was mostly on the mark. Mistakes had been made. Mineralic, purely inorganic objects had been misinterpreted as fossil. Better and more evidence, carefully gathered and dispassionately considered, was much needed. But his dismissive rejection of *Cryptozoon* and his bold assertion that "we can hardly expect to find in Pre-Cambrian rocks any actual proof of the existence of bacteria" turned out to be misguided. Yet his influence was pervasive. It took another 30 years and a bit of serendipity to put the field back on track.

EMERGENCE OF A NEW FIELD OF SCIENCE

In the mid-1960s—a full century after Darwin broached the problem of the missing early fossil record—the hunt for early life began to stir, and in the following two decades the flood-gates would finally swing wide open. But this surge, too, had harbingers, now dating from the 1950s.

A Benchmark Discovery by an Unsung Hero

The worker who above all others set the course for modern studies of ancient life was Stanley A. Tyler (1906–1963) of the University of Wisconsin, the geologist who in 1953 discovered the now famous mid-Precambrian (2,100-million-year-old) microbial assemblage petrified in carbonaceous cherts of the Gunflint Formation of Ontario, Canada. A year later, together with Harvard paleobotanist Elso S. Barghoorn (1915–1984), Tyler published a short note announcing the discovery (Tyler and Barghoorn, 1954), a rather sketchy report that on the basis of study of petrographic thin sections documents that the fossils are indigenous to the deposit but fails to note either the exact provenance of the find or that the fossils are present within, and were actually the microbial builders of, large *Cryptozoon*-like stromatolites (an association that, once recognized, would prove key to the development of the field). Substantive, full-fledged reports would come later—although not until after Tyler's untimely death, an event that cheated him from receiving the great credit he deserved—but this initial short article on "the oldest structurally preserved organisms that clearly exhibit cellular differentiation and original carbon complexes which have yet been discovered in pre-Cambrian sediments" (Tyler and Barghoorn, 1954) was a benchmark, a monumental "first."

Contributions of Soviet Science

At about the same time, in the mid-1950s, a series of articles by Boris Vasil'evich Timofeev (1916–1982) and his colleagues at the Institute of Precambrian Geochronology in Leningrad (St. Petersburg) reported discovery of microscopic fossil spores in Precambrian siltstones of the Soviet Union. In thin sections, like those studied by Tyler and Barghoorn, fossils are detected within the rock, entombed in the mineral matrix, so the possibility of laboratory contamination can be ruled out. But preparation of thin sections requires special equipment, and their microscopic study is tedious and time-consuming. A faster technique, pioneered for Precambrian studies in Timofeev's lab, is to dissolve a rock in mineral acid and concentrate the organic-walled microfossils in the resulting sludgelike residue. This maceration technique, however, is notoriously subject to error-causing contamination—and because during these years there was as yet no established early fossil record with which to compare new finds, mistakes were easy to make. Although Timofeev's laboratory was not immune, much of his work has since proved sound (Schopf, 1992), and the technique he pioneered to ferret-out microfossils in Precambrian shaley rocks is now in use worldwide.

Famous Figures Enter the Field

Early in the 1960s, the fledgling field was joined by two geologic heavyweights, an American, Preston Cloud (1912–1991), and an Australian, Martin Glaessner (1906–1989), both attracted by questions posed by the abrupt appearance and explosive evolution of shelly invertebrate animals that marks the start of the Phanerozoic Eon.

A feisty leader in the development of Precambrian paleobiology, Cloud was full of energy, ideas, opinions, and good hard work. His Precambrian interests were first evident in the late-1940s, when he argued in print that although the known Early Cambrian fossil record is woefully incomplete it is the court of last resort and, ultimately, the only court that matters (Cloud, 1948). By the 1960s, he had become active in the field, authoring a paper that to many certified the authenticity of the Tyler-Barghoorn Gunflint microfossils (Cloud, 1965) and, later, a series of reports adding new knowledge of the early microbial fossil record (Cloud and Licari, 1968; Cloud *et al.*, 1975; Cloud and Morrison, 1980). But above all, he was a gifted synthesist, showing his mettle in a masterful article of 1972 that set the stage for modern understanding of the interrelated atmospheric-geologic-biologic history of the Precambrian planet (Cloud, 1972).

In the early 1960s, a second prime player entered this now fast-unfolding field, Martin Glaessner (1906–1989), of the University of Adelaide in South

Australia. A scholarly, courtly, old-school professor, Glaessner was the first to make major inroads toward understanding the (very latest) Precambrian record of multicelled animal life (Radhakrishna, 1991; McGowran, 1994).

In 1947, three years before Glaessner joined the faculty at Adelaide, Reginald C. Sprigg announced his discovery of fossils of primitive soft-bodied animals, chiefly imprints of saucer-sized jellyfish, at Ediacara, South Australia (Sprigg, 1947). Although Sprigg thought the fossil-bearing beds were Cambrian in age, Glaessner showed them to be Precambrian (albeit marginally so), making the Ediacaran fossils the oldest animals known. Together with his colleague, Mary Wade, Glaessner spent much of the rest of his life working on this benchmark fauna (see, e.g., Glaessner and Wade, 1966, 1971), bringing it to international attention in the early 1960s (Glaessner, 1962) and, later, in a splendid monograph (Glaessner, 1984).

With Glaessner in the fold, the stage was set. Like a small jazz band—Tyler and Barghoorn trumpeting microfossils in cherts, Timofeev beating on fossils in siltstones, Cloud strumming the early environment, Glaessner the earliest animals—great music was about to be played. At long last, the curtain was to rise on the missing record of Precambrian life.

Breakthrough to the Present

My own involvement dates from 1960, when as a sophomore in college I became enamored with the problem of the missing Precambrian fossil record, an interest that was to become firmly rooted during the following few years, when I was the first of Barghoorn's graduate students to focus on early life. I have recently recounted in some detail my recollections of those heady days (Schopf, 1999) and need not reiterate the story here. Suffice it to note that virtually nothing had been published on the now-famous Gunflint fossils (Fig. 3) in the nearly 10 years that had passed between the Tyler-Barghoorn 1954 announcement of the find and my entry into graduate school in June, 1963. Then, quite unexpectedly, in October of that year, Stanley Tyler passed away at the age of 57, never to see the ripened fruits of his long-term labor reach the published page. Within a year thereafter, a series of events that would shape the field began to unfold, set off first by a squabble between Barghoorn and Cloud as to who would scoop whom in a battle for credit over the Gunflint fossils (Schopf, 1999). By late 1964 this spat had been settled, with Cloud electing to hold off publication of his paper "illustrating some conspicuous Gunflint nannofossils and discussing their implications until Barghoorn could complete his part of a descriptive paper with the by-then deceased Tyler" (Cloud, 1983, p. 23). The two articles appeared in *Science* in 1965, first Barghoorn and Tyler's "Microorganisms from the Gunflint

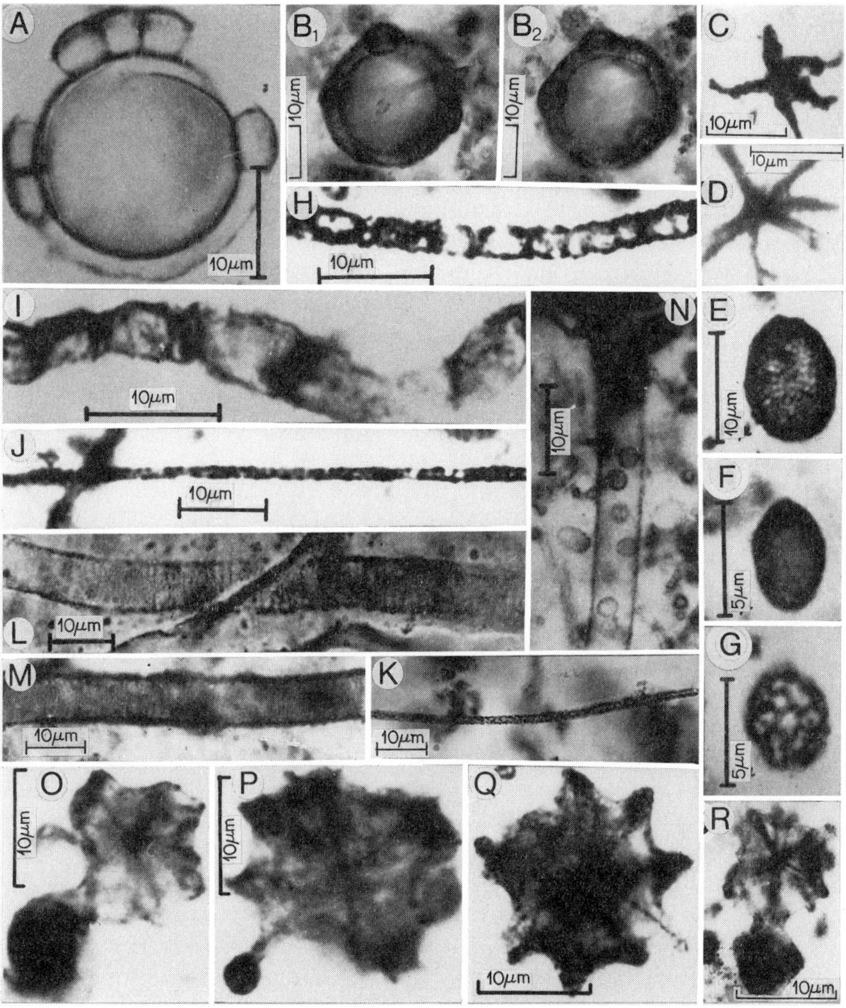

FIGURE 3. Microfossils of the Paleoproterozoic (≈2,100-million-year-old) Gunflint chert of southern Canada. (A and B) Eosphaera, in B shown in two views of the same specimen. (C and D) Eoastrion. (E–G) Huroniospora. (H–K) Gunflintia. (L and M) Animikiea. (N) Entosphaeroides. (O–R) Kakabekia.

chert" (Barghoorn and Tyler, 1965), followed a few weeks later by Cloud's contribution, "Significance of the Gunflint (Precambrian) microflora" (Cloud, 1965). Landmark papers they were!

Unlike the 1954 Tyler-Barghoorn announcement of discovery of the Gunflint fossils, which had gone largely unnoticed, the Barghoorn-Tyler 1965 article—backed by Cloud's affirmation of its significance—generated enormous interest. Yet it soon became apparent that acceptance of ancient life would come only grudgingly. The well had been poisoned by Dawson's debacle, the *Cryptozoon* controversy, Seward's criticism—object lessons that had been handed down from professor to student, generation to generation, to become part of accepted academic lore. Moreover, it was all too obvious that the Gunflint organisms stood alone. Marooned in the remote Precambrian, they were isolated by nearly a billion and half years from all other fossils known to science, a gap in the known fossil record nearly three times longer than the entire previously documented history of life. Skepticism abounded. Conventional wisdom was not to be easily dissuaded. The question was asked repeatedly: "Couldn't this whole business be some sort of fluke, some hugely embarrassing awful mistake?"

As luck would have it, the doubts soon could be laid to rest. During field work the previous year (and stemming from a chance conversation with a local oil company geologist by the name of Helmut Wopfner), Barghoorn had collected a few hand-sized specimens of Precambrian stromatolitic black chert in the vicinity of Alice Springs, deep in the Australian outback. Once the Gunflint paper had been completed, I was assigned to work on the samples, which quite fortunately contained a remarkable cache of new microscopic fossils, most nearly indistinguishable from extant cyanobacteria and almost all decidedly better preserved than the Gunflint microbes (Fig. 4). Although the age of the deposit (the Bitter Springs Formation) was known only approximately, it seemed likely to be about 1,000 million years, roughly half as old as the Gunflint chert.

Barghoorn and I soon sent a short report to *Science* (Barghoorn and Schopf, 1965), publication of which—viewed in light of the earlier articles on the Gunflint organisms—not only served to dispell lingering doubts about whether Precambrian fossils might be some sort of fluke, but seemed to show that the early fossil record was surprisingly richer and easier to unearth than anyone had dared imagine. Indeed, it now appears that the only truly odd thing about the Gunflint and Bitter Springs fossils is that similar finds had not been made even earlier. Walcott had started the train down the right track only for it to be derailed by the conventional wisdom that the early history of life was unknown and evidently unknowable, a view founded on the assumption that the tried and true techniques of the Phanerozoic hunt for large fossils would prove equally rewarding in the Precambrian. Plainly put, this was wrong.

LESSONS FROM THE HUNT

The Gunflint and Bitter Springs articles of 1965 charted a new course, showing for the first time that a search strategy centered on the peculiari-

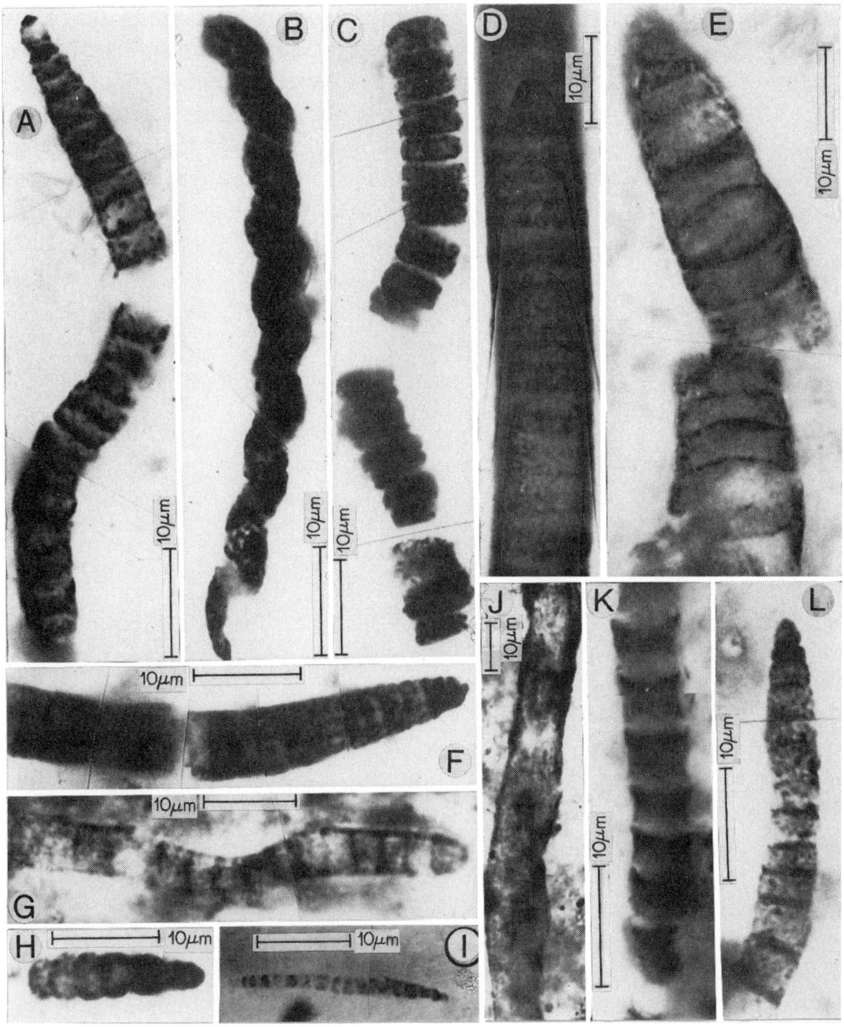

FIGURE 4. Filamentous microfossils of the Neoproterozoic (≈850-million-year-old) Bitter Springs chert of central Australia. Because the petrified microbes are three-dimensional and sinuous, composite photos have been used to show the specimens A–G, I, K, and L. (A, F, I, and L) Cephalophytarion. (B) Helioconema. (C and G) Oscillatoriopsis. (D) Unnamed cyanobacterium. (E) Obconicophycus. (H) Filiconstrictosus. (J) Siphonophycus. (K) Halythrix.

ties of the Precambrian fossil record would pay off. The four keys of the strategy, as valid today as they were three decades ago, are to search for (*i*) microscopic fossils in (*ii*) black cherts that are (*iii*) fine-grained and (*iv*) associated with *Cryptozoon*-like structures. Each part plays a role.

(*i*) Megascopic eukaryotes, the large organisms of the Phanerozoic, are now known not to have appeared until shortly before the beginning of the Cambrian—except in immediately sub-Cambrian strata, the hunt for large body fossils in Precambrian rocks was doomed from the outset.

(*ii*) The blackness of a chert commonly gives a good indication of its organic carbon content—like fossil-bearing coal deposits, cherts rich in petrified organic-walled microfossils are usually a deep jet black color.

(*iii*) The fineness of the quartz grains making up a chert provides another hint of its fossil-bearing potential—cherts subjected to the heat and pressure of geologic metamorphism are often composed of recrystallized large grains that give them a sugary appearance whereas cherts that have escaped fossil-destroying processes are made up of cryptocrystalline quartz and have a waxy glasslike luster.

(*iv*) *Cryptozoon*-like structures (stromatolites) are now known to have been produced by flourishing microbial communities, layer upon layer of microscopic organisms that make up localized biocoenoses. Stromatolites permineralized by fine-grained chert early during diagenesis represent promising hunting grounds for the fossilized remnants of the microorganisms that built them.

Measured by virtually any criterion one might propose (Fig. 5), studies of Precambrian life have burst forth since the mid-1960s to culminate in recent years in discovery of the oldest fossils known, petrified cellular microbes nearly 3,500 million years old, more than three-quarters the age of the Earth (Schopf, 1993). Precambrian paleobiology is thriving—the vast majority of all scientists who have ever investigated the early fossil record are alive and working today; new discoveries are being made at an ever quickening clip—progress set in motion by the few bold scientists who blazed this trail in the 1950s and 1960s, just as their course was charted by the Dawsons, Walcotts, and Sewards, the pioneering pathfinders of the field. And the collective legacy of all who have played a role dates to Darwin and the dilemma of the missing Precambrian fossil record he first posed. After more than a century of trial and error, of search and final discovery, those of us who wonder about life's early history can be thankful that what was once "inexplicable" to Darwin is no longer so to us.

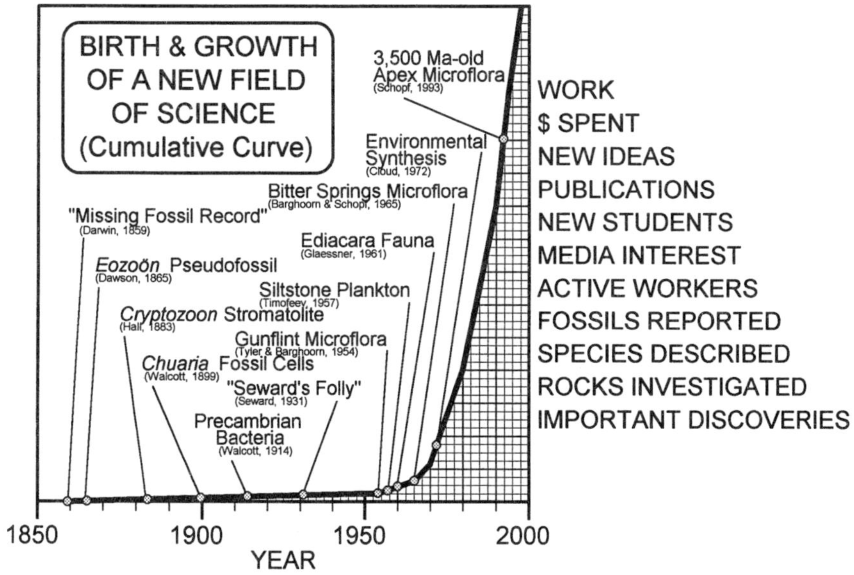

FIGURE 5. Birth and growth of Precambrian paleobiology.

REFERENCES

Barghoorn, E. S. & Schopf, J. W. (1965) Microorganisms from the Late Precambrian of central Australia. *Science* 150, 337–339.

Barghoorn, E. S. & Tyler, S. A. (1965) Microorganisms from the Gunflint chert. *Science* 147, 563–577.

Cloud, P. (1948) Some problems and patterns of evolution exemplified by fossil invertebrates. *Evol.* 2, 322–350.

Cloud, P. (1965) Significance of the Gunflint (Precambrian) microflora. *Science* 148, 27–45.

Cloud, P. (1972) A working model of the primitive Earth. *Am. J. Sci.* 272, 537–548.

Cloud, P. (1983) Early biogeologic history: The emergence of a paradigm. In *Earth's Earliest Biosphere, An Interdisciplinary Study*, ed. Schopf, J. W. (Princeton Univ. Press, Princeton, NJ), pp. 14–31.

Cloud, P & Licari, G. R. (1968) Microbiotas of the banded iron formations. *Proc. Nat. Acad. Sci. USA* 61, 779–786.

Cloud, P. & Morrison, K. (1980) New microbial fossils from 2 Gyr old rocks in northern Michigan. *Geomicrobiol. J.* 2, 161–178.

Cloud, P., Moorman, M., & Pierce, D. (1975) Sporulation and ultrastructure in a late Proterozoic cyanophyte: Some implications for taxonomy and plant phylogeny. *Quart. Rev. Biol.* 50, 131–150.

Darwin, C. (1859) *On the Origin of Species by Means of Natural Selection, or the Preservation of Favoured Races in the Struggle for Life*, facsimile of the 1st edition of 1859 (Harvard Univ. Press, Cambridge, MA, 1964); facsimile of the 6th (and last) edition of 1872 (John Murray, London, 1902).

Dawson, J. W. (1875) *The Dawn of Life* (Hodder & Stoughton, London).

Glaessner, M. F. (1962) Pre-Cambrian fossils. *Biol. Rev. Cambridge Phil. Soc.* 37, 467–494.

Glaessner, M. F. (1984) *The Dawn of Animal Life, A Biohistorical Study* (Cambridge Univ. Press, Cambridge, UK).

Glaessner, M. F. & Wade, M. (1966) The late Precambrian fossils from Ediacara, South Australia. *Palaeontol.* 9, 599-628.

Glaessner, M. F. & Wade, M. (1971) *Precambridium*—a primitive arthropod. *Lethaia* 4, 71–77.

Gould, S. J. (1989) *Wonderful Life* (Norton, New York).

King, W. & Rowney, T. H. (1866) On the so-called "Eozoonal Rock." *Quart. J. Geol. Soc. London* 22, 215–222.

McGowran, B. (1994) Martin Fritz Glaessner 1906–1989. *Hist. Rec. Austral. Sci.* 10, 61–81.

O'Brien, C. F. (1970) *Eozoön canadense*, "the dawn animal of Canada." *Isis* 61, 206–223.

O'Brien, C. F. (1971) Sir William Dawson, a life in science and religion. *Mem. Am. Phil. Soc.* 84, 1–207.

Radhakrishna, B. P., ed. (1991) *The World of Martin F. Glaessner* (Geol. Soc. India, Bangalore), Mem. 20, pp. iii–xxiv.

Schopf, J. W. (1970) Precambrian micro-organisms and evolutionary events prior to the origin of vascular plants. *Biol. Rev. Cambridge Phil. Soc.* 45, 319–352.

Schopf, J. W. (1992) Historical development of Proterozoic micropaleontology. In *The Proterozoic Biosphere, A Multidisciplinary Study* eds. Schopf, J. W. & Klein, C. (Cambridge Univ. Press, New York), pp. 179–183.

Schopf, J. W. (1993) Microfossils of the Early Archean Apex chert: New evidence of the antiquity of life. *Science* 260, 640–646.

Schopf, J. W. (1999) *Cradle of Life, The Discovery of Earth's Earliest Fossils* (Princeton Univ. Press, Princeton, NJ).

Seward, A. C. (1931) *Plant Life Through the Ages* (Cambridge, Univ. Press, Cambridge, UK).

Sprigg, R. C. (1947) Early Cambrian (?) jellyfishes from the Flinders Ranges, South Australia. *Trans. Roy. Soc. S. Austral.* 71, 212–224.

Stebbins, G. L. (1950) *Variation and Evolution in Plants* (Columbia Univ. Press, New York).

Tyler, S. A. & Barghoorn, E. S. (1954) Occurrence of structurally preserved plants in pre-Cambrian rocks of the Canadian shield. *Science* 119, 606–608.

Walcott, C. D. (1883) Pre-Carboniferous strata in the Grand Canyon of the Colorado, Arizona. *Am. J. Sci.* 26, 437–442.

Walcott,C. D. (1899) Pre-Cambrian fossiliferous formations. *Bull. Geol. Soc. Am.* 10, 199–214.

Walcott,C. D. (1906) Algonkian formations of northwestern Montana. *Bull. Geol. Soc. Am.* 17, 1–28.

Walcott,C. D. (1911) Report of the Director, 1910. *Smithsonian Inst. Ann. Rept. 1910*, 1–39.

Walcott,C. D. (1915) Discovery of Algonkian bacteria. *Proc. Nat. Acad. Sci. USA* 1, 256–257.

Yochelson, E. L. (1967) Charles Doolittle Walcott 1850-1927. *Biograph. Mem., Nat. Acad. Sci.USA* 39, 471–540.

Yochelson, E. L. (1997) *Charles Doolittle Walcott, Paleontologist* (Kent State Univ. Press, Kent, OH).

3

The Chimeric Eukaryote: Origin of the Nucleus from the Karyomastigont in Amitochondriate Protists

LYNN MARGULIS*, MICHAEL F. DOLAN*,
and RICARDO GUERRERO‡

We present a testable model for the origin of the nucleus, the membrane-bounded organelle that defines eukaryotes. A chimeric cell evolved via symbiogenesis by syntrophic merger between an archaebacterium and a eubacterium. The archaebacterium, a thermoacidophil resembling extant *Thermoplasma,* generated hydrogen sulfide to protect the eubacterium, a heterotrophic swimmer comparable to *Spirochaeta* or *Hollandina* that oxidized sulfide to sulfur. Selection pressure for speed swimming and oxygen avoidance led to an ancient analogue of the extant cosmopolitan bacterial consortium *"Thiodendron latens."* By eubacterial-archaebacterial genetic integration, the chimera, an amitochondriate heterotroph, evolved. This "earliest branching protist" that formed by permanent DNA recombination generated the nucleus as a component of the karyomastigont, an intracellular complex that assured genetic continuity of the former symbionts. The karyomastigont organellar system, common in extant amito-

*Department of Geosciences, Organismic and Evolutionary Biology Graduate Program, University of Massachusetts, Amherst, MA 01003; and ‡Department of Microbiology, and Special Research Center Complex Systems (Microbiology Group), University of Barcelona, 08028 Barcelona, Spain

This paper was presented at the National Academy of Sciences colloquium "Variation and Evolution in Plants and Microorganisms: Toward a New Synthesis 50 Years After Stebbins," held January 27–29, 2000, at the Arnold and Mabel Beckman Center in Irvine, CA.

chondriate protists as well as in presumed mitochondriate ancestors, minimally consists of a single nucleus, a single kinetosome and their protein connector. As predecessor of standard mitosis, the karyomastigont preceded free (unattached) nuclei. The nucleus evolved in karyomastigont ancestors by detachment at least five times (archamoebae, calonymphids, chlorophyte green algae, ciliates, foraminifera). This specific model of syntrophic chimeric fusion can be proved by sequence comparison of functional domains of motility proteins isolated from candidate taxa.

Archaeprotists / spirochetes / sulfur syntrophy / *Thiodendron* / trichomonad

TWO DOMAINS, NOT THREE

All living beings are composed of cells and are unambiguously classifiable into one of two categories: prokaryote (bacteria) or eukaryote (nucleated organisms). Here we outline the origin of the nucleus, the membrane-bounded organelle that defines eukaryotes. The common ancestor of all eukaryotes by genome fusion of two or more different prokaryotes became "chimeras" via symbiogenesis (Gupta and Golding, 1995). Long term physical association between metabolically dependent consortia bacteria led, by genetic fusion, to this chimera. The chimera originated when an archaebacterium (a thermoacidophil) and a motile eubacterium emerged under selective pressure: oxygen threat and scarcity both of carbon compounds and electron acceptors. The nucleus evolved in the chimera. The earliest descendant of this momentous merger, if alive today, would be recognized as an amitochondriate protist. An advantage of our model includes its simultaneous consistency in the evolutionary scenario across fields of science: cell biology, developmental biology, ecology, genetics, microbiology, molecular evolution, paleontology, protistology. Environmentally plausible habitats and modern taxa are easily comprehensible as legacies of the fusion event. The scheme that generates predictions demonstrable by molecular biology, especially motile protein sequence comparisons (Chapman *et al.*, 2000), provides insight into the structure, physiology, and classification of microorganisms.

Our analysis requires the two- (Bacteria/Eukarya) not the three- (Archaea/Eubacteria/Eukarya) domain system (Woese *et al.*, 1990). The prokaryote vs. eukaryote that replaced the animal vs. plant dichotomy so far has resisted every challenge. Microbiologist's molecular biology-based threat to the prokaryote vs. eukaryote evolutionary distinction seems idle (Mayr, 1998). In a history of contradictory classifications of microorganisms since 1820, Scamardella (1999) noted that Woese's entirely nonmorphological system ignores symbioses. But bacterial consortia and protist endosymbioses irreducibly underlie evolutionary transitions from prokaryotes to eukaryotes. Although some prokaryotes [certain Gram-positive bacteria (Gupta, 1998a)] are intermediate between eubacteria and archaebacteria, no organisms intermediate between prokaryotes and eukaryotes exist. These facts render the 16S rRNA and other nonmorphological taxonomies of Woese and others inadequate. Only all-inclusive taxonomy, based on the work of thousands of investigators over more than 200 years on live organisms (Margulis and Schwartz, 1998), suffices for detailed evolutionary reconstruction (Mayr, 1998).

When Woese (1998) insists "there are actually three, not two, primary phylogenetic groupings of organisms on this planet" and claims that they, the "Archaebacteria" (or, in his term that tries to deny their bacterial nature, the "Archaea") and the "Eubacteria" are "each no more like the other than they are like eukaryotes," he denies intracellular motility, including that of the mitotic nucleus. He minimizes these and other cell biological data, sexual life histories including cyclical cell fusion, fossil record correlation (Margulis, 1996), and protein-based molecular comparisons (Gupta, 1998a, b). The tacit, uninformed assumption of Woese and other molecular biologists that all heredity resides in nuclear genes is patently contradicted by embryological, cytological, and cytoplasmic heredity literature (Sapp, 1999). The tubulin-actin motility systems of feeding and sexual cell fusion facilitate frequent viable incorporation of heterologous nucleic acid. Many eukaryotes, but no prokaryotes, regularly ingest entire cells, including, of course, their genomes, in a single phagocytotic event. This invalidates any single measure alone, including ribosomal RNA gene sequences, to represent the evolutionary history of a lineage.

As chimeras, eukaryotes that evolved by integration of more than a single prokaryotic genome (Gupta, 1998b) differ qualitatively from prokaryotes. Because prokaryotes are not directly comparable to symbiotically generated eukaryotes, we must reject Woese's three-domain interpretation. Yet our model greatly appreciates his archaebacterial-eubacterial distinction: the very first anaerobic eukaryotes derived from both of these prokaryotic lineages. The enzymes of protein synthesis in eukaryotes come primarily from archaebacteria whereas in the motility system (microtubules and their organizing centers), many soluble heat-shock and other proteins originated from eubacteria (Margulis, 1996). Here we apply Gupta's idea (from protein sequences) (1998a) to comparative protist data (Dolan *et al.*, 2000) to show how two kinds of prokaryotes made the first chimeric eukaryote. We reconstruct the fusion event that produced the nucleus.

THE CHIMERA: ARCHAEBACTERIUM\EUBACTERIUM MERGER

Study of conserved protein sequences [a far larger data set than that used by Woese *et al.* (1990)] led Gupta (1998a) to conclude "all eukaryotic cells, including amitochondriate and aplastidic cells received major genetic contributions to the nuclear genome from both an archaebacterium (very probably of the eocyte, i.e., thermoacidophil group and a Gram-negative bacterium . . . [t]he ancestral eukaryotic cell never directly descended from archaebacteria but instead was a chimera formed by fusion and integration of the genomes of an archaebacerium and a Gram-negative bacterium" (p. 1487). The eubacterium ancestor has yet to be identified; Gupta rejects our spirochete hypothesis. In answer to which microbe provided the eubacterial contribution, he claims: "the sequence data . . . . suggest that the archaebacteria are polyphyletic and are close relatives of the Gram-positive bacteria" (p. 1485). The archaebacterial sequences, we posit, following Searcy (1992), come from a *Thermoplasma acidophilum*-like thermoacidophilic (eocyte) prokaryote. This archaebacterial ancestor lived in warm, acidic, and sporadically sulfurous waters, where it used either elemental sulfur (generating H_2S) or less than 5% oxygen (generating H_2O) as terminal electron acceptor. As does its extant descendant, the ancient archaebacterium survived acid-hydrolysis environmental conditions by nucleosome-style histone-like protein coating of its DNA (Searcy, 1992) and actin-like stress-protein synthesis (Searcy and Delange, 1980). The wall-less archaebacterium was remarkably pleiomorphic; it tended into tight physical association with globules of elemental sulfur by use of its rudimentary cytoskeletal system (Searcy and Hixon, 1994). The second member of the consortium, an obligate anaerobe, required for growth the highly reduced conditions provided by sulfur and sulfate reduction to hydrogen sulfide. Degradation of carbohydrate (e.g., starch, sugars such as cellobiose) and oxidation of the sulfide to elemental sulfur by the eubacterium generated carbon-rich fermentation products and electron acceptors for the archaebacterium. When swimming eubacteria attached to the archaebacterium, the likelihood that the consortium efficiently reached its carbon sources was enhanced. This hypothetical consortium, before the integration to form a chimera (Fig. 1), differs little from the widespread and geochemically important *"Thiodendron"* (Dubinina *et al.*, 1993a, b).

THE *"THIODENDRON"* STAGE

The *"Thiodendron"* stage refers to an extant bacterial consortium that models our idea of an archaebacteria-eubacteria sulfur syntrophic motility symbiosis. The partners in our view merged to become the chimeric predecessor to archaeprotists. The membrane-bounded nucleus, by hypoth-

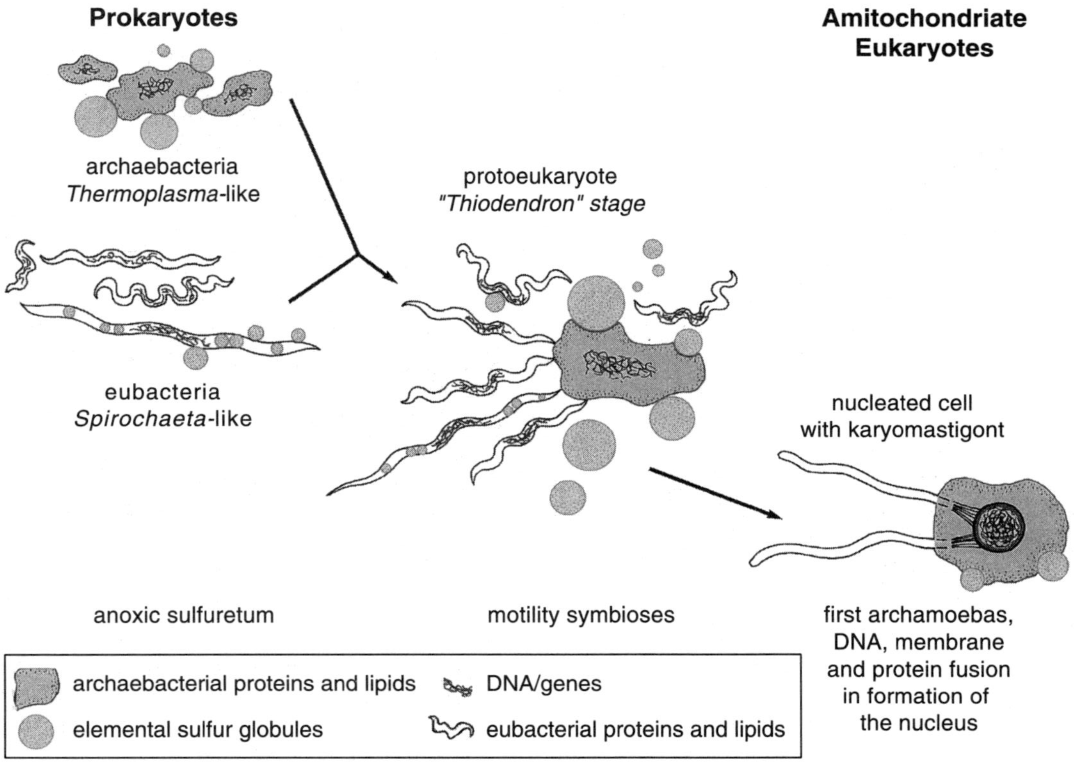

FIGURE 1. Origin of the chimeric eukaryote with karyomastigonts from a motile sulfur-bacteria consortium.

esis, is the morphological manifestation of the chimera genetic system that evolved from a *Thiodendron*-type consortium. Each phenomenon we suggest, from free-living bacteria to integrated association, enjoys extant natural analogues.

Study of marine microbial mats revealed relevant bacterial consortia in more than six geographically separate locations. Isolations from Staraya Russa mineral spring 8, mineral spring Serebryani, Lake Nizhnee, mudbaths; littoral zone at the White Sea strait near Veliky Island, Gulf of Nilma; Pacific Ocean hydrothermal habitats at the Kurile Islands and Kraternaya Bay; Matupi Harbor Bay, Papua New Guinea, etc. (Dubinina *et al.*, 1993a) all yielded "*Thiodendron latens*" or very similar bacteria. Samples were taken from just below oxygen-sulfide interface in anoxic waters (Dubinina *et al.*, 1993a, b). Laboratory work showed it necessary to abolish the genus *Thiodendron* because it is a sulfur syntrophy. A stable ectosymbiotic association of two bacterial types grows as an anaerobic consortium between 4 and 32°C at marine pH values and salinities. Starch, cellobiose, and other carbohydrates (not cellulose, amino acids, organic acids, or alcohol) supplemented by heterotrophic CO_2 fixation provide it carbon. *Thiodendron* appears as bluish-white spherical gelatinous colonies, concentric in structure within a slimy matrix produced by the consortium bacteria. The dominant partner invariably is a distinctive strain of pleiomorphic spirochetes: they vary from the typical walled *Spirochaeta* 1:2:1 morphology to large membranous spheres, sulfur-studded threads, gliding or nonmotile cells of variable width (0.09–0.45 μm) and lengths to millimeters. The other partner, a small, morphologically stable vibrioid, *Desulfobacter* sp., requires organic carbon, primarily acetate, from spirochetal carbohydrate degradation. The spirochetal *Escherichia coli*-like formic acid fermentation generates energy and food. *Desulfobacter* sp. cells that reduce both sulfate and sulfur to sulfide are always present in the natural consortium but in far less abundance than the spirochetes. We envision the *Thiodendron* consortium of "free-living spirochetes in geochemical sulfur cycle" (Dubinina *et al.*, 1993b, p. 456) and spirochete motility symbioses (Margulis, 1993) as preadaptations for chimera evolution. *Thiodendron* differs from the archaebacterium-eubacterium association we hypothesize; the marine *Desulfobacter* would have been replaced with a pleiomorphic wall-less, sulfuric-acid tolerant soil *Thermoplasma*-like archaebacterium. New thermoplasmas are under study. We predict strains that participate in spirochete consortia in less saline, more acidic, and higher temperature sulfurous habitats than *Thiodendron* will be found.

When "pure cultures" that survived low oxygen were first described [by B. V. Perfil'ev in 1969, in Russian (see Dubinina *et al.*, 1993a, b)] a complex life history of vibrioids, spheroids, threads and helices was attributed to "*Thiodendron latens.*" We now know these morphologies are artifacts of environmental selection pressure: Dubinina *et al.* (1993a, p. 435), reported that "the pattern of bacterial growth changes drastically when the redox potential of the medium is brought down by addition of 500 mg/l of sodium sulfide." The differential growth of the two tightly associated partners in the consortium imitates the purported *Thiodendron* bacterial developmental patterns. The syntrophy is maintained by lowering the level of oxygen enough for spirochete growth. The processes of sulfur oxidation-reduction and oxygen removal from oxygen-sensitive enzymes, we suggest, were internalized by the chimera and retained by their protist descendants as developmental cues.

Metabolic interaction, in particular syntrophy under anoxia, retained the integrated prokaryotes as emphasized by Martin and Müller (1998). However, we reject their concept, for which no evidence exists, that the archaebacterial partner was a methanogen. Our sulfur syntrophy idea, by contrast, is bolstered by observations that hydrogen sulfide is still generated in amitochondriate, anucleate eukaryotic cells (mammalian erythrocytes) (Searcy and Lee, 1998).

T. acidophilum in pure culture attach to suspended elemental sulfur. When sulfur is available, they generate hydrogen sulfide (Searcy and Hixon, 1994). Although severely hindered by ambient oxygen, they are microaerophilic in the presence of small quantities (<5%) of oxygen. The *Thermoplasma* partner thus would be expected to produce sulfide and scrub small quantities of oxygen to maintain low redox potential in the spirochete association. The syntrophic predecessors to the chimera is metabolically analogous to *Thiodendron* where *Desulfobacter* reduces sulfur and sulfate producing sulfide at levels that permit the spirochetes to grow. We simply suggest the replacement of the marine sulfidogen with *Thermoplasma*. In both the theoretical and actual case, the spirochetes would supply oxidized sulfur as terminal electron acceptor to the sulfidogen.

The DNA of the *Thermoplasma*-like archaebacterium permanently recombined with that of the eubacterial swimmer. A precedent exists for our suggestion that membrane hypertrophies around DNA to form a stable vesicle in some prokaryotes: the membrane-bounded nucleoid in the eubacterium *Gemmata obscuriglobus* (Fuerst and Webb, 1991). The joint *Thermoplasma*-like archaebacterial DNA package that began as the consortium nucleoid became the chimera's nucleus.

The two unlike prokaryotes together produced a persistent protein exudate package. This step in the origin of the nucleus—the genetic integration of the two-membered consortium to form the chimera—is traceable by its morphological legacy: the karyomastigont. The attached swimmer partner, precursor to mitotic microtubule system, belonged to genera like the nearly ubiquitous consortium-former *Spirochaeta* or the cytoplasmic tubule-maker *Hollandina* (Margulis, 1993). The swimmer's attachment structures hypertrophied as typically they do in extant motility symbioses (Margulis, 1993). The archaebacterium-eubacterium swimmer attachment system became the karyomastigont. The proteinaceous karyomastigont that united partner DNA in a membrane-bounded, jointly produced package, assured stability to the chimera. All of the DNA of the former prokaryotes recombined inside the membrane to become nuclear DNA while the protein-based motility system of the eubacterium, from the moment of fusion until the present, segregated the chimeric DNA. During the lower Proterozoic eon (2,500–1,800 million years ago), many interactions inside the chimera generated protists in which mitosis and eventually meiotic sexuality evolved. The key concept here is that the karyomastigont, retained by amitochondriate protists and later by their mitochondriate descendants, is the morphological manifestation of the original archaebacterial-eubacterial fused genetic system. Free (unattached) nuclei evolved many times by disassociation from the rest of the karyomastigont. The karyomastigont, therefore, was the first microtubule-organizing center.

KARYOMASTIGONTS PRECEDED NUCLEI

The term "karyomastigont" was coined by Janicki (1915) to refer to a conspicuous organellar system he observed in certain protists: the mastigont ("cell whip," eukaryotic flagellum, or undulipodium, the [9 (2) + (2)] microtubular axoneme underlain by its [9 (3) + 0)] kinetosome) attached by a "nuclear connector" or "rhizoplast" to a nucleus. The need for a term came from Janicki's work on highly motile trichomonad symbionts in the intestines of termites where karyomastigonts dominate the cells. When kinetosomes, nuclear connector, and other components were present but the nucleus was absent from its predictable position, Janicki called the organelle system an "akaryomastigont." In the Calonymphidae, one family of entirely multinucleate trichomonads, numerous karyomastigonts, and akaryomastigonts are simultaneously present in the same cell (e.g., *Calonympha grassii*) (Kirby and Margulis, 1994).

The karyomastigont, an ancestral feature of eukaryotes, is present in "early branching protists" (Dacks and Redfield, 1998; Delgado-Viscogliosi *et al.*, 2000; Edgcomb *et al.*, 1998). Archaeprotists, a large inclusive taxon (phylum of Kingdom Protoctista) (Margulis and Schwartz, 1998) are heterotrophic unicells that inhabit anoxic environments. All lack mitochondria. At least 28 families are placed in the phylum Archaeprotista. Examples include archaemoebae (*Pelomyxa* and *Mastigamoeba*), metamonads (*Retortamonas*), diplomonads (*Giardia*), oxymonads (*Pyrsonympha*), and the two orders of Parabasalia: Trichomonadida [*Devescovina*, *Mixotricha*, *Monocercomonas*, *Trichomonas*, and calonymphids (*Coronympha*, *Snyderella*)] and Hypermastigida (*Lophomonas*, *Staurojoenina*, and *Trichonympha*). These cells either bear karyomastigonts or derive by differential organelle reproduction (simple morphological steps) from those that do (Table 1). When, during evolution of these protists, nuclei were severed from their karyomastigonts, akaryomastigonts were generated (Kirby, 1949). Nuclei, unattached, at least temporarily, to undulipodia were freed to proliferate and occupy central positions in cells. Undulipodia, also freed to proliferate, generated larger, faster-swimming cells in the same evolutionary step.

The karyomastigont is the conspicuous central cytoskeleton in basal members of virtually all archaeprotist lineages [three classes: Archamoeba, Metamonads, and Parabasalia (Brugerolle, 1991)] (Fig. 2). In trichomonads, the karyomastigont, which includes a parabasal body (Golgi complex), coordinates the placement of hydrogenosomes (membrane-bounded bacterial-sized cell inclusions that generate hydrogen). The karyomastigont reproduces as a unit structure. Typically, four attached kinetosomes with rolled sheets of microtubules (the axostyle and its extension the pelta) reproduce as their morphological relationships are retained. Kinetosomes reproduce first, the nucleus divides, and the two groups of kinetosomes separate at the poles of a thin microtubule spindle called the paradesmose. Kinetosomes and associated structures are partitioned to one of the two new karyomastigonts. The other produces components it lacks such as the Golgi complex and axostyle.

Nuclear α-proteobacterial genes were interpreted to have originated from lost or degenerate mitochondria in at least two archaeprotist species [*Giardia lamblia* (Roger *et al.*, 1998); *Trichomonas vaginalis* (Roger *et al.*, 1996; Germont *et al.*, 1996)] and in a microsporidian (Sogin, 1997). Hydrogenosomes, at least some types, share common origin with mitochondria. In the hydrogen hypothesis (Martin and Müller, 1998), hydrogenosomes are claimed to be the source of eubacterial genes in amitochondriates. That mitochondria were never acquired in the ancestors we consider more likely than that they were lost in every species of these anaerobic protists. Eubacterial genes in the nucleus that are not from the original spirochete probably were acquired in amitochondriate protists from proteobacterial symbionts other than those of the mitochondrial lineage. Gram-negative bacteria, some of which may be related to ancestors of hydrogenosomes, are rampant as epibionts, endobionts, and even endonuclear symbionts—for example, in *Caduceia versatilis* (d'Ambrosio *et al.*, 1999).

Karyomastigonts freed (detached from) nuclei independently in many lineages both before and after the acquisition of mitochondria. Calonymphid ancestors of *Snyderella* released free nuclei before the mitochondrial symbiosis (Dolan *et al.*, 2000), and *Chlamydomonas*-like ancestors of other chlorophytes such as *Acetabularia* released the nuclei after the lineage was fully aerobic (Hall and Luck, 1995). In trophic forms of protists that lack mastigote stages, the karyomastigont is generally absent. An exception is *Histomonas*, an amoeboid trichomonad cell that lacks an axoneme but bears enough of the remnant karyomastigont structure to permit its classification with parabasalids rather than with rhizopod amoebae (Dyer, 1990). This organellar system appears in the zoospores, motile trophic forms, or sperm of many organisms, suggesting the relative ease of karyomastigont development. The karyomastigont, apparently in some cells, is easily lost, suppressed, and regained. In many taxa of multinucleate or multicellular protists (foraminifera, green algae) and even in plants, the karyomastigont persists only in the zoospores or gametes.

In yeast, nematode, insect, and mammalian cells, nonkaryomastigont microtubule-organizing centers are "required to position nuclei at specific locations in the cytoplasm" (Raff, 1999). The link between the microtubule organizing center and the nuclei "is mysterious" (Raff, 1999). To

TABLE 1. Karyomastigont distribution in unicellular protoctists

	Archaeprotista*		
Class	Karyomastigont	Kinetosome	Nucleus
Pelobiontids	+†	–	+
Metamonads	+	+	+/–
Parabasalids	+	+	+
Trichomonads	+	+	+/–
Hypermastigids	–	+	–
	Chlorophyta		
Genus	Karyomastigont	Kinetosome	Nucleus
Chlamydomonas	+	+	–
Chlorella			
Acetabularia	+	+	+
	Ciliophora		
Subphyla	Karyomastigont	Kinetosome	Nucleus
Postciliodesmatophora	–	+	+
Rabdophora	–	+	+
Cyrtophora	–	+	+
	Discomitochondria		
Class	Karyomastigont	Kinetosome	Nucleus
Amoebomastigotes	+	–	+/–
Kinetoplastids	–	–	–
Euglenids	–/?	+	–
Pseudociliates	–	+	–
	Granuloreticulosa		
Class	Karyomastigont	Kinetosome	Nucleus
Reticulomixids	–/+	–	+
Foraminiferans	+	–	+

us, the link is an evolutionary legacy, a remnant of the original archaebacterial-eubacterial connector. The modern organelles (i.e., centriole-kinetosomes, untethered nuclei, Golgi, and axostyles) derive from what first ensured genetic continuity of the chimera's components: the karyomastigont, a structure that would have been much more conspicuous to Proterozoic investigators than to us.

We thank our colleagues Ray Bradley, Michael Chapman, Floyd Craft, Kathryn Delisle (for figures), Ugo d'Ambrosio, Donna Reppard, Dennis Searcy, and Andrew Wier. We acknowledge research assistance from the University of Massachusetts Graduate School via Linda Slakey, Dean of Natural Science and Mathematics, from the Richard Lounsbery Foundation, and from the American Museum of Natural History Department of Invertebrates (New York). Our research is supported by National Aeronautics and Space Administration Space Sciences and Comision Interministerial de Ciencia y Tecnologia Project No. AMB98-0338 (to R.G.).

Hemimastigophora			
Genus	Karyomastigont	Kinetosome	Nucleus
Stereonema	–	+	+/–
Spironema	–	+	–
Hemimastix	–	+	–
Zoomastigota			
Class	Karyomastigont	Kinetosome	Nucleus
Jakobids	?	–	–
Bicosoecids	+/?	+	–
Proteromonads	+	–	–
Opalinids	–	+	+
Choanomastigotes	+	–	–

*Bold entries are protoctist phyla. All species of Archaeprotists lack mitochondria. "Karyomastigont," "kinetosome," and "nucleus," refer to relative proliferation of these organelles. Members of the phylum Archaeprotista group into one of three classes: Pelobiontid giant amoebae; Metamonads, which include three subclasses: Diplomonads (*Giardia*), Retortamonads (*Retortamonas*), and Oxymonads (such as *Pyrsonympha* and *Saccinobaculus*); and Parabasalia. The Class Parabasalia unites trichomonads, devescovinids, calonymphids, and hypermastigotes such as *Trichonympha*. The phylum Discomitochondria includes amoebomastigotes, kinetoplastids (*Trypanosoma*), euglenids, and pseudociliates (*Stephanopogon*). The Hemimastigophora comprise a new southern-hemisphere phylum of free-living mitochondriate protists (Foissner *et al.* 1988). Hemimastigophorans probably evolved from members of the kinetoplastid-euglenid taxon (Foissner and Foissner, 1993). If so, they represent a seventh example of release of the nucleus from the karyomastigont and subsequent kinetosome proliferation. The phylum Granuloreticulosa includes the shelled (Class Foraminifera) and unshelled (Class Reticulomyxa) foraminiferans. The phylum Zoomastigota includes five classes of single-celled, free-living and symbiotrophic mitochondriate protists; Jakobids, Bicosoecids, Proteromonads, Opalinids, and Choanomastigotes. Details of the biology are in the work by Margulis *et al.* (1993). A current phylogeny is depicted in Figure 2.
†Structure known but not demonstrated for all species at the electron microscopic level.

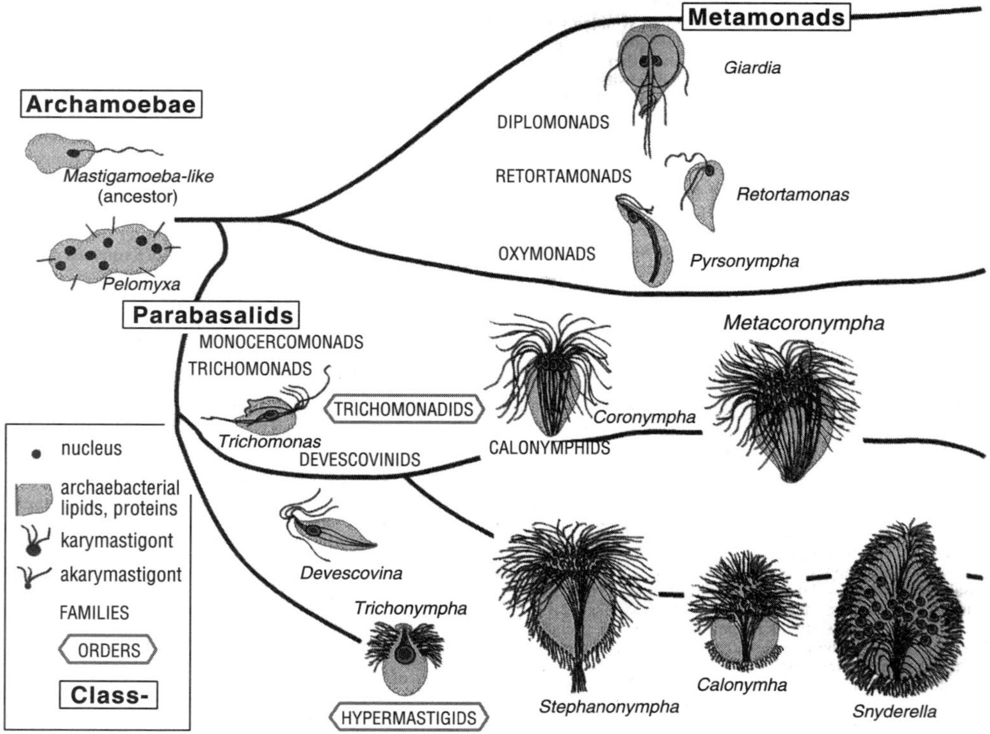

FIGURE 2. Biological phylogeny of chimeric eukaryotes taken to be primitively amitochondriate.

REFERENCES

Brugerolle, G. (1991) Flagellar and cytoskeletal systems in amitochondriate flagellates: Archamoeba, Metamonada and Parabasalia. *Protoplasma* 164, 70–90.

Chapman, M. Dolan, M. F. & Margulis, L. (2000) Centriole–kinetosomes: Form, function and evolution *Quart. Rev. Biol.* 75, 409-420.

Dacks, J. B. & Redfield, R. (1998) Phylogenetic placement of *Trichonympha. J. Euk. Microbiol.* 45, 445–447.

d'Ambrosio, U., Dolan, M., Wier, A. & Margulis, L. (1999) Devescovinid trichomonad with axostyle–based rotary motor ("Rubberneckia"): taxonomic assignment as *Caduceia versatilis* sp. nov. *Europ. J. Protistol.* 35, 327–337.

Delgado-Viscogliosi, P., Viscogliosi, E., Gerbod, D., Kuldo, J., Sogin, M. L. & Edgcomb, V. (2000) Molecular phylogeny of Parabasalids based on small subunit rRNA sequences, with emphasis on the Trichomonadinae subfamily. *J. Euk. Microbiol.* 47, 70–75.

Dolan, M. F., d'Ambrosio, U., Wier, A. & Margulis, L. (2000) Surface kinetosomes and disconnected nuclei of a calonymphid: ultrastructure and evolutionary significance of *Snyderella tabogae. Acta Protozool.* **39,** 135–141.

Dubinina, G. A., Leshcheva, N. V. & Grabovich, M. Yu. (1993a) The colorless sulfur bacterium *Thiodendron* is actually a symbiotic association of spirochetes and sulfidogens *Mikrobiologiya* 62, 717–732 (translated, Plenum Publ., pp. 432–444)

Dubinina, G. A., Grabovich, M. Yu. & Lesheva, N. V. (1993b) Occurrence, structure and metabolic activity of "*Thiodendron*" sulfur mats in various saltwater environments. *Mikrobiologiya* 62, 740–750 (translated, Plenum Publ., pp. 450–456)

Dyer, B. (1990) Phylum Zoomastigina Class Parabasalia. In *Handbook of Protoctista,* eds. Margulis, L., Corliss, J. O., Melkonian, M. & Chapman, D. J. (Jones and Bartlett. Boston), pp. 252–258.

Edgcomb, V., Viscogliosi, E., Simpson, A. G. B., Delgado-Viscogliosi, P., Roger, A. J. & Sogin, M. L. (1998) New insights into the phylogeny of trichomonads inferred from small subunit rRNA sequences. *Protist* 149, 359–366.

Foissner, W. Blatterer, H. & Foissner, I. (1988) The Hemimastigophora (*Hemimastix amphikineta* nov. gen., nov. sp.), a new protistan phylum from Gondwanian soils. *Europ. J. Protistol.* 23, 361–383.

Foissner, W. & Foissner, I. (1993) Revision of the Family Spironemidae Doflein (Protista, Hemimastigophora), with description of two new species, *Spironema terricola,* nov. sp. and *Stereonema geiseri* nov. gen., nov. sp. *J. Euk. Microbiol.* 40, 422–438.

Fuerst. J. A. & Webb, R. I. (1991) Membrane–bounded nucleoid in the eubacterium *Gemmata obscuriglobus. Proc. Natl. Acad. Sci. USA* 88, 8184–8188.

Germont, A., Philippe, H. & Le Guyader, H. (1996) Presence of a mitochondrial–type 70K Da heat shock protein in *Trichomonas vaginalis* suggest a very early mitochondrial endosymbiosis in eukaryotes. *Proc. Natl. Acad. Sci. USA* 93, 14614–14617.

Golding, G. B. & Gupta, R. S. (1995) Protein–based phylogenies support a chimeric origin for the eukaryotic genome. *Molec. Biol. Evol.* 12, 1–6.

Gupta, R. S. (1998a) Protein phylogenies and signature sequences: A reappraisal of evolutionary relationships among archaebacteria, eubacteria and eukaryotes. *Microbiol. Molec. Biol. Rev.* 62, 1435–1491.

Gupta, R. S. (1998b) What are archaebacteria: life's third domain or monoderm prokaryotes related to Gram–positive bacteria? A new proposal for the classification of prokaryotic organisms. *Molec. Microbiol.* 29, 695–708.

Gupta, R. S. (1998c) Life's third domain (Archaea): an established fact or an endangered paradigm? A new proposal for classification of organisms based on protein sequences and cell structure. *Theor. Popul. Biol.* 54, 91–104.

Hall, J. & Luck, D. J. L. (1995) Basal body–centriolar DNA: in situ studies on *Chlamydomonas reinhardtii. Proc. Natl. Acad. Sci. USA* 92, 5129–5133.

Kirby, H. (1949) Systematic differentiation and evolution of flagellates in termites. *Revista de la Sociedad Mexicana de Historia Natural* 10, 57–79.

Kirby, H. & Margulis, L. (1994) Harold Kirby's symbionts of termites: karyomastigont reproduction and calonymphid taxonomy. *Symbiosis* 16, 7–63.

Janicki, C. (1915) Unterschugen an parasitichen Flagellaten. II. Teil: Die Gattungen *Devescovina, Parajoenina, Stephanonympha, Calonympha.* Ueber den Parabasalapparat. Ueber Kernkonstitution und Kernteilung. *Zeitschrift wissen Zoologie* 112, 573–691.

Margulis, L. (1993) *Symbiosis in Cell Evolution* (W. H. Freeman, New York), 2nd Ed.

Margulis, L., McKhann, H. I. & Olendzenski, L., eds. (1993) *Illustrated Glossary of the Protoctista* (Jones and Bartlett Publishers, Sudbury, MA.).

Margulis, L. (1996) Archaeal–eubacterial mergers in the origin of Eukarya: Phylogenetic classification of life. *Proc. Natl. Acad. Sci. USA* 93, 1071–1076.

Margulis, L. & Schwartz, K. V. (1998) *Five Kingdoms, An illustrated Guide to the Phyla of Life on Earth* (W. H. Freeman, New York), 3rd Ed.

Martin, W. & Müller, M. (1998) The hydrogen hypothesis for the first eukaryote. *Nature* 392, 37–41.

Mayr, E. (1998) Two Empires or three? *Proc. Natl. Acad. Sci. USA* 95, 9720–9723.

Raff, J. W. (1999) Nuclear migration: The missing L(UNC)? *Curr. Biol.* 9, R708–R710.

Roger, A. J., Clark, C. G. & Doolittle, W. M. (1996) A possible mitochondrial gene in the early-branching amitochondriate protist *Trichomonas vaginalis. Proc. Natl. Acad. Sci. USA* 93, 14618–14677.

Roger, A. J., Srard, S. G., Tovar, J., Clark, C. G., Smith, M. W., Gillin, F. D. & Sogin, M. L. (1998) A mitochondrial-link chaperonin 60 gene in *Giardia lamblia*: Evidence that diplomonads once harbored an endosymbiont related to the progenitor of mitochondria *Proc. Natl. Acad. Sci. USA* 95, 229–234.

Sapp J. (1999) Free–wheeling centrioles. *History and Philosophy of the Life Sciences* 20, 3–38.

Scamardella, J. M. (1999) Protist, Protozoa and Protoctista: Not plants or animals: A brief history of the origin of Kingdoms Protozoa, Protista and Protoctista. *Internatl. Microbiol.* 2, 207–216.

Searcy, D. G. (1992) Origins of mitochondria and chloroplasts from sulfur–based symbioses. In *The Origin and Evolution of the Cell*, eds. Hartman, H. & Matsuno, K. (World Scientific, Singapore).

Searcy, D. G. & Delange, R. J. (1980) *Thermoplasma acidophilum* histone–like protein: partial amino acid sequence suggestive of homology to eukaryotic histones. *Biochim. Biophys. Acta* 609, 197–200.

Searcy, D. & Hixon, W. G. (1994) Cytoskeletal origins in sulfur-metabolizing archaebacteria. *BioSystems* 10, 19–28.

Searcy, D. & Lee, S. H. (1998) Sulfur reduction by human erythrocytes. *J. Exp. Zool.* 282, 310–322.

Sogin, M. L. (1997) History assignment: When was the mitochondrion founded? *Current Opinion in Genetics and Development* 7, 792–799.

Woese, C. R., Kandler, O. & Wheelis, M. L. (1990) Towards a natural system of organisms: Proposals for the domains Archaea, Bacteria, and Eukarya. *Proc. Natl. Acad. Sci. USA* 87, 4576–4579.

Woese, C. R. (1998) Default taxonomy: Ernst Mayr's view of the microbial world. *Proc. Natl. Acad. Sci. USA* 95, 11043–11046.

4
Dynamic Evolution of Plant Mitochondrial Genomes: Mobile Genes and Introns and Highly Variable Mutation Rates

JEFFREY D. PALMER, KEITH L. ADAMS, YANGRAE CHO[†], CHRISTOPHER L. PARKINSON[‡], YIN-LONG QIU[§], AND KEMING SONG[¶]

We summarize our recent studies showing that angiosperm mitochondrial (mt) genomes have experienced remarkably high rates of gene loss and concomitant transfer to the nucleus and of intron acquisition by horizontal transfer. Moreover, we find substantial lineage-specific variation in rates of these structural mutations and also point mutations. These findings mostly arise from a Southern blot survey of gene and intron distribution in 281 diverse angiosperms. These blots reveal numerous losses of mt ribosomal protein genes but, with one exception, only rare loss of respiratory genes. Some lineages of angiosperms have kept all of their mt ribosomal protein genes whereas others have lost

Department of Biology, Indiana University, Bloomington, IN 47405

This paper was presented at the National Academy of Sciences colloquium "Variation and Evolution in Plants and Microorganisms: Toward a New Synthesis 50 Years After Stebbins," held January 27–29, 2000, at the Arnold and Mabel Beckman Center in Irvine, CA.

Abbreviation: mt, mitochondrial.

[†]Present address: Stanford Genome Center, 855 California Avenue, Palo Alto, CA 94304.

[‡]Present address: Department of Biology, University of Central Florida, Orlando, FL 32816-2368.

[§]Present address: Department of Biology, University of Massachusetts, Amherst, MA 01003-5810.

[¶]Present Address: Sigma Chemical Company, 3300 South Second Street, St. Louis, MO 63118.

most of them. These many losses appear to reflect remarkably high (and variable) rates of functional transfer of mt ribosomal protein genes to the nucleus in angiosperms. The recent transfer of cox2 to the nucleus in legumes provides both an example of interorganellar gene transfer in action and a starting point for discussion of the roles of mechanistic and selective forces in determining the distribution of genetic labor between organellar and nuclear genomes. Plant mt genomes also acquire sequences by horizontal transfer. A striking example of this is a homing group I intron in the mt cox1 gene. This extraordinarily invasive mobile element has probably been acquired over 1,000 times separately during angiosperm evolution via a recent wave of cross-species horizontal transfers. Finally, whereas all previously examined angiosperm mtDNAs have low rates of synonymous substitutions, mtDNAs of two distantly related angiosperms have highly accelerated substitution rates.

The evolutionary dynamics of plant mitochondrial (mt) genomes have long been known to be unusual compared with those of animals and most other eukaryotes at both the sequence level (exceptionally low rate of point mutations) and structural level (high rates of rearrangement, duplication, genome growth and shrinkage, and incorporation of foreign DNA). The rate of synonymous substitutions (a useful approximation of the neutral point mutation rate) was shown in the 1980s to be lower in angiosperm mitochondria than in any other characterized genome, and fully 50–100 times lower than in vertebrate mitochondria (Wolfe *et al.*, 1987; Palmer and Herbon, 1988). This gulf largely persists despite the more recent discovery of modest substitutional rate heterogeneity within angiosperms (Eyre-Walker and Gaut, 1997; Laroche *et al.*, 1997) and vertebrates (Martin *et al.*, 1992; Waddell *et al.*, 1999).

Angiosperms have by far the largest mtDNAs, at least 200 kb to over 2,000 kb in size (larger than some bacterial genomes) (Palmer 1990, 1992). These genomes grow and shrink relatively rapidly; for example, within the cucumber family, mt genome size varies by more than six-fold (Ward *et al.*, 1981). Plant mitochondria rival the eukaryotic nucleus (especially the plant nucleus) in terms of the C-value paradox they present: i.e., larger plant mt genomes do not appear to contain more genes than smaller ones, but simply have more spacer DNA (intron content and size also do not vary significantly across angiosperms). This paradox extends to plant/animal comparisons. For example, the one sequenced angiosperm mt genome (from *Arabidopsis*; Unseld *et al.*, 1997; Marienfeld *et al.*, 1999) is 367 kb in size yet contains only one more RNA gene and twice the number of protein genes (27 vs. 13) than our own mt genome, which is over 20 times smaller (16.6 kb). Angiosperm mtDNAs are large in part because of frequent duplications. These most commonly result in small (2–10 members) repeat families whose elements range up to a few hundred base pairs in size, although large duplications and triplications of up to 20 kb are almost always found at least once and sometimes several times within a genome. Angiosperm mtDNAs also grow promiscuously via the relatively frequent capture of sequences from the chloroplast and nucleus (Palmer, 1992; Unseld *et al.*, 1997; Marienfeld *et al.*, 1999). The functional significance of this foreign DNA seems entirely limited to chloroplast-derived tRNA genes, which provide many of the tRNAs used in plant mt protein synthesis (Miyata *et al.*, 1998).

Recombination between the small and large repeats scattered throughout angiosperm mtDNAs creates a very dynamic genome structure, both evolutionarily and in real time. Recombination between repeats of about 2 kb and larger is so frequent as to create a dynamic equilibrium in which an individual plant's mtDNA exists as a nearly equimolar mixture of recombinational isomers differing only in the relative orientation of the single copy sequences flanking the rapidly recombining repeats (Palmer, 1990; Mackenzie *et al.*, 1994). Plants such as maize, with many different sets of these large, usually direct "recombination repeats" somehow manage to perpetuate their mt genomes despite their dissolution into a bewildering complexity of subgenomic molecules via repeat-mediated deletion events (Mackenzie *et al.*, 1994; Fauron *et al.*, 1995). Recombination between smaller repeats appears to occur less frequently, although perhaps frequently enough to help maintain a reservoir of low-level, rearranged forms of the genome (termed "sublimons") that persist together with the main mt genome. On an evolutionary time-scale, recombination between short dispersed inverted repeats generates large inversions frequently enough to scramble gene order almost completely, even among relatively close members of the same genus (Palmer and Herbon, 1988; Fauron *et al.*, 1995). The combined forces of frequent duplication and inversion have led to the fairly common creation of novel, chimeric genes in plant mitochondria. A number of these chimeric genes lead to cytoplasmic-nuclear incompatibilities manifest as cytoplasmic male sterility (Hanson, 1991; Mackenzie *et al.*, 1994).

The above picture, encapsulated as the title of a 1988 paper (Palmer and Herbon, 1988)—"Plant mitochondrial DNA evolves rapidly in structure, but slowly in sequence"—with its corollary that animal mtDNA evolves oppositely in all respects, was largely complete by the late 1980s (Palmer, 1990). This picture was derived from comparison, at both fine- and broad-scale taxonomic levels, of a relatively small number of angiosperms, belonging primarily to but five economically important families [the crucifers (Brassicaceae), the cucurbits (Cucurbitaceae), the legumes (Fabaceae), the grasses (Poaceae), and the nightshades (Solanaceae)], and with two genera (*Brassica* and *Zea*) serving as exemplars (Palmer and Herbon, 1988; Fauron *et al.*, 1995). The completion of the *Arabidopsis* mt genome sequence in 1997 (Unseld *et al.*, 1997; Marienfeld *et al.*, 1999) gave a more fine-grained and comprehensive picture of the extent to which foreign DNA uptake and internal duplication have influenced the size, structure, and evolutionary potential of a particular mt genome, but without significantly changing prior notions of the structural dynamics of plant mt genomes.

A few years ago, we set up a large-scale Southern blot survey, of 281 diverse angiosperms, to better elucidate the evolution of two fundamental classes of mt features—their content of genes and of introns—that were poorly characterized relative to the traits whose evolution is described above. Our hope was that by surveying hundreds of diverse plants, we could discern the basic tempo and pattern of gene and intron loss and gain and, furthermore, identify especially attractive candidates for follow-up study to learn about the processes and mechanisms underlying these kinds of structural changes. An additional, utilitarian goal, whose achievements (see, e.g., Qiu *et al.*, 1998) will not be described in this report, was to use these presence/absence characters to help unravel the phylogeny of angiosperms and other land plants. This paper will summarize our results, both recently published and unpublished, on gene content and intron content evolution in angiosperm evolution. We show that rates of gene loss, of accompanying gene transfer to the nucleus, and of intron acquisition by cross-species horizontal transfer can be remarkably high for particular classes of these genetic elements, and that these rates also vary substantially across lineages of flowering plants. As a completely unexpected bonus, these surveys have also led to the discovery of two exceptional groups of plants with vastly elevated rates of synonymous substitutions.

SOUTHERN BLOT SURVEY FOR CHANGES IN MT GENE AND INTRON CONTENT

We went to some lengths to sample angiosperm diversity, extracting total DNAs from 281 species that represent 278 genera and 169 families of angiosperms (species listed at www.bio.indiana.edu/~palmerlab). Twelve sets of pseudoreplicate filter-blots were made, each set containing one digest (with either *Bam*HI or *Hin*dIII) of each of the 281 DNAs. The digested DNAs were arranged according to the presumptive phylogenetic relationships of their cognate plants as understood about 5 years ago. To date, the sets of blots have been sequentially hybridized with nearly 100 different probes, mostly for segments of various mt genes and introns, but also for several chloroplast genes and introns.

Virtually all mt genes and introns tested hybridized well across the full spectrum of angiosperms examined, and some even hybridized well across additional blots containing the full diversity of land plants, a roughly 450-million-year time span (see, e.g., Qiu *et al.*, 1998). The success of these hybridizations, carried out at moderate stringencies [washes at 60°C in 2x standard saline citrate (SSC)/0.1% SDS], across such large timespans testifies to the very low substitution rates of the great majority of plant mtDNAs. A probe sequence was inferred to be absent from the mt genome of a particular filter-bound preparation of total DNA if there was no detectable hybridization on an overexposed autoradiograph against two layers of controls: good hybridization to the DNA in question using other mt probes and good hybridization to other DNAs with the probe in question.

Fig. 1 shows examples of the three general categories of results obtained with the various mt probes used. Many probes, such as rRNA probes and the *cox1* exon probe used in the middle panel of Fig. 1, hybridized strongly to essentially all DNAs tested; i.e., the lane-to-lane variations in hybridization intensity were reproducible across all probes in this category. We interpret these variations as primarily reflecting differences in amount of mtDNA loaded per lane, and conclude that each mtDNA probably contains an intact copy of the sequence probed (see the penultimate section for an explanation of the weak *cox1* hybridization in lane 4 of the middle panel). Many other probes, such as the *rps7* gene probe used in the top panel of Fig. 1, while hybridizing strongly to many lanes

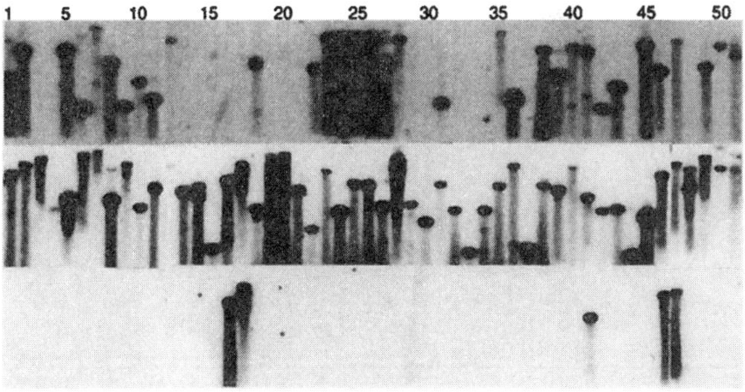

FIGURE 1. Southern blot survey illustrating three distinct presence/absence patterns of mitochondrial genes and introns. BamHI-cut total DNAs from 51 of 281 angiosperms surveyed were arranged according to presumptive phylogenetic relationships, were electrophoresed, and were blotted and hybridized with probes internal to the rps7 gene (Upper; gene mostly present, with several losses evident), the cox1 coding sequence (Middle; gene universally present), and the cox1 intron (Lower; intron rarely present, each presence thought to reflect an independent acquisition by horizontal transfer).

(normalizing intensities to the category 1 results described above), hybridized at best only very weakly (again normalizing) and often not at all to many other lanes (i.e., lanes 3–4, 13–17, 19–21, 29–30, 32–34, 37, 44, 48). We conclude that most or all of the *rps7* probe region is probably missing from these mt genomes. The third category of results was obtained with but a single probe, the *cox1* intron shown in Fig. 1 *Bottom*. This probe gave a singularly patchy, sporadic hybridization pattern, whose molecular basis will be explained in a later section.

RIBOSOMAL PROTEIN GENES ARE LOST FREQUENTLY, RESPIRATORY GENES ONLY RARELY

The survey blots were hybridized with probes for each of the 14 ribosomal protein genes known from angiosperm mt genomes, and with probes for 11 of the 21 known respiratory genes. The cases of inferred gene absences were plotted onto a multigene phylogeny of the surveyed angiosperms (Soltis *et al.*, 1999) to estimate the number of phylogenetically separate gene losses. A total of only two losses were inferred among 10 of the respiratory genes; these include the previously described loss of mt *cox2* in the legume *Vigna* (Nugent and Palmer, 1991, 1993; Covello and Gray, 1992; Adams *et al.*, 1999) and the loss of *nad3* in the Piperaceae. The small respiratory gene *sdh4* was an exception, with about 10 separate losses inferred (K.A., Y.-L.Q., and J.D.P., unpublished data). In striking contrast to the respiratory genes as a group, probes for all 14 ribosomal protein gene probes failed to hybridize to mtDNAs of many, disparately related angiosperms (see, e.g., Fig. 1 *Upper*), suggesting numerous gene losses (at least 10 losses for most genes, over 200 losses in total). The losses vary in phylogenetic depth, with most being limited to one or two related families, whereas several encompass many related families or even orders (K.A., Y.-L.Q., and J.D.P., unpublished data). Probes for *rps2* and *rps11* did not hybridize to the lanes of most higher eudicots (a group comprising 182 of the 281 angiosperms in our survey), suggesting relatively ancient losses early in the evolution of eudicots. Both *rps2* and *rps11* have been isolated from the nucleus of *Arabidopsis* (Perrotta *et al.*, 1998; *rps11* expressed sequence tags from the GenBank database), suggesting that gene loss followed functional transfer to the nucleus. The relatively few losses of these two genes (4 and 6, respectively) reflect the reduced potential for many ("subsequent") losses when such an ancient loss occurs.

Our blot surveys will not detect mt pseudogenes unless much or all of the probe region is missing, and thus our survey probably underestimates the number of gene losses. Several ribosomal protein pseudogenes have been reported in angiosperm mt genomes, for example, of *rps14* and *rps19* in *Arabidopsis* (Sanchez *et al.*, 1996; Unseld *et al.*, 1997; Marienfeld *et al.*, 1999). In some cases, weak hybridization by a probe was observed to DNA from a species known to contain only a fragment of a gene in the mitochondrion (e.g., *rps12* in *Oenothera*).

The hybridization results show that ribosomal protein gene content in angiosperm mitochondrial genomes varies considerably, as suggested previously (Nugent and Palmer, 1993; Unseld *et al.*, 1997), although the magnitude of the variation and frequency of gene loss are unexpectedly high. The high number of ribosomal protein gene losses compared with the low number of respiratory gene losses is also striking. In some angiosperm lineages, the rate of ribosomal protein gene loss appears to be comparable to, or even higher than, the silent substitution rate (K.A., Y.-L.Q., and J.D.P., unpublished work), whereas in many other lineages, including the most ancient ones, there has been no loss at all. These latter lineages have retained all 14 ribosomal proteins that were present in the common ancestor of angiosperms, whereas several fairly recently arisen angiosperm lineages have lost 10 or more of the 14 genes (in one case, apparently all 14). The rate of ribosomal protein gene loss thus varies enormously across angiosperms lineages; some factor(s) must have triggered a rapid rate of loss in certain recent lineages.

The mt gene losses detected by our survey could be explained by functional transfer of the gene to the nucleus, by functional substitution by another protein (see, e.g., Sanchez *et al.*, 1996), or by the protein being dispensable in certain plants. Six ribosomal protein genes (Grohmann *et al.*, 1992; Wischmann and Schuster, 1995; Kadowaki *et al.*, 1996; Sanchez *et al.*, 1996; Perrotta *et al.*, 1998; Figueroa *et al.*, 1999a; Kubo *et al.*, 1999) that are present in the mitochondrion of many flowering plants, along with the respiratory gene *cox2* (Nugent and Palmer, 1991, 1993; Covello and Gray, 1992; Adams *et al.*, 1999), have been reported to have been transferred to the nucleus. Thus, the most likely explanation for loss of a gene from the mitochondrion is transfer to the nucleus.

DO MULTIPLE GENE LOSSES REFLECT MULTIPLE GENE TRANSFERS?

Assuming that most of the genes lost from the mitochondrion have been transferred to the nucleus, then the many separate losses of each mt ribosomal protein gene and of *sdh4* could reflect either an equivalent number of separate transfers, each more or less coincident in time with the loss, or a smaller number of earlier transfers (as few as one for each gene), with several or all losses stemming from the same ancestral transfer of a particular gene. The latter, early-transfer/multiple-dependent-loss model predicts a prolonged period of retention of dual intact and expressed genes in both compartments after gene transfer. This model seems inconsistent with both theory (Marshall *et al.*, 1994; Herrman, 1997) and with empirical results (see the next section) indicating fairly rapid loss of one compartment's gene or the other after gene transfer/duplication. On the other hand, the one-loss = one-transfer model and other multiple transfer models seem unlikely considering the complex series of events required for each successful functional transfer (i.e., reverse transcription, movement to the nucleus, chromosomal integration, and functional activation, which in almost all cases requires acquisition of sequences conferring both proper expression and also targeting of the now cytoplasmically synthesized protein to the mitochondrion).

Nuclear sequences for each of three mitochondrially derived ribosomal protein genes have been reported from two separate lineages of mt gene loss as defined by our blot survey, while we have been studying the transferred *rps10* gene in a number of angiosperms. These genes provide a useful starting point for investigating the number and timing of gene transfers during angiosperm evolution. In the *rps14* loss lineage that includes maize and rice, the transferred gene is located within an intron of the *sdh2* gene, and the *sdh2* targeting sequence is alternatively spliced to *rps14* transcripts (Figueroa *et al.*, 1999a; Kubo *et al.*, 1999), whereas nuclear *rps14* in *Arabidopsis* shares none of these features (Figueroa *et al.*, 1999b). Rice *rps11* was duplicated in the nucleus after transfer but before targeting sequence acquisition (Kadowaki *et al.*, 1996), with the two *rps11* genes having acquired their targeting sequences from two different nuclear genes for mt proteins (*atpB* and *coxVb*), whereas the targeting sequence of pea *rps11* (Kubo *et al.*, 1998) has no similarity to any sequences currently in the databases. Finally, nuclear *rps19* genes of *Arabidopsis* (Sanchez *et al.*, 1996) and soybean (expressed sequence tags in the GenBank database), both rosids, also have unrelated targeting sequences and structures. The dissimilar structural features of members of each pair of these three genes strongly suggests that each was derived from a separate activation event. Because activation probably occurs relatively soon after transfer, before the nuclear gene is permanently disabled by mutation (Thorsness and Weber, 1996; Adams *et al.*, 1999; see next section), we think that these ribosomal protein genes were not only independently activated but also independently transferred to the nucleus. Considering that *rps14*, *rps11*, and *rps19* have been lost from the mt genomes of many different angiosperm lineages, as revealed by our Southern blot survey, it is possible that each gene has been independently transferred many times, not just twice. Indeed, we have recently obtained evidence for many, recent independent transfers of *rps10*, which has been lost from the mt genome over 20 times among the 281 angiosperms surveyed by blots (K.A., D. Daley, Y.-L.Q., J. Whelan, and J.D.P., unpublished work). It is increasingly evident, therefore, that functional transfer of mt genes to the nucleus occurs at a surprisingly high frequency in angiosperm evolution, especially, it would appear, in some groups of flowering plants (see preceding section).

MITOCHONDRIAL GENE TRANSFER IN ACTION: RECENT TRANSFER OF *COX2* TO THE NUCLEUS IN LEGUMES

The most extensively studied example of recent mt gene transfer in flowering plants (or any group of eukaryotes) is the cytochrome oxidase subunit 2 gene (*cox2*) in legumes (Nugent and Palmer, 1991, 1993; Covello and Gray, 1992; Adams *et al.*, 1999). *Cox2*, present in the mitochondrion of virtually all plants, was transferred to the nucleus during recent legume evolution (Adams *et al.*, 1999). Examination of nuclear and mt *cox2* presence and expression in over 25 legume genera has revealed a wide range of intermediate stages in the process of mt gene transfer, providing a portrayal of the gene transfer process in action (Fig. 2; Adams *et al.*, 1999). *Cox2* was transferred to the nucleus via an edited RNA intermediate (Nugent and Palmer, 1991; Covello and Gray, 1992). Once nuclear *cox2* was activated, a state of dual intact and expressed genes—of transcompartmental functional redundancy—was established; this transition stage persists most fully (i.e., with both compartments' *cox2* genes highly expressed in terms of steady-stated, properly processed RNAs; COX2 protein levels have not been assayed) only in *Dumasia* among the many studied legumes. Four other, phylogenetically disparate legumes also retain intact and expressed copies of *cox2* in both compartments, but with only one of the two genes expressed at a high enough level, in the one tissue type examined thus far, to presumably support respiration (Fig. 2). Silencing of either nuclear or mt *cox2* has occurred multiple times and in a variety of ways, including disruptive insertions or deletions, cessation of transcription or RNA editing, and partial to complete gene loss (Adams *et al.*, 1999). Based on phylogenetic evidence, we infer that mt *cox2* and nuclear *cox2* have been silenced approximately three to five times each during the evolution of the studied legumes (Fig. 2). Although *cox2* in legumes is the only known example of gene inactivation after recent transfer and activation in the nucleus, a comparative phylogenetic approach might reveal that the nuclear copy of other recently transferred organelle genes has become inactivated in one or more species related to the single plant studied so far.

ROLES OF SELECTION AND CHANCE IN MITOCHONDRIAL GENE TRANSFERS DURING ANGIOSPERM EVOLUTION

All but the last step (gene loss) in the complicated and evolutionarily unidirectional process by which mt genes move to the nucleus and disap-

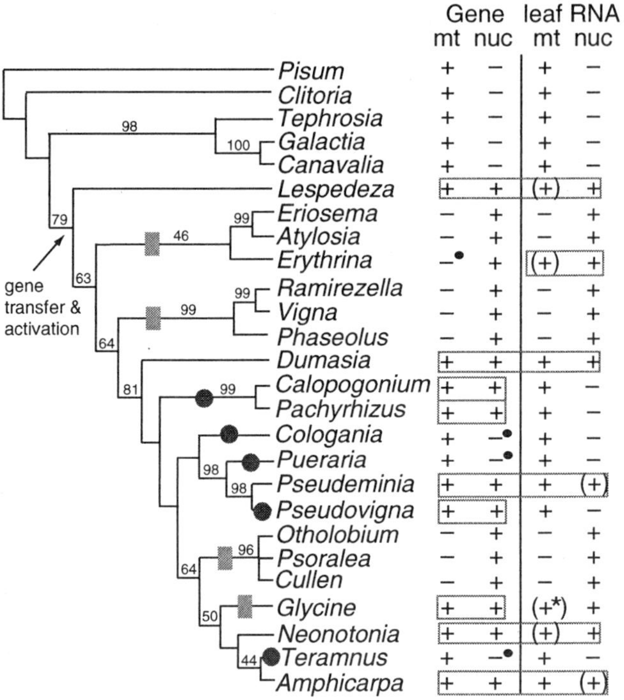

FIGURE 2. Summary of legume cox2 gene distribution and expression data in a phylogenetic context. The left two columns indicate the presence (+) or absence (–) of an intact cox2 gene in the mitochondrion (mt) or nucleus (nuc) of the indicated species. Bullets indicate genes containing small insertions or deletions that disrupt the reading frame or intron splicing. The right two columns indicate the presence (+) or absence (–) of detectable mitochondrial and nuclear cox2 transcripts in young leaves. Parentheses indicate transcripts present at low levels; the asterisk indicates transcripts that are not properly edited. Boxing highlights dual intact genes and/or dual transcription and proper processing (of dual cox2 genes, intact or not) in a given plant. Light rectangles and dark circles indicate loss or silencing of mt and nuclear cox2, respectively. The phylogenetic tree is one of three equally parsimonious trees obtained from parsimony analysis of a data set consisting of 2,154 bp of two chloroplast gene sequences (rbcL and ndhF) and 557 chloroplast restriction sites. Bootstrap values above 40% are shown. The figure is modified from Adams *et al.* (1999).

pear from the mitochondrion may be driven largely by mechanistic forces and chance mutations. These prior steps include reverse transcription (which could also occur after either of the next two steps), exit from the mitochondrion, entry into the nucleus, integration into the nuclear genome, gain of a nuclear promoter and other elements conferring properly regulated expression, and gain of a mt targeting sequence. Nucleic acids could escape from the mitochondrion by several mechanisms, as thoroughly discussed by Thorsness and Weber (1996). The rate of mtDNA escape and uptake by the nucleus has been estimated to be relatively high (Thorsness and Fox, 1990; Thorsness and Weber, 1996). Once inside the nucleus, nucleic acids can integrate into the nuclear genome by double-strand break repair, as shown recently for yeast (Ricchetti *et al.*, 1999), and perhaps by other mechanisms. Clues have been revealed as to the mechanisms of gene activation and targeting sequence acquisition, including gain of a targeting sequence (and perhaps upstream promoter and other regulatory elements) from a preexisting gene for a mt protein by a shuffling process (Kadowaki *et al.*, 1996; our unpublished data) and by integration into a preexisting gene for a mt protein (Figueroa *et al.*, 1999; Kubo *et al.*, 1999).

After the nuclear copy of a transferred gene is activated and gains a mt targeting sequence, both genes must be expressed at least transiently, as described above for *cox2* in legumes (Fig. 2; Adams *et al.*, 1999). It is possible that the genes in both genomes could become fixed. Both nuclear and mt *atp9* genes have been retained in *Neurospora crassa* (van den Boogaart *et al.*, 1982) and *Aspergillus nidulans* (Brown *et al.*, 1984; Ward and Turner, 1986), and both are functional at certain times during the life cycle of *Neurospora* (Bittner-Eddy *et al.*, 1994). However, the most commonly observed outcome is that one gene (usually the mt gene, although this is influenced by sampling biases) becomes silenced and lost. Inactivation of the nuclear copy of a transferred gene results in a failed transfer, but the opportunity for repeated "attempts" at transfer can create a gene transfer ratchet (Doolittle, 1998). Both selection and chance factors may play a role in determining which gene is retained and which gene is inactivated. Our finding of approximately equal numbers of *cox2* silencings in legume mt and nuclear genomes raises the possibility that it is largely random as to which gene became silenced in a given species, with disabling mutations inactivating either *cox2* gene at comparable frequencies. Alternatively, if the rate of production of disabling mutations is higher in one genome or the other, then the equal numbers of silencings would reflect selection favoring the gene's retention in the high-rate genome. This is difficult to assess because, although we know that substitution rates are much higher in legume nuclear than mt genes (Wolfe *et al.*, 1987), we do not know what the overall rate of disabling mutations is in either genome.

Several hypotheses have been proposed for selection favoring retention of the nuclear copy of a transferred and activated organelle gene and loss of the organellar copy. Presence of a gene in the nucleus allows for crossing over during meiosis, perhaps enabling beneficial mutations to be fixed more rapidly than in asexual organelle genomes (Blanchard and

Lynch, 2000). Alternatively, the progressive accumulation of detrimental mutations in asexual mt genomes by Muller's ratchet may favor transfer of genes to the nucleus. Evidence for Muller's ratchet has been found in tRNA genes of animal mitochondria (Lynch, 1996) and in the genomes of endosymbiotic bacteria (Moran, 1996). However, the rate of nucleotide substitutions is very low in plant mitochondria (about 10-fold lower than in the nucleus), which should counterbalance the effects of Muller's ratchet (Martin and Herrman, 1998; Race *et al.*, 1999) and negate (for plants) the hypothesis (Allen and Raven, 1996) that a nuclear location is favored because it provides relief from the effects of oxygen free radical damage incurred by organellar genes. Selection for a small, compact genome, although perhaps operating in other eukaryotes, is unlikely to be a factor favoring continued gene transfer in plants, because plant mt genomes readily incorporate and retain foreign DNA (Nugent and Palmer, 1988; Palmer, 1992; Unseld *et al.*, 1997; Cho *et al.*, 1998; Marienfeld *et al.*, 1999) and are very large and mostly noncoding (Ward *et al.*, 1981; Palmer, 1990, 1992; Unseld *et al.*, 1997; Marienfeld *et al.*, 1999). Finally, there is the possibility that genes for some organellar proteins may be better regulated in the nucleus (Thorsness and Weber, 1996). Although this possibility is intriguing, we are unaware of any evidence to support or refute it.

Why are a few protein genes preferentially retained by mt genomes across all or most eukaryotes? One view is that the products of these genes, all of which function in respiration, are highly hydrophobic and difficult to both import into mitochondria and properly insert (post-translationally) into the inner mt membrane (see, e.g., Popot and de Vitry, 1990; Thorsness and Weber, 1996; Palmer, 1997). Evidence for this includes experiments in which cytoplasmically synthesized cytochrome *b*, a highly hydrophobic protein with eight transmembrane helices, could not be imported in its entirety, with successful import limited to regions comprising only three to four transmembrane domains (Claros *et al.*, 1995). In general, genes whose products have many hydrophobic transmembrane domains are usually located in the mitochondrion whereas genes whose products have few such domains are more often transferred to the nucleus (Popot and de Vitry, 1990; Gray *et al.*, 1998). Indeed, the only two protein genes contained in all of the many completely sequenced mt genomes encode what are by some criteria (Claros *et al.*, 1995) the two most hydrophobic proteins present in the mitochondrion, cytochrome *b* and subunit 1 of cytochrome oxidase (Gray *et al.*, 1998; Gray, 1999). Although the hydrophobicity hypothesis cannot account for the distribution of every mt gene in every eukaryote, it seems likely to be a factor favoring the retention of certain respiratory genes. A second hypothesis for retention of certain genes in organelles is that their products are toxic when present in the cytosol or in some other, inappropriate cellular compartment to which they might be misrouted after cytosolic synthesis (Martin and Schnarrenberger, 1997). Although this toxicity hypothesis is eminently testable, we are unaware of any empirical evidence for it. A third hypothesis for organellar gene retention is to allow their expression to be directly and quickly regulated by the redox state of the organelle (Allen, 1993; reviewed in Race *et al.*, 1999). Evidence for redox regulation of organellar gene expression has been reported for chloroplasts (Pfannschmidt *et al.*, 1999) but not, to our knowledge, for mitochondria.

Although selective factors may be responsible for the transfer of some genes to the nucleus and the retention of others in the mitochondrion, chance factors may also be at work. At the stage of dual expression, whether the nuclear or mt copy of a transferred gene is retained may in some cases depend solely on the roll of the evolutionary dice—on which gene first sustains a gene-inactivating mutation, or a mutation that is either deleterious or beneficial to the gene product's function. In the latter two cases, selection would be involved in the sense that it would act to either fix the gene with the beneficial mutation or eliminate the gene with the deleterious mutation.

We are left with a picture of organelle gene transfer as a complex, historically contingent process whose outcome undoubtedly depends on a combination of mechanistically driven factors and chance mutations, together with selective forces. The process seems to be driven by the high rate of physical duplication of organelle genes into the nucleus (which appears to be true for all eukaryotes, regardless of whether functional gene transfer is still occurring), and proceeds seemingly exclusively in one direction: from organelles to nucleus. Indeed, with one disputed exception, a *mutS* homolog in coral mt DNA (Pont-Kingdon *et al.*, 1995, 1998), there are no examples known of the reverse process, of functional genes moving from the nucleus to the mitochondrion or chloroplast.

Why has gene transfer been so pervasively unidirectional? Flowering plant mitochondria are certainly able to accept foreign sequences: Numerous examples are known of the uptake of chloroplast DNA (Nugent and Palmer, 1988; Palmer, 1992; Unseld *et al.*, 1997; Marienfeld *et al.*, 1999), nuclear DNA (Knoop *et al.*, 1996; Unseld *et al.*, 1997; Marienfeld *et al.*, 1999), and sequences from other organisms (Vaughn *et al.*, 1995; Cho *et al.*, 1998; see below), and a few chloroplast-derived genes are expressed in the mitochondrion (Joyce and Gray, 1989; Kanno *et al.*, 1997; Miyata *et al.*, 1998). Nonetheless, the initial driving force (the rate of physical transfer/duplication of sequences from one genome into the other) may be much stronger toward the nucleus than in the reverse direction; certainly this seems to be the case for yeast by several orders of magnitude (Thorsness and Fox, 1990; Thorsness and Weber, 1996). Compounding this, each mt gene physically transferred to the nucleus can potentially result in functional transfer whereas only a small fraction of nuclear genes could play a useful role if transferred to the mitochondrion. The pervasively unidirectional flow of mt genes to the nucleus may, therefore, be driven largely, perhaps even entirely, by a huge imbalance in the relative likelihood of gene movement and potential functionality in one direction versus the other.

EXPLOSIVE INVASION OF PLANT MITOCHONDRIA BY A GROUP I INTRON

Thus far, we have discussed intracellular horizontal evolution entirely as a means of relocating plant mt genes to the nucleus. As mentioned in the introduction, plant mt genomes are also well known to acquire foreign sequences by intracellular gene transfer, from both the chloroplast and nucleus. We have recently described (Cho *et al.*, 1998; Cho and Palmer, 1999), and will briefly review here, a case of horizontal evolution that stands out in three respects: (*i*) It is the first case of cross-species acquisition of DNA by plant mt genomes; (*ii*) it is unparalleled with respect to how frequently the same piece of DNA has been acquired, over and over again, during angiosperm evolution; and (*iii*) all of these many invasions have occurred very recently, as an explosive wave within the last 10 million years or so.

The piece of DNA in question here is a homing group I intron. These introns encode site-specific endonucleases with relatively long target sites that catalyze their efficient spread from intron-containing to intron-lacking alleles of the same gene in genetic crosses. A few cases of the evolutionary spread of these introns by horizontal homing between species were known when, in 1995, we in collaboration with Jack Vaughn's group reported (Vaughn *et al.*, 1995) that the angiosperm *Peperomia* had acquired, quite recently (Adams *et al.*, 1998), a group I intron in its mt *cox1* gene by long-distance horizontal transfer, most likely from a fungus. We subsequently discovered a closely related form of this intron, located at the same position in *cox1*, in a very distantly related angiosperm, *Veronica*. This stimulated us to use the *Veronica* intron as a probe against our survey blots of 281 angiosperm DNAs. As shown in Fig. 1 *Lower*, the intron probe hybridized strongly to relatively few DNAs, in an unusually patchy manner phylogenetically [and always to the same band as a *cox1* exon probe (Fig. 1 *Middle*), indicating that the hybridizing region is always located in the same gene]. All told, 48 of the 281 angiosperm DNAs on the blots, scattered across most of the major groups represented, hybridized well to the intron (Cho *et al.*, 1998).

The exceptionally patchy phylogenetic distribution of the intron (see Fig. 2 of Cho *et al.*, 1998) caused us to sequence the intron from 30 diverse intron-containing taxa. We then compared the congruence of intron (Fig. 3*A*) and "organismal" (Fig. 3*B*) phylogenies to assess the relative contributions of vertical and horizontal transmission to the intron's evolutionary history in angiosperms. These phylogenies are highly incongruent. From this, we concluded that the intron had been independently acquired, by cross-species horizontal transfer, many times separately among the examined plants. For example, consider the closely related rosids *Bursera* and *Melia*, whose intron-hybridizing DNAs are in adjacent lanes in Fig. 1 (lanes 46 and 47; recall that DNAs are arranged according to presumptive phylogenetic order in these blots) and which group with 100% bootstrap support in the organismal tree of Fig. 3*B*. Their *cox1* intron sequences do not, however, group together (Fig. 3*A*), suggesting that *Bursera* and *Melia* acquired their introns independently of one another. Three, more convincing pairs of examples of phylogenetic evidence for independent acquisition consist of *Ilex*/*Hydrocotyle*, *Symplocus*/*Diospyros*, and *Maranta*/*Hedychium*. Each pair again receives 100% bootstrap support in Fig. 3*B*, and in each case the two members of the pair are now separated by multiple, well supported nodes in the intron tree (Fig. 3*A*).

All told, we inferred at least 32 separate cases of intron gain to account for the intron's presence in the 48 angiosperms revealed to contain the intron by the 281-taxa Southern blot survey (Cho *et al.*, 1998). Some 25 of these cases are marked on Fig. 3*A* by plus signs, while 7 additional gains were inferred by criterion *ii*, as we shall now describe. Overall, the inferences of independent intron gain were based on four criteria: (*i*) the many incongruencies, some strongly supported, some less so, between intron and organismal phylogenies (Fig. 3); (*ii*) the highly disjunct phylogenetic distribution of intron-containing plants; (*iii*) different lengths of co-conversion tracts among otherwise related introns (Fig. 3); and (*iv*) the existence of ancestrally intron-lacking taxa within families containing the intron. This last form of evidence also relates to co-conversion, the process by which donor exonic sequences flanking the intron replace recipient exonic sequences when the intron is inserted into the cox1 gene. Space limitations preclude any meaningful discussion of the complicated logic behind the two criteria that are largely or entirely based on co-conversion tract evidence; the interested reader is instead referred to Cho *et al.* (1998) and Cho and Palmer (1999).

More extensive sampling within the monocot family Araceae showed that 6 of the 14 Araceae sampled contain the intron and that these 6 taxa probably acquired their introns by at least 3 and quite possibly 5 separate horizontal transfers (Cho *et al.*, 1998; Cho and Palmer, 1999). In addition, unpublished studies from our lab and that of Claude dePamphilis reveal many more cases of independent gain of this promiscuous group I intron. Given that we have still sampled only a tiny fraction of the >300,000

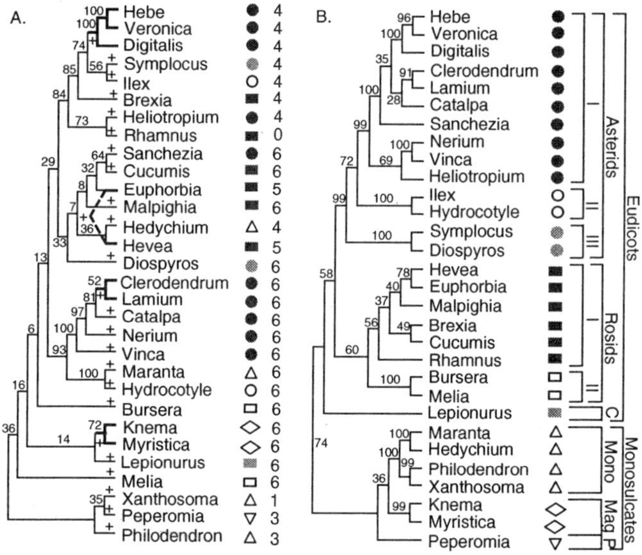

FIGURE 3. Phylogenetic evidence for horizontal transfer of cox1 introns. (A) Maximum-likelihood tree of 30 angiosperm cox1 introns. Numbers on the tree are bootstrap values. Plus signs on the tree mark 25 inferred gains of the intron among these taxa. Symbols to the immediate right of names are as in Fig. 3B. Numbers at far right indicate number of 3′-flanking nucleotides changed by co-conversion (see text). Bold branches mark four small clades of introns thought to have originated from the same intron gain event. (B) Organismal tree from a maximum-likelihood analysis of a combined data set of chloroplast rbcL and mt cox1 coding sequences. Numbers are bootstrap values. Symbols mark the nine major groups of angiosperms represented in this analysis. The figure is modified from Cho *et al.* (1998).

species of angiosperms, we are confident that the intron has been horizontally acquired at least hundreds of times during angiosperm evolution and probably over 1,000 times. Equally remarkably, all of these transfers seem to have occurred very recently, in the last 10 million years or so of angiosperm evolution.

Many fascinating questions can be asked about the evolution of this wildly invasive group I intron. What does the inevitably complex historical network of horizontal transfers look like; i.e., who is the donor and who is the recipient in each specific instance of intron transfer? Phylogenetic evidence suggests at least one, perhaps initiating long-distance transfer of the intron from a fungus to a flowering plant (Cho *et al.*, 1998). Have many or most of the transfers occurred via this long-distance route, in which case all of the fungi must themselves be closely related [because all of the plant introns are (Cho *et al.*, 1998)]? Or have most transfers, perhaps all but the first, occurred via "short"-distance transfer, i.e., from angiosperm to angiosperm? These two alternative models make contrasting phylogenetic predictions as elaborated elsewhere (Cho *et al.*, 1998). Has transfer, especially if largely plant-to-plant, been mediated by vectoring agents, and if so which ones (e.g., viruses, bacteria, aphids, mycorrhizal fungi, etc.)? Or has it occurred by transformation-like uptake of DNA from the environment or by the occasional direct fusion (perhaps pollen-mediated) of two unrelated plants? To some extent, the answers are probably yes, yes, and yes; considering the large number of independent transfers, each a unique and rare (except on the evolutionary timescale) historical event, almost any imaginable kind of vector and method of intron transfer could have been used at least once. Why has the intron burst on the angiosperm scene in such a rampant manner only so recently? Has this recent wave of lateral transfers been triggered by some key shift in the intron's invasiveness within angiosperms, and if so, what has changed? We hope to provide at least partial answers to some of these fascinating but challenging questions over the coming years.

In passing, we note that the overwhelmingly horizontal evolution of this remarkable group I intron is in striking contrast to the vertical pattern of evolution of the 23 other introns in angiosperm mt genomes, all of which are group II introns (Unseld *et al.*, 1997). We have used probes for 11 of these introns in our Southern blots surveys (Qiu *et al.*, 1998; Y.-L.Q. and J.D.P., unpublished data). All 11 are present in most or all major groups of angiosperms, and in many other groups of vascular plants, and thus were clearly present in the common ancestor of all angiosperms. These group II introns appear to have been transmitted in a strictly vertical manner, including occasional to frequent losses.

HIGHLY ACCELERATED SUBSTITUTION RATES IN TWO LINEAGES OF PLANTS

As already emphasized, our wide-scale Southern blot survey for presence or absence of mt genes and introns is predicated entirely on the uncommonly low rates of nucleotide substitutions observed to date in plant mitochondria. If a lineage of plants were to sustain for very long a radically higher substitution rate (say at the 50- to 100-fold higher level characteristic of mammalian mitochondria), then all of its mt genes might hybridize poorly or not at all, as if the genome no longer existed. Poor hybridization with all mt probes tested was observed for two of the 281 angiosperms on our blots, *Pelargonium hortorum* (the common garden geranium) (Fig. 1, lane 4) and *Plantago rugelii* (plantain, a common lawn weed); this was despite normal loadings of total DNA in these two lanes and strong hybridization with all chloroplast probes used. To explore this, portions of several mt protein and rRNA genes were PCR-amplified from both taxa and sequenced (Y.C., C.L.P., Y.-L.Q., and J.D.P., unpublished work). In all cases, the genes are exceptionally divergent. Most critically, in the case of the protein genes, most of the enhanced divergence is confined to synonymous sites. This indicates that the neutral point mutation rate, the rate of occurrence of nucleotide substitutions irrespective of selection, is markedly enhanced in both genera. To illustrate this effect for *cob* and *cox2*, Fig. 4 shows phylogenetic trees constructed with third codon positions only. Most third position changes are silent, and therefore the extremely long branch lengths leading to the *Pelargonium* and *Plantago* sequences in each tree graphically illustrate the high point mutation rate in these two distantly unrelated angiosperms.

Analysis of chloroplast and nuclear gene sequences from both plants indicates that these two genomes are not undergoing accelerated evolution (consistent with the strong hybridization of chloroplast probes mentioned above). Thus, the mutation rate increases in these two plants are restricted to their mitochondrial genomes. This distinguishes these cases of rate variation from those reported by Eyre-Walker and Gaut (1997), in which all three genomes of grasses were shown to exhibit higher rates of synonymous substitution than in palms. Another distinction is the magnitude of the rate variation: Grasses and palms differ only several-fold in their (plant-wide) substitution rates, whereas *Pelargonium* and *Plantago*

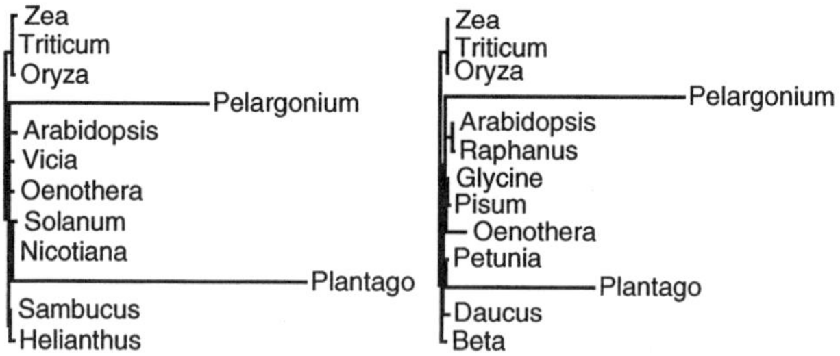

FIGURE 4. Accelerated evolution of mitochondrial cob (left tree) and cox2 (right tree) genes in P. hortorum and P. rugelii. Trees are from a maximum likelihood analysis of third codon positions only (344 positions for cob and 116 for cox2). Tree topologies were constrained to match current views of phylogenetic relationships for the organisms whose genes are analyzed. Trees are shown at the same scale; scale bar indicates 0.1 substitution/nucleotide. Pelargonium and Plantago sequences are from our unpublished data, and the rest are from the GenBank database.

show elevated (mt-specific) rates some 50–100 times higher than normal, putting them on a par with the very rapidly evolving mt genomes of mammals. Furthermore, analysis of several other species from the *Pelargonium* (Geraniaceae) and *Plantago* (Plantaginaceae) families shows a range of enhanced mt divergences in both families, as if sequential increases in the mt mutation rate had occurred during their evolution (Y.C., C.L.P., and J.D.P., unpublished work).

The magnitude and recency of these mutation rate shifts appear to be unprecedented for evolutionary lineages of species (as opposed to the well known but ephemeral mutator strains of laboratory mutant cultures, of wild strains of bacteria, or of human colon cancers). It remains to be seen whether the same sorts of underlying mechanisms are involved, such as changes in the fidelity and efficacy of DNA replication and mismatch repair (Modrich and Lahue, 1996).

POSTSCRIPT

Plant mt genomes continue to spring marvelous evolutionary surprises. The discovery that certain angiosperm groups are rapidly moving a large set of mt ribosomal proteins to the nucleus seems remarkable in two contexts: first, that they still have these so "easily transferred" genes left to transfer after a roughly 2-billion-year period of mt existence; second, that animals lost all of these ribosomal protein genes at least 0.6 billion years ago (i.e., before they became animals) and that there has been absolutely no functional gene transfer within the long period of metazoan evolution (plants still do it, animals don't!). For reasons as yet unfathomable, rates of functional gene transfer appear to vary hugely across lineages and over time. The very recent and explosive burst of *cox1* intron invasions into angiosperm mt genomes and the discovery of unprecedentedly large increases in the mt point mutation rate in two groups of angiosperms also speak to the surprising fluidity of the forces that control the rates of all manner of classes of mutations. These discoveries pave the way for more reductionist studies aimed at elucidating molecular mechanisms underlying these striking evolutionary patterns and rate changes. They also point to the opportunity afforded by microarray technology to mine new veins of molecular evolutionary gold by scaling up by orders of magnitude the Southern blot approach so successfully used thus far.

ACKNOWLEDGMENTS

We thank Jeff Doyle, Jane Doyle, Peter Kuhlman, Jackie Nugent, Phil Roessler, and Andy Shirk for various contributions and Jeff Blanchard, Dan Daley, Will Fischer, Patrick Keeling, and Jim Whelan for helpful discussions. This study was supported by National Institutes of Health Research Grant GM-35087 to J.D.P., U.S. Department of Agriculture Plant Biotechnology Fellowship 95-38420-2214 to K.L.A., and National Institutes of Health Postdoctoral Fellowships GM-17923 and GM-19225 to Y.L.Q. and C.L.P., respectively.

REFERENCES

Adams, K. L., Clements, M. J. & Vaughn, J. C. (1998) The Peperomia mitochondrial coxI group I intron: Timing of horizontal transfer and subsequent evolution of the intron. *J. Mol. Evol.* 46, 689–696.

Adams, K. L., Song, K., Roessler, P.G., Nugent, J. M., Doyle, J. L., Doyle, J. J. & Palmer, J. D. (1999) Intracellular gene transfer in action: Dual transcription and multiple silencings of nuclear and mitochondrial cox2 genes in legumes. *Proc. Natl. Acad. Sci. USA* 96, 13863–13868.

Allen, J. F. (1993) Control of gene expression by redox potential and the requirement for chloroplast and mitochondrial genomes. *J. Theor. Biol.* 165, 609–631.

Allen, J. F. & Raven, J. A. (1996) Free-radical-induced mutation vs redox regulation: Costs and benefits of genes in organelles. *J. Mol. Evol.* 42, 482–492.

Bittner-Eddy, P., Monroy, A. F. & Brambl, R. (1994) Expression of mitochondrial genes in the germinating conidia of Neurospora crassa. *J. Mol. Biol.* 235, 881–897.

Blanchard, J. & Lynch, M. (2000) Organellar genes. Why do they end up in the nucleus? *Trends Genet.* **16,** 315–320.

Brown, T. A., Ray, J. A., Waring, R. B., Scazzocchio, C. & Davies, R. W. (1984) A mitochondrial reading frame which may code for a second form of ATPase subunit 9 in Aspergillus nidulans. *Curr. Genet.* 8, 489–492.

Cho, Y., Qiu, Y.-L., Kuhlman, P. & Palmer, J. D. (1998) Explosive invasion of plant mitochondria by a group I intron. *Proc. Natl. Acad. Sci. USA* 95, 14244–14249.

Cho, Y. & Palmer, J. D. (1999) Multiple acquisitions via horizontal transfer of a group I intron in the mitochondrial cox1 gene during evolution of the Araceae family. *Mol. Biol. Evol.* 16, 1155–1165.

Claros, M. G., Perea, J., Shu, Y., Samatey, F. A., Popot, J.-L. & Jacq, C. (1995) Limitations to in vivo import of hydrophobic proteins into yeast mitochondria. The case of a cytoplasmically synthesized apocytochrome b. *Eur. J. Biochem.* 228, 762–771.

Covello, P. S. & Gray, M. W. (1992) Silent mitochondrial and active nuclear genes for subunit 2 of cytochorme c oxidase (cox2) in soybean: evidence for RNA-mediated gene transfer. *EMBO J.* 22, 3815–3820.

Doolittle, W. F. (1998) You are what you eat: a gene transfer ratchet could account for bacterial genes in eukaryotic nuclear genomes. *Trends Genet.* 14, 307–311.

Eyre-Walker, A. & Gaut, B. S. (1997) Correlated rates of synonymous site evolution across plant genomes. *Mol. Biol. Evol.* 14, 455–460.

Fauron, C. M.-R., Moore, B. & Casper M. (1995) Maize as a model of higher plant mitochondrial genome plasticity. *Plant Science* 112, 11–32.

Figueroa, P., Gomez, I., Holuigue, L., Araya, A. & Jordana, X. (1999a) Transfer of rps14 from the mitochondrion to the nucleus in maize implied integration within a gene encoding the iron-sulphur subunit of succinate dehydrogenase and expression by alternative splicing. *Plant J.* 18, 601–609.

Figueroa, P., Gomez, I., Carmona, R., Holuigue, L., Araya, A. & Jordana, X. (1999b) The gene for mitochondrial ribosomal protein S14 has been transferred to the nucleus in Arabidopsis thaliana. *Mol. Gen. Genet.* 262, 139–144.

Gray, M. W. (1999) Evolution of organellar genomes. *Curr. Opin. Genet. Dev.* 9, 678–687.

Gray, M. W., Lang, B. F., Cedergren, R., Golding, G. B., Lemieux, C., Sankoff, D., Turmel, M., Brossard, N., Delage, E., Littlejohn, T. G., Plante, I., Rious, P., Saint-Louis, D., Zhu, T. & Burger, G. (1998) Genome structure and gene content in protist mitochondrial DNAs. *Nucleic Acids Res.* 26, 865–878.

Grohmann, L., Brennicke, A. & Schuster, W. (1992) The mitochondrial gene encoding ribosomal protein S12 has been translocated to the nuclear genome in Oenothera. *Nucleic Acids Res.* 20, 5641–5646.

Hanson, M. R. (1991) Plant mitochondrial mutations and male sterility. *Annu. Rev. Genet.* 25, 461–486.

Herrmann, R. G. (1997) Eukaryotism, towards a new interpretation. In *Eukaryotism and Symbiosis*, ed. Schenk, H. E. A., Herrmann, R. G., Jeon, K. W., Müller, N. E. & Schwemmler, W. (Springer-Verlag, Vienna), pp. 73–118.

Joyce, P. B. & Gray, M. W. (1989) Chloroplast-like transfer RNA genes expressed in wheat mitochondria. *Nucleic Acids Res.* 17, 5461–5476.

Kadowaki, K., Kubo, N., Ozawa, K. & Hirai, A. (1996) Targeting presequence acquisition after mitochondrial gene transfer to the nucleus occurs by duplication of existing targeting signals. *EMBO J.* 15, 6652–6661.

Kanno, A., Nakazono, M., Hirai, A. & Kameya, T. (1997) A chloroplast derived trnH gene is expressed in the mitochondrial genome of gramineous plants. *Plant Mol. Biol.* 34, 353–356.

Knoop, V., Unseld, M., Marienfeld, J., Brandt, P., Sunkel, S., Ullrich, H. & Brennicke, A. (1996) copia-, gypsy- and LINE-like retrotransposon fragments in the mitochondrial genome of Arabidopsis thaliana. *Genetics* 142, 579–585.

Kubo, N., Harada, K. & Kadowaki, K. (1998) Transfer of an rps11 gene from mitochondrion to the nucleus in pea. In *Plant Mitochondria: From Gene to Function*, eds. Moller, I. M., Gardestrom, P., Glimelius, K. & Glaser, E. (Backhuys, Leiden), pp. 25–27.

Kubo, N., Harada, K., Hirai, A. & Kadowaki, K. (1999) A single nuclear transcript encoding mitochondrial RPS14 and SDHB of rice is processed by alternative splicing: Common use of the same mitochondrial targeting signal for different proteins. *Proc. Natl. Acad. Sci. USA* 96, 9207–9211.

Laroche, J., Li, P., Maggia, L. & Bousquet, J. (1997) Molecular evolution of angiosperm mitochondrial introns and exons. *Proc. Natl. Acad. Sci. USA* 94, 5722–5727.

Lynch, M. (1996) Mutation accumulation in transfer RNAs: molecular evidence for Muller's ratchet in mitochondrial genomes. *Mol. Biol. Evol.* 13, 209–220.

Mackenzie, S., He, S. & Lyznik, A. (1994) The elusive plant mitochondrion as a genetic system. *Plant Physiol.* 105, 775–780.

Marienfeld, J., Unseld, M. & Brennicke, A. (1999) The mitochondrial genome of Arabidopsis is composed of both native and immigrant information. *Trends Plant Sci.* 4, 495–502.

Marshall, C. R., Raff, E. C. & Raff, R. A. (1994) Dollo's law and the death and resurrection of genes. *Proc. Natl. Acad. Sci. USA* 91, 12283–12287.

Martin, A. P., Naylor, G. J. P. & Palumbi, S. R. (1992) Rates of mitochondrial DNA evolution in sharks are slow compared with mammals. *Nature* 357, 153–155.

Martin, W. & Schnarrenberger, C. (1997) The evolution of the Calvin cycle from prokaryotic to eukaryotic chromosomes: a case study of functional redundancy in ancient pathways through endosymbiosis. *Curr. Genet.* 32, 1–18.

Martin, W. & Herrmann, R. G. (1998) Gene transfer from organelles to the nucleus: How much, what happens, and why? *Plant Physiol.* 118, 9–17.

Miyata, S., Nakazono, M. & Hirai, A. (1998) Transcription of plastid-derived tRNA genes in rice mitochondria. *Curr. Genet.* 34, 216–220.

Modrich, P. & Lahue, R. (1996) Mismatch repair in replication fidelity, genetic recombination, and cancer biology. *Annu. Rev. Biochem.* 65, 101–133.

Moran, N. (1996) Accelerated evolution and Muller's rachet in endosymbiotic bacteria. *Proc. Natl. Acad. Sci. USA* 93, 2873–2878.

Nugent, J. M. & Palmer, J. D. (1988) Location, identity, amount and serial entry of chloroplast DNA sequences in crucifer mitochondrial DNAs. *Curr. Genet.* 14, 501–509.

Nugent, J. M. & Palmer, J. D. (1991) RNA-mediated transfer of the gene coxII from the mitochondrion to the nucleus during flowering plant evolution. *Cell* 66, 473–481.

Nugent, J. M. & Palmer, J. D. (1993) Evolution of gene content and gene organization in flowering plant mitochondrial DNA: A general survey and further studies on coxII gene transfer to the nucleus. In *Plant Mitochondria*, eds. Brennicke, A. & Kuck, U. (VCH Publishers, New York), pp. 163–170.

Palmer, J. D. (1990) Contrasting modes and tempos of genome evolution in land plant organelles. *Trends. Genet.* 6, 115–120.

Palmer, J. D. (1992) Chloroplast and mitochondrial genome evolution in land plants. In *Plant Gene Research: Cell Organelles*, ed. Herrmann, R. G. (Springer-Verlag, Vienna), pp. 99–133.

Palmer, J. D. (1997) Organelle genomes: going, going, gone! *Science* 275, 790–791.

Palmer, J. D. & Herbon, L. A. (1988) Plant mitochondrial DNA evolves rapidly in structure, but slowly in sequence. *J. Mol. Evol.* 28, 87–97.

Perrotta, G., Grienenberger, J. M. & Gualberto, J. M. (1998) Plant mitochondrial rps2 genes code for proteins with a C-terminal extension that apparently is processed. In *Plant Mitochondria: From Gene to Function*, eds. Moller, I. M., Gardestrom, P., Glimelius, K. & Glaser, E. (Backhuys, Leiden), pp. 37–41.

Pfannschmidt, T., Nilsson, A. & Allen, J. F. (1999) Photosynthetic control of chloroplast gene expression. *Nature* 397, 625–628.

Pont-Kingdon, G. A., Okada, N. A., Macfarlane, J. L., Beagley, C. T., Wolstenholme, D. R., Cavalier-Smith, T., & Clark-Walker, G. D. (1995) A coral mitochondrial mutS gene. *Nature* 375, 109–111.

Pont-Kingdon, G., Okada, N. A., Macfarlane, J. L., Beagley, C. T, Watkins-Sims, C. D., Cavalier-Smith, T., Clark-Walker, G. D. & Wolstenholme, D. R. (1998) Mitochondrial DNA of the coral Sarcophyton glaucum contains a gene for a homologue of bacterial mutS: A possible case of gene transfer from the nucleus to the mitochondrion. *J. Mol. Evol.* 46, 419–431.

Popot, J.-L. & de Vitry, C. (1990) On the microassembly of integral membrane proteins. *Annu. Rev. Biophys. Biophys. Chem.* 19, 369–403.

Qiu, Y.-L., Cho, Y., Cox, J. C. & Palmer, J. D. (1998) The gain of three mitochondrial introns identifies liverworts as the earliest land plants. *Nature* 394, 671–674.

Race, H. L., Herrmann, R. G. & Martin, W. (1999) Why have organelles retained genomes? *Trends Genet.* 15, 364–370.

Ricchetti, M., Fairhead, C. & Dujon, B. (1999) Mitochondrial DNA repairs double-strand breaks in yeast chromosomes. *Nature* 402, 96–100.

Sanchez, H., Fester, T., Kloska, S., Schroder, W. & Schuster, W. (1996) Transfer of rps19 to the nucleus involves the gain of an RNP-binding motif which may functionally replace RPS13 in Arabidopsis mitochondria. *EMBO J.* 15, 2138–2149.

Soltis, P. S., Soltis, D. E. & Chase, M. W. (1999) Angiosperm phylogeny inferred from multiple genes as a tool for comparative biology. *Nature* 402, 402–404.

Thorsness, P. E. & Fox, T. D. (1990) Escape of DNA from mitochondria to the nucleus in Saccharomyces cerevisiae. *Nature* 346, 376–379.

Thorsness, P. E. & Weber, E. R. (1996) Escape and migration of nucleic acids between chloroplasts, mitochondria, and the nucleus. *Intl. Rev. Cytol.* 165, 207–233.

Unseld, M., Marienfeld, J. R., Brandt, P. & Brennicke, A. (1997) The mitochondrial genome of Arabidopsis thaliana contains 57 genes in 366, 924 nucleotides. *Nature Genet.* 15, 57–61.

van den Boogaart, P., Samallo, J. & Agsteribbe, E. (1982) Similar genes for a mitochondrial ATPase subunit in the nuclear and mitochondrial genomes of Neurospora crassa. *Nature* 298, 187–189.

Vaughn, J. C., Mason, M. T., Sper-Whitis, G. L., Kuhlman, P. & Palmer, J. D. (1995) Fungal origin by horizontal transfer of a plant mitochondrial group I intron in the chimeric coxI gene of Peperomia. *J. Mol. Evol.* 41, 563–572.

Waddell, P. J., Cao, Y., Hasegawa, M. & Mindell, D. P. (1999) Assessing the cretaceous superordinal divergence times within birds and placental mammals by using whole mitochondrial protein sequences and an extended statistical framework. *Syst. Biol.* 48, 119–137.

Ward, B. L., Anderson, R. S. & Bendich, A. J. (1981) The mitochondrial genome is large and variable in a family of plants (Cucurbitaceae). *Cell* 25, 793–803.

Ward, M. & Turner, G. (1986) The ATP synthase subunit 9 gene of Aspergillus nidulans: sequence and transcription. *Mol. Gen. Genet.* 205, 331–338.

Wischmann, C. & Schuster, W. (1995) Transfer of rps10 from the mitochondrion to the nucleus in Arabidopsis thaliana: evidence for RNA-mediated transfer and exon shuffling at the integration site. *FEBS Letters* 375, 152–156.

Wolfe, K. H., Li, W.-H. & Sharp, P. M. (1987) Rates of nucleotide substitution vary greatly among plant mitochondrial, chloroplast, and nuclear DNAs. *Proc. Natl. Acad. Sci. USA* 84, 9054–9058.

Part II

VIRAL AND BACTERIAL MODELS

Andrés Moya and colleagues point out advantages offered by RNA viruses for the experimental investigation of evolution ("The Evolution of RNA Viruses: A Population Genetics View," Chapter 5); notably, the phenotypic features ("phenotypic space") map fairly directly onto the "genetic space." In other organisms, from bacteria to humans, the expression of the genetic make up in the phenotype is mediated, to a lesser or greater degree, but always importantly, by complex interactions between genes, between cells, and with the environment. These authors' model is the vesicular stomatitis virus (VSV), a rhabdovirus, containing 11.2 kb of RNA encoding five proteins. The authors grow different viral clones under variable demographic and environmental conditions, and measure the evolution of fitness in these clones by competition with a control clone. Fitness generally decreases through the serial viral transfers from culture to culture, particularly when bottlenecks associated with transfers are small. Fitness may, however, increase when the transmission rates are high, although the response varies from clone to clone. Moya *et al.* conclude with an examination of the advantages and disadvantages of traditional population genetics theory for the description of viral evolution vis-à-vis the quasi-species concept, which proposes that the target of natural selection is not a single genotype but rather a cloud of mutants distributed around a most frequent one, the "master sequence."

Robin M. Bush and colleagues ("Effects of Passage History and Sampling Bias on Phylogenetic Reconstruction of Human Influenza A Evolution," Chapter 6) had noticed in their earlier reconstruction of the phylogeny of influenza A virus, based on the hemagglutinin gene, an excess of non-silent nucleotide substitutions in the terminal branches of the tree. They explore two likely hypotheses to account for this excess: (1) that these nucleotide replacements are host-mediated mutations that have appeared or substantially increased in frequency during passage of the virus in the embryonated eggs in which they are cultured—this hypothesis can account at most for 59 (7.9%) of the 745 non-silent substitutions observed; (2) sampling bias, induced by the preference of investigators for sequencing antigenetically dissimilar strains for the purpose of identifying new variants that might call for updating the vaccine—which seems to be the main factor accounting for the replacement excess in terminal branches. The authors point out that the matter is of consequence in vaccine development, and that host-mediated mutations should be removed before making decisions about influenza evolution.

Bruce R. Levin and Carl T. Bergstrom ("Bacteria are Different: Observations, Interpretations, Speculations, and Opinions about the Mechanisms of Adaptive Evolution in Prokaryotes," Chapter 7) note that adaptive evolution in bacteria compared to plants and animals is different in three respects. The two most important factors are (1) the frequency of homologous recombination, which is low in bacteria but high in sexual eukaryotes; and (2) the phylogenetic range of gene exchange, which is broad in bacteria but narrow (typically, intraspecific) in eukaryotes. A third factor is that the role of viruses, plasmids, and other infectiously transmitted genetic elements is nontrivial in the adaptive evolution of bacteria, while it is negligible in eukaryotes.

5

The Evolution of RNA Viruses: A Population Genetics View

ANDRÉS MOYA, SANTIAGO F. ELENA, ALMA BRACHO, ROSARIO MIRALLES, AND ELADIO BARRIO

RNA viruses are excellent experimental models for studying evolution under the theoretical framework of population genetics. For a proper justification of this thesis we have introduced some properties of RNA viruses that are relevant for studying evolution. On the other hand, population genetics is a reductionistic theory of evolution. It does not consider or make simplistic assumptions on the transformation laws within and between genotypic and phenotypic spaces. However, such laws are minimized in the case of RNA viruses because the phenotypic space maps onto the genotypic space in a much more linear way than on higher DNA-based organisms. Under experimental conditions, we have tested the role of deleterious and beneficial mutations in the degree of adaptation of vesicular stomatitis virus (VSV), a nonsegmented virus of negative strand. We also have studied how effective population size, initial genetic variability in populations, and environmental heterogeneity shapes the impact of mutations in the evolution of vesicular stomatitis virus. Finally, in an integrative

Institut Cavanilles de Biodiversitat i Biología Evolutiva and Departament de Genètica, Universitat de València, Apartat 2085, 46071 València, Spain

This paper was presented at the National Academy of Sciences colloquium "Variation and Evolution in Plants and Microorganisms: Toward a New Synthesis 50 Years After Stebbins," held January 27–29, 2000, at the Arnold and Mabel Beckman Center in Irvine, CA.

Abbreviations: VSV, vesicular stomatitis virus; BHK, baby hamster kidney; MARM, monoclonal antibodies mutant.

attempt, we discuss pros and cons of the quasispecies theory compared with classic population genetics models for haploid organisms to explain the evolution of RNA viruses.

RNA VIRUSES: BIOLOGICAL AND POPULATION PROPERTIES

Despite their great functional and structural diversity, all RNA viruses share the following properties (Domingo and Holland, 1997): (*i*) Cell–virus junction is mediated by means of specific membrane receptors. (*ii*) A viral particle penetrates the cell, loses its capsid, and releases its nucleic acids within the cell. (*iii*) The replication of the viral genome is regulated by the expression of viral genes (i.e., RNA replicase is encoded by the virus genome). (*iv*) The component parts of the viruses are assembled and released as virions out of the cell. In addition, these properties are complemented with four others that are relevant to understanding the evolution of RNA viruses. (*v*) The number of viral particles in a given infected organism may be as high as 10^{12} (Domingo and Holland, 1997). Such population sizes, several orders of magnitude larger than any population size for DNA-based organisms, are related to viral generation time. (*vi*) In fact, a single infectious particle can produce, on average, 100,000 copies in 10 h (Domingo and Holland, 1997). If the replication machinery is working optimally, a new RNA genome is produced every 0.4 s. (*vii*) Genome sizes range between 3 and 30 kb. Accordingly, the number of genes per genome is also very small. (*viii*) Finally, RNA viruses show extremely high mutation rates (Drake and Holland, 1999). Because of the lack of proofreading by their replicases, RNA viruses show the highest mutation rates among living beings (Drake and Holland, 1999), on the order of one mutation per genome and replication round.

The above-mentioned properties of large population size, high replication rate, and short generation time are responsible, in general, for the extremely high genetic variability of RNA viral populations. Recombination and segmentation also may play an important role in generating new genetic variability (Domingo and Holland, 1997). In any case, the extent of genetic variability per generation time of any RNA virus is usually much higher than that corresponding to any DNA-based organism, providing an excellent opportunity for studying ongoing evolution in accessible terms for human observers.

RNA VIRUSES MEET THE POPULATION GENETICS THEORY OF EVOLUTION: THEORETICAL BACKGROUND

According to population genetics, evolution is the change in the genetic properties of populations. Changes considered to be evolutionarily relevant are those inherited via the genetic material. Population genetics, in a more formal sense, is the study of those variables that are responsible for changes in the frequency of alleles in populations. In essence, the theory is reductionistic because it makes simplified assumptions on the transformation laws, especially those related to development, that are operating within and between genotypic and phenotypic spaces (Lewontin, 1974). Such simplifications allow the estimation of allele frequencies of a given generation as a function (probabilistic and/or deterministic) of frequencies from previous generations, as well as a set of state variables, mostly including mutation, selection, migration, and random drift. The vast majority of these transformation laws are not considered by population genetics or, if considered, they are incorporated as linear transformations, i.e., the genes of the genotypic space map linearly on the phenotypic space. When applied, for instance, to a phenotypic trait such as fitness, linearity means that a change in a gene promotes a certain fitness change, and that the genotype is an array of independent units contributing to the fitness in an additive way. Mendelian laws are the only transformation laws formally incorporated to the core of the population genetics theory (Lewontin, 1974).

It seems likely that populations of organisms governed by simple transformation laws (specially epigenetic) will meet the theoretical predictions of population genetics much better than those organisms governed by unknown, but probably very complex, transformation laws. Therefore, RNA viruses should better meet the theoretical predictions of population genetics. As the number of their genes is small, the type and number of epistatic interactions among their products should be of minor relevance compared with organisms with larger genomes (Elena, 1999). Thus, epistasis is expected to be of minor importance in simple-genome RNA viruses (Elena, 1999).

The environment of an RNA virus has several components, all of which could have different effects on its adaptive process. The closest environmental component is formed by cytoplasmic components of the infected cell, but intercellular spaces within tissues, tissues within individual hosts, and the ecological environment where host species are living are other components that modulate the adaptive response of RNA viruses. The theoretical framework to understand the dynamics of haploid organisms can be found in a series of classic papers that appeared many years ago (Moran, 1957; Robson, 1957; Muller, 1964; Drobník and Dlouhá, 1966; Felsenstein, 1971; Karlin and McGregor, 1971; Cook and Nassar, 1972; Gillespie, 1973; Gladstein, 1973; Trajstman, 1973; Cannings, 1974, 1975; Haigh, 1978; Emigh, 1979a, b; Strobeck, 1979). These studies, together with more recent statistical procedures for testing the presence of positive Darwinian selection or neutrality at the nucleotide level

(Kimura, 1983; Gillespie, 1991; Li, 1997), constitute the main body of population genetics to account for the dynamics of RNA viral populations.

EXPERIMENTAL VIRUS MODEL AND FITNESS ASSAYS

Vesicular stomatitis virus (VSV) is the prototype of the well-defined Rhabdoviridae family. It has a wide host range of vertebrates and arthropods. This virus is identified by its elongated bullet-shaped form of approximately 180 x 70 nm in size, with a nucleocapsid covered by a lipid-rich envelope. The RNA of the virion is complementary in its sequence to the mRNA for the viral proteins. The Rhabdoviridae are the simplest of the so-called minus-strand viruses. The VSV genome contains approximately 11.2 kb and transcribes into five mRNAs coding for five proteins (Fig. 1). All of the studies described here are *in vitro* experiments with VSV, in which viruses were grown in different types of cell cultures depending on the experiments, although BHK cells (from baby hamster kidney) were the usual ones.

The study of fitness effects of state variables and/or experimental regimes on viral populations is a three-step process. First, before any experimental treatment, relative fitness assays of two VSV competing clones (stock clones) were carried out as described in the third step. The first competing clone was a surrogate wild type, and the second was one of the following four different resistant to monoclonal antibodies mutant (MARM) clones. MARM C is an approximately neutral variant (with fitness 1.02 ± 0.03 relative to wild type) that contains an $Asp^{259} \rightarrow$ Ala substitution in the surface glycoprotein (G, Fig. 1). This amino acid substitution allows the mutant to replicate under I_1 mAb concentration levels that neutralize the wild-type clone (VandePol *et al.*, 1986). MARM *R* clone was isolated after repeated plaque-to-plaque transfers of MARM C and showed a lower fitness than the parental virus (0.87 ± 0.05). The I_1 MAb phenotype of MARM X is conferred by an $Asp^{257} \rightarrow$ Val substitution in the G protein. It has a much higher fitness (2.52 ± 0.16) relative to the wild

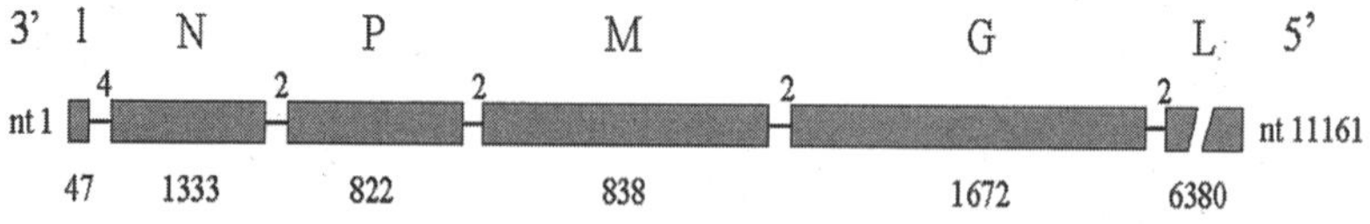

FIGURE 1. Schematic representation of the VSV genome, with its nontranscribed leader (l), and five consecutively transcribed mRNAs (gray blocks). Each letter refers to the proteins. From left to right: N, nucleocapsid; P, phosphoprotein; M, matrix; G, glycoprotein; L, viral replicase. Numbers below each block and above lines are lengths in nts of the genes and nontranscribed sequence, respectively.

type, acquired after 61 consecutive transfers of large virus populations on BHK cells. Finally, MARM *F* is an extremely low fitness clone (0.00015 ± 0.00001) obtained after 20 plaque-to-plaque transfers from MARM *X*. MARM *C* and *X* originally were isolated from the wild-type virus by picking spontaneous I_1 mAb-resistant clones (VandePol *et al.*, 1986). Therefore, they were isogenic with wild type with the only exception of the above mutations responsible for the resistance.

The second step corresponds with the experiment itself (see next four sections), normally carried out with one of the four MARM clones, and in which viral populations experienced different demographic regimes and environmental conditions.

Third, fitness of the evolved viral populations was evaluated by competition assays with the ancestral wild-type clone in the following form. The evolved MARM population was mixed with a known amount of the wild-type clone. A differential quantitation of MARM clone, compared with the total virus, was done by parallel plating of the virus with and without I_1 mAb. These virus mixtures then were used to initiate replicate serial competition passages. After each competition passage, the resulting virus mixture was 10^4-fold diluted and used to initiate the next competition transfer by infection of a fresh monolayer. The number of competition passages varied between two and a maximum of five, depending on the speed with which one competitor displaced the other. The antilogarithm of the slope of the regression $\ln[p_t/(1-p_t)] = \ln[p_0/(1-p_0)] + t\ln\bar{W}$ is taken as an estimate of the mean fitness of the corresponding MARM population relative to the wild type, where p_t and $1 - p_t$ are the proportions at passage number t of MARM and wild type, respectively (Duarte *et al.*, 1992; Clarke *et al.*, 1993; Elena *et al.*, 1996, 1998).

THE DYNAMICS OF DELETERIOUS MUTATIONS IN FINITE POPULATIONS

When finite populations with high mutation rates are considered, a significant proportion of the mutants should be deleterious. If populations are asexual and small in size, mutation-free individuals become rare and can be lost by random genetic drift. In that case a kind of irreversible ratchet mechanism gradually will decrease the mean fitness of the populations (Muller, 1964). Chao (1990) provided the first experimental evidence for the action of Muller's ratchet in RNA viruses. As can be observed in Table 1, there is a common pattern of fitness decline, but the magnitude of decline strongly depended on the virus studied. For instance, in the case of VSV (Duarte *et al.*, 1992; Clarke *et al.*, 1993; Duarte *et al.*, 1993, 1994) we performed genetic bottleneck passages (plaque-to-plaque transfers) and quantified the relative fitness of bottlenecked clones

TABLE 1. Percentage of fitness decline, with respect to the corresponding initial viral clone, in experiments with different RNA viruses subjected to a different number and continuous bottleneck transfers

Virus	No. of bottleneck passages	% of fitness decrease		Reference
		Average	Range	
φ6	40	22	12–33	29
VSV	20	18	0.1–99	25, 26, 30, 31
FMDV	30	60	14–99	32
HIV-1	15	94	89–99	33
MS2	20	17	16–18	34

FMDV, foot-and-mouth disease virus.

by allowing direct competition in mixed infections as described above. We documented variable fitness drops after 20 or more plaque-to-plaque transfers of VSV. The relevance of these findings for evolutionary biology is clear: whenever bottlenecks occur, fitness decreases.

In relation to the accumulation of deleterious mutations we carried out two additional types of studies. First, we explored which is the model that better accounts for the distribution of deleterious mutational effects. Second, we addressed the question of how large the effective population size should be to overcome the Muller's ratchet effect.

After a series of independent experiments of plaque-to-plaque passages of approximately 120 generations for each of the three VSV mutant clones (MARM *C*, MARM *R*, and MARM *X*, with 24, 16, and nine independent mutation-accumulation lines, respectively), relative mean fitness was estimated and the nature of its distribution was studied (Elena and Moya, 1999). Because fitness effects were not normally distributed, we fitted the observed distribution to alternative models. Table 2 shows the basic statistics of the fitting of three different models. The first model tested was the negative exponential (Mukai *et al.*, 1972). The only parameter of this model, α, is the inverse of the expected fitness ($1/\bar{W}$). As can be observed in Table 2, although significant to each mutant data set, R^2 values are not good. The model shows the property that mutations with small effects are more common than mutations with larger effects. The second model tested was the gamma distribution (Keightley, 1994). This model has two parameters, α and β, related with fitness as: $\bar{W} = \alpha/\beta$. This model fits significantly better to the data than the negative exponential (Elena and Moya, 1999). However, it still fails to explain those cases with large fitness effects for MARM *X* and MARM *C* mutants (Table 2; for more details see Elena and Moya, 1999). For a better description of larger deleterious effects, we considered a third model, which is a linear combination of a uniform and a gamma distribution. In this model a fraction p

TABLE 2. Fitting the deleterious fitness effects distribution of MARM *X*, *C*, and *R* to three theoretical models

MARM	Mean fitness	Model	Predicted fitness	Parameters α	β	p	P*	R^2
X	0.43 ± 0.28	Exponential	0.49	2.05 ± 0.24	—	—		0.74
		Gamma	0.40	48.23 ± 17–86	19–39 ± 7.19	—	0.0001	0.92
		Compound	0.44	400.05 ± 245.91	158.32 ± 96.89	0.41 ± 0.08	0.0031	0.96
C	0.74 ± 0.26	Exponential	0.93	1.08 ± 0.12	—	—		0.60
		Gamma	0.82	38.28 ± 11.22	31.23 ± 9.28	—	<0.0001	0.89
		Compound	0.72	168.78 ± 25.25	147.97 ± 22.33	0.42 ± 0.02	<0.0001	0.99
R	0.70 ± 0.28	Exponential	0.78	1.28 ± 0.20	—	—		0.73
		Gamma	0.67	5–90 ± 1.68	3.97 ± 1.07	—	0.0021	0.94
		Compound	0.66	6.72 ± 3.96	5.25 ± 3.64	0.44 ± 0.37	0.3510	0.95

See text for more details.
*P, significance level for the partial F-test used for testing the reduction in the error sum of squares for models with an increasing number of parameters.

TABLE 3. Mean estimates, with standard error, of $E(\Delta W)$, the mean fitness change per generation, as well as U, the deleterious mutation rate and, $E(s)$, its average effect under a constant model of effect distribution

MARM	$E(\Delta W)$	U	$E(s)$
X	−0.00311 ± 0.00020	1.5753 ± 0.5676	−0.0022 ± 0.0008
C	−0.00210 ± 0.00004	0.8839 ± 0.0541	−0.0024 ± 0.0001
R	−0.00252 ± 0.00004	1.1610 ± 0.0519	−0.0022 ± 0.0001

of the mutants are drawn from a uniform distribution and $1 - p$ from a gamma distribution (Elena and Moya, 1999; Elena *et al.*, 1998). MARM *X* and MARM *C* fit better than the gamma model. The fit for MARM *R* is not better than that obtained with the gamma model, but this could be caused by the low number of replicates for this clone.

The distribution of deleterious effects, on the other hand, strongly depended on which one of the three VSV mutant clones was used in the experiment. Table 3 shows $E(\Delta W)$, the mean fitness reduction per generation, for the three MARM clones. They are statistically different from zero and proportional to the initial fitness (Elena *et al.*, 1996). Applying Mukai–Bateman's method of a constant distribution of mutational fitness effects (Elena and Moya, 1999), we obtained (see Table 3) the estimates of U, the deleterious mutation rate per genome and generation, and $E(s)$, the expected selection coefficient against deleterious mutations. The average mean fitness reduction for the entire data set was –0.26% per generation, which is compatible with the idea that fitness was reduced by the accumulation of many mutations ($U \approx 1$) of small effect (–0.1% each). In other words, the huge decline in fitness reported above was the consequence of the accumulation of many mutations of small effect. The effect of deleterious mutations on long-term survival of viral population should be greater if we consider large numbers of mutations of small average effect compared with few changes of large effect because, in the latter case, selection should be more efficient in purifying deleterious load.

The population size necessary to stop Muller's ratchet not only depended on the demographic experimental regimes of the viral populations, but also on the genetic composition and the fitness of the initial populations. In a series of bottleneck experiments performed with MARM *X*, MARM *C*, MARM *U*, and MARM *N* populations (for biological properties of the last two mutants not mentioned in this review see Novella *et al.*, 1995), initial fitness and fitness after 20 2-to-2, 5-to-5, and 30-to-30 clone passages was estimated. Table 4 gives a summary of the results obtained. As can be observed, after 20 2-to-2 clone passages the initial low-fitness MARM *N* clones showed slight fitness changes. When the virus transmission rate was raised to five, fitness gain was observed in the six experiments

TABLE 4. Mean fitness, with standard error (based on six different experiments), after different population dynamics for different MARM clonal populations

MARM	Fitness before passage	Passage dynamics	Mean fitness after passage
X	3.05 ± 0.03	5-to-5	1.70 ± 0.03***
		30-to-30	3.00 ± 0.40^{ns}
U	1.00 ± 0.20	5-to-5	1.30 ± 0.20^{ns}
C	0.91 ± 0.03	5-to-5	1.20 ± 0.20^{ns}
N	0.38 ± 0.01	2-to-2	0.38 ± 0.01^{ns}
		5-to-5	0.55 ± 0.05**

ns, not significant; **, $P < 0.01$; ***, $P < 0.001$.

performed. The neutral MARM *C* and *U* mutants, with passage transmission rates of five, gave similar mean fitness results to the MARM *N* mutant at a transmission rate of two (i.e., no significant change in fitness). Finally, the high fitness MARM *X* clone showed no fitness changes with a transmission rate of 30, but fitness decrease at a transmission rate of five clones (Novella *et al.*, 1995). The new fitness values were similar after 20 more passages with 5-to-5 plaque transfers in MARM *X*, MARM *U*, and MARM *N* clones (Novella *et al.*, 1996). The results obtained can be explained by the sampling of infectious particles of lower, equal, or greater fitness than the average fitness showed by the original nonbottlenecked population (see Table 4). Mutants of equal or greater fitness should be sampled more frequently from populations with low fitness, because highly fit populations have lower probability of undergoing further mutational improvement. One discrepancy with this explanation was observed when similar experiments were done with the extremely debilitated MARM *F* clone (Elena *et al.*, 1998). We did not expect any fitness recovery after 20 single plaque-to-plaque (1-to-1) passages, but we observed significant fitness recoveries, ranging from 0.11 to 0.94 (Elena *et al.*, 1998). A partial explanation for this may be that competition is taking place between variants within a single plaque. In the context of a highly mutable population, a beneficial mutation that confers faster replication may occur during the growth of a single plaque. Then, the probability of picking this fitter clone from the plaque is very high because of its higher frequency.

ADAPTATION: COMPETITION IN CONSTANT, CHANGING, AND SUBDIVIDED ENVIRONMENTS

Until now, we have shown experimental evidence for the accumulation of many deleterious mutations during the evolution of RNA viral populations, especially those subjected to strong bottlenecking, a common feature of natural virus populations. But from time to time, the appearance of mutations with beneficial effect on fitness cannot be ruled out. This, and the next two sections, will provide experimental evidence for the presence of such mutations and its effect on the long-term evolution of RNA viruses. All of the topics treated can be reduced to the following two: (*i*) is there evolution of fitness in competing populations under different environmental situations, and (*ii*) is there any limitation to the rate of viral adaptation?

In population biology, the competitive exclusion principle (Hardin, 1960) states that in the absence of niche differentiation, one competing species always will eliminate or exclude the other. The very high error frequency during RNA genome replication and, hence, its rapid evolution, may render the prolonged coexistence of two (or more) genetically distinct viral populations unlikely. Different genomes that may have different replication abilities and/or encapsidation rates are constantly being generated. Then, the appearance of one or more mutations in one of the competing populations may confer enough selective advantage to its carrier as to facilitate its fixation. However, Clarke *et al.* (1994) showed that two viral competing populations of approximately initial equal fitness could coexist through numerous generations during prolonged replication in a constant seasonal environment. A mechanism that may explain the long preservation of both populations is frequency-dependent selection. Evidence of that also has been obtained by Elena *et al.* (1997). The biological scenario where frequency-dependent selection might be present is as follows: low-frequency genotype faces a new and sparsely populated biological niche, whereas the niche of the most common genotype may approach saturation. This scenario requires certain niche heterogeneity. For the cell–culture environment we can imagine factors like age, density stage of the cellular cycle, location, etc. It is then possible that each viral genotype is using this smooth variability in a constant environment in different ways.

To explain the relevance of competition in evolution, Van Valen (1973) proposed the Red Queen's hypothesis according to which each species is competing in a zero-sum game against others; each game is a dynamic equilibrium between competing species where, although both are increasingly adapted, no species can ever win. This is exactly what we observed (Clarke *et al.*, 1994). Fig. 2 shows the relative fitness estimates of both winners and losers of two competition series (H and G) compared with the fitness of the initial competing populations. As can be observed, both winners and losers show a higher fitness than the initial ones.

The process of adaptation cannot continue forever for a constant environment. Once the virus becomes adapted to the new environment, no further improvement is expected unless the environment changes. How-

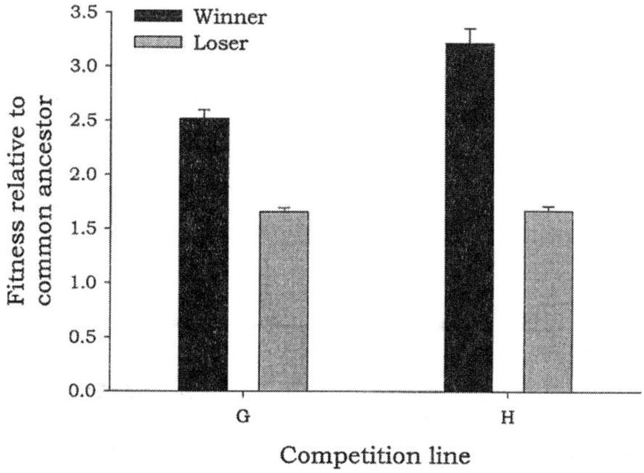

FIGURE 2. Relative fitness to the ancestral populations of winners and losers in long-term competition experiments (G and H). Initial populations were neutral. Modified from Clarke *et al.* (1994).

ever, what type of trajectory is followed by fitness during the adaptation of a viral population? In a study of the fitness recovery of the MARM *F* clone under different demographic regimes (Elena *et al.*, 1998) we observed that, for the case of large and consecutive population expansions, a maximum fitness value close to 1 was reached after few passages in four independent experiments. Moreover, the rate of fitness increases slowed down. Although the issue is still controversial, a hyperbolic function on a logarithmic scale [$\ln\bar{W}_t = \ln\bar{W}_0 + at/(b + t)$,] where t is the passage number, a is the asymptotic log-fitness value, and b is the passage number at which the log fitness is equal to half the maximum value, seems to provide a good fit to the observed trajectories (Table 5). It catches some relevant

TABLE 5. Estimated parameters for the hyperbolic model of fitness, with standard errors, in four independent experiments of fitness recovery of MARM *F* clone during 40 large and consecutive population expansions

Replicate	$\ln \bar{W}_0$	a	b	R^2	$\bar{W}$
I	−9.3 ± 0.7	9.3 ± 0.8	0.4 ± 0.2	0.95	1.1 ± 0.0
II	−7.9 ± 0.6	7.9 ± 0.7	0.3 ± 0.1	0.95	1.2 ± 0.1
III	−9.0 ± 0.6	9.3 ± 0.8	0.8 ± 0.3	0.99	1.2 ± 0.2
IV	−10.0 ± 0.3	10.2 ± 0.4	1.2 ± 0.2	1.00	1.0 ± 0.1

Last column corresponds to fitness estimates at passage 40. For more details see text and Elena *et al.* (1998).

properties of the fitness trajectories. The model has two main properties: mean fitness evolves rapidly during the first passages and asymptotically approaches a maximum value. The reason for the deceleration in the rate of adaptation has to do with the availability of beneficial mutations and the magnitude associated with each available mutation. At the beginning of the process, when the viral population is far from the optimum, any beneficial mutation will increase fitness by driving it upward in a fitness landscape (Orr, 1998). However, as the population approaches the fitness peak, only fine-tuning mutations, less abundant than the former, will be needed. Once the virus becomes adapted to its new environment, no further improvement is expected unless the environment changes.

The issue regarding fitness changes also should be considered when viruses shift among environments. Three types of *in vitro* experiments have been done in relation to environmental changes. First, experiments of adaptation of VSV to new hosts (Holland *et al.*, 1991; Novella *et al.*, 1999) has been done. VSV clones, previously adapted to BHK cells, gained fitness after few passages on alternative new host cells. It is controversial, however, if VSV also concurrently increased fitness on ancestral BHK cells. This question needs to be addressed under appropriate and more replicated experiments. Second, the fitness changes associated with adaptation to fluctuating environments also were explored. Experiments have been done with VSV and Eastern equine encephalitis virus (Novella *et al.*, 1999; Weaver *et al.*, 1999), where viruses were grown in two new cell types, changing daily. Both viruses were selected with increased fitness in both novel environments. Adaptation seems, however, to be reversible. VSV clones adapted to persistently replicate in LL5 insect cells showed low fitness on BHK cells and in mouse brain (Novella *et al.*, 1995), but attenuated virus soon recovered fitness and virulence after a few passages back in BHK. Third, *in vitro* experiments were performed with antiviral agents added to the media. When cells were treated with α-interferon, the size of VSV viral populations experienced a dramatic reduction (99.9%). Those viruses that developed α-interferon resistance quickly increased their fitness in the presence of the interferon (Novella *et al.*, 1996). However, fitness decreased when interferon was removed from the media.

Another question that could be of great importance to viral populations is its eventual differentiation into subpopulations (for instance different organs or tissues within an individual), and then subsequent migration among subpopulations. We have experimentally modeled this situation by means of an *in vitro* system to simulate migration of VSV among isolated homogeneous host cell populations (see Miralles *et al.*, 1999 for a detailed description of the experimental protocol). The results clearly demonstrated a positive correlation between migration rate and the magnitude of the mean fitness across subpopulations and, with less support, a decrease in the fitness difference among subpopulations with the magnitude of the migration. The results, in full agreement with population genetics theory, can be explained by the spread of beneficial mutations, originated in single isolated populations, through the entire population.

CLONAL INTERFERENCE IMPOSES A LIMIT ON THE RATE OF VIRUS ADAPTATION

Viral populations adapt through the appearance and fixation of beneficial mutations. In large asexual populations such as of VSV, where mutation rate and population size are the key state variables, beneficial mutations may arise frequently enough that two or more are coexisting. When this happens, among other properties of the clonal interference model (Guerrish and Lenski, 1992), we expect, for increasing population sizes, that (*i*) the fitness effect associated with a fixed beneficial mutation will be larger, and (*ii*) the rate of adaptation will tend toward a limit. These two predictions were demonstrated to hold in experimental populations of VSV (Miralles *et al.*, 1999). As already described, we carried out experiments with two competing populations of initially similar fitness (i.e., MARM C and wild type). Experiments were done at seven (five times replicated) different effective population sizes ranging from 100 to 10^8 viral particles. Once one of two variants became fixed, we measured its fitness relative to its nonevolved counterpart. The major results are summarized in Fig. 3. The larger the population size, the bigger the effect associated with a beneficial mutation that becomes fixed in the population (Fig. 3*A*). In addition (Fig. 3*B*), the rate of adaptation slows down with increasing effective population size as a consequence of longer times required for fixation of beneficial mutations: in other words, the winning clone must outblock more alternative less-beneficial genotypes.

From data shown in Fig. 3 it is possible to estimate the rate at which beneficial mutations are generated, as well as its average effect on fitness (Guerrish and Lenski, 1992). The estimated value for the beneficial mutation rate (Miralles *et al.*, 1999) was 6.4×10^{-8} beneficial mutations per genome and generation. The effective population sizes studied, all higher than 10^9, warranted the possibility of appearance of a beneficial mutation, giving support to the assumption that a single beneficial mutation in each lineage is responsible for the fitness increase (Fig. 3). Comparing our estimate with that of the total mutation rate in VSV (2–3.5 substitutions per genome and generation), we can infer that one of 2×10^8 mutations produced in VSV can be considered as beneficial. On the other hand, the maximum-likelihood estimate of the mean selective advantage of all ben-

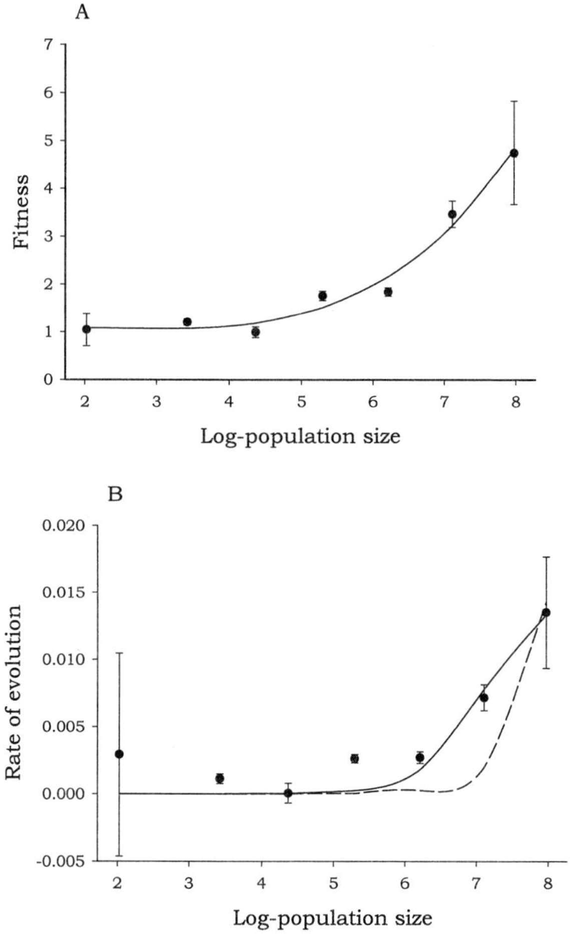

FIGURE 3. (A) Influence of effective population size on the fitness effect of beneficial mutations that are fixed. The model of the fitted line is described in Miralles *et al.* (1999). (B) Effect of the effective population size on the rate of evolution. See text for more details. The dashed line corresponds to the fit of a liner model to the data. The solid line is the fit to a hyperbolic model. Both curves appear to be exponential because of the logarithmic scale of x axis.

eficial mutations produced in the population (see Miralles *et al.*, 1999 for details) was 31% per day. Most mutations have small deleterious effect on fitness, whereas the few beneficial ones that appear in the population will have a large effect on it.

NUCLEOTIDE DIVERSITY AND FITNESS RECOVERY IN THE EVOLUTION OF A HIGHLY DEBILITATED VSV EXPERIMENTAL POPULATION: THE SAMPLING PROBLEM

As mentioned above, we have studied fitness recovery in the evolution of a highly sick VSV viral population clone (MARM *F*) obtained from 20 consecutive plaque-to-plaque passages (Elena *et al.*, 1998). The MARM *F* clone was diluted and plated on a monolayer of BHK cells, and four well-isolated plaques were collected. Fitness of each subclone was estimated and the relative mean fitness was 0.00015 relative to wild type (Elena *et al.*, 1998). After a few large population passages (Elena *et al.*, 1998, regime E), the four subclones recovered fitness several orders of magnitude, and by passage 40 it reached a mean value of 1.1 ± 0.0, not significantly different from the wild type. Then, we estimated fitness recovery in one of the subclones at three more times (after 28 h, 2 days, and 5 days of large consecutive passages) (Table 6). In addition, two VSV genome regions, representing ≈10% of the VSV genome were sequenced. One of them, 514 nt long, includes part of the G gene, and the other one, 510 nt long, comprises part of the P and M genes (see Fig. 1), including a 56-nt noncoding region of the P gene. This noncoding region contains a highly conserved sequence of 11 nt (3′-AUACUUUUUUU-5′) in the vesiculovirus genus, which is involved in the synthesis of the poly(A) tail during transcription process (Conzelmann, 1998). The poly(U) tract of this sequence suffers frequent insertions and deletions that could involve a fitness reduction. Thus, in VSV, removal of a single U residue abolished the synthesis of the monocistronic upstream transcript completely (Conzelmann, 1998). A less dramatic effect showed the addition of U residues to the VSV signal. We sequenced 20 clones of each region from isolates taken at six time points of the experiment: 0 h (the original subclone), 28 h, 2 days, 5 days, 8 days, and after 40 days of daily mass expansion (Bracho *et al.*, 1998; Elena *et al.*, 1998). Some of the results obtained are shown in Table 6. As can be observed, the number of synonymous and nonsynonymous polymorphic sites was low for both regions at any time. These numbers, as well as nucleotide diversity, were expected to increase from 0 h to 40 days. They were derived from a single plaque with an extremely low initial fitness. Such tendency was observed only when comparing the sample taken at day 40 with any of the other five moments for both regions (average between sampled corrected distances of 0.0024 ± 0.0003 for region G, and 0.0037 ± 0.0004 for regions M–P). However, the low numbers of mutations detected in this study forced us to consider two key points related with the estimation of genetic variability of RNA viral populations and its relationship with fitness change. First, how correct are the available estimates of the mutation rate for VSV of 10^{-3} to 10^{-4} substitu-

TABLE 6. Mean fitness, with standard error, number of polymorphic, nonsynonymous (NS) and synonymous (S) sites as well as nucleotide diversity, π (estimated as number of substitutions per site), with standard error, of regions G (510 nt) and M-P (514 nt)

		G region			M-P region			
Time	Fitness	NS	S	π ($\times 10^3$)	NS	S	π ($\times 10^3$)	Indels
0 h	0.00041 ± 0.00004	0	0	0.00 ± 0.00	0	0	0.00 ± 0.00	0(14), + 1 (4),+ 2(1), –1(1)
28 h	0.22 ± 0.08	0	0	0.00 ± 0.00	0	1	0.20 ± 0.32	0(17), + 1(3)
2 d	0.22 ± 0.07	0	0	0.00 ± 0.00	2	0	0.39 ± 0.24	0(14), + 1(4), + 2(1), +13(l)
5 d	—	0	0	0.00 ± 0.00	2	0	0.39 ± 0.24	0(19), + 2(l)
8 d	0.40 ± 0.10	1	0	0.20 ± 0.30	0	0	0.00 ± 0.00	0(19), + 6(l)
40 d	1.10 ± 0.00	3	2	2.12 ± 0.32	1	1	0.39 ± 0.24	0(19), + 2(l)

In the studied M-P noncoding region of the VSV genome there is a highly conserved stretch of seven Us (see text). Indel distribution corresponds to the number (between parenthesis) of zero, added (+), or deleted (–) Us found in a 20-clone sample. The values have been obtained in six time points of an experiment of fitness recovery of a highly debilitated MARM clone.

tions per nucleotide and per replication round (Drake and Holland, 1999)? If that figure is valid, then how large should both nucleotide and population samples be to detect a significant proportion of its genetic variability? It is expected that the fitness recovery of an expanding population of a highly debilitated clone should be characterized at the molecular level by repeated nonsynonymous substitution fixation, a pattern that has not been observed in the regions studied. If we consider that the estimated number of beneficial mutations in VSV expanding populations is low, then it might be possible that the two regions selected have not been the appropriate ones. This statement points toward a much higher nucleotide sampling. However, as can be seen in Table 6, there is a reduction of the frequency of VSV molecules showing indels in the U-stretch of the intergenic region. As mentioned above, indels in this region have an effect on the levels of transcription of the adjacent genes and can be responsible for the fitness reduction of the MARM *F* clone. A recent study (Escarmís *et al.*, 1999) on the evolution of highly debilitated foot-and-mouth disease virus clones under continuous population expansions give support to this idea. Escarmís *et al.* (1999) observed that the original debilitated clone had six-point mutations spread over the genome in addition to an elongated internal polyadenylate tract immediately preceding the second initiation AUG codon (Escarmís *et al.*, 1996). The point mutations were replaced and the polyadenylate tract disappeared after a large number of passages (Escarmís *et al.*, 1999).

QUASISPECIES AND POPULATION GENETICS THEORIES OF THE EVOLUTION OF RNA VIRUSES

There is abundant theoretical literature related to the dynamics of populations in which mutation is a frequent event. This is the case of Eigen and Schuster's notion of quasispecies (Eigen and Schuster, 1977) where the target of selection is no longer a single fittest genotype but rather a cloud of mutants distributed around a most frequent one quoted as master sequence. Following those authors, population genetics theory is a good descriptor for those populations where mutations are rare events and purifying selection is the main evolutionary force generating a homogeneous population, in addition to possible neutral variation fixed by random drift (Nowak and Schuster, 1989). They also suggested that the quasispecies theory is able to handle all situations, from small to large mutation rates. If true, then we have two different theories with different explanatory power, formally being the quasispecies theory of a wider application range than population genetics, in a way similar to the statement that general relativity is more general than Newton's mechanics. However, the theoretical models of population genetics are not compromised at all by the assumption of small mutation rates, as it can be appreciated when studying any formal presentation of the theory, and even much less when considering the evolution of haploid organisms. Consequently, in terms of mutation rates we have at least two competing theories of similar application range.

One important issue of the quasispecies theory is the notion that the target of selection is not just the fastest growing (i.e., the fittest) replicator, but a broad spectrum of mutants produced by erroneous copying of the original sequence. Nowak (1992) and Eigen (1996) defined the quasispecies in precise mathematical terms as the "dominant eigenvector which belongs to the largest eigenvalue of the replication matrix." This replication matrix contains the replication rates (i.e., selective values) of each mutant class, as well as mutation probabilities. However, because neutral mutants are not considered and population size is infinite, genetic drift formally seems not to operate in the quasispecies framework. The dynamics of a population of infinite size under continuous selection pressure will be determined by selective advantage of the fastest replicating variant. However, the frequency of a given mutant also depends on the probability with which it is generated by mutation from other closely related variants and their frequencies. The consequence of this scenario (Nowak, 1992) is that the individual sequence no longer serves as the unit (or target) of selection. On the contrary, the entire quasispecies should be this target.

The thesis of the present work is that both population genetics and quasispecies theories, when applied to the evolution of RNA viral populations, might have different explanatory power and/or application range. Historically, they represent two research traditions, and both have their own theoretical tools to explain the evolutionary dynamics of highly mutating populations of infinite size in which genetic drift is not present and mutants have always different selection coefficients. However, population genetics theory (*i*) does not require selective imposition of the fittest genotype, as the proponents of quasispecies theory seem to suggest, and (*ii*) formally contemplates the action of random drift and eventual fixation of some neutral mutants.

Most of the evidence giving support to the quasispecies theory comes from RNA viruses because they show high mutation rates and reach very high population numbers in a short time. In fact, virus populations normally consist of a widely dispersed mutant distribution rather than a homogeneous one formed by a single, most-fit, wild-type sequence (Frank, 1996). However, the realization of high levels of genetic variability in viral populations does not constitute *per se* an observation giving exclusive support to the quasispecies theory or to the theory of population genetics. A pattern of extremely high genetic variability also can be derived from a model exclusively based on mutation and selection on single replicons, a result derivable from the theoretical background on haploid organisms.

One central issue that could discriminate between both theories is related to the unit of selection of highly mutating replicons. According to the quasispecies theory, as previously stated, the unit is not the single replicon but a set of related mutant sequences, whose degree of kinship can be expressed by a genetic distance. Such a set of closely related sequences forms an evolving cloud that has a higher evolutionary (i.e., adaptive) plasticity than an evolving population exclusively based on independent replicons. Population genetics is not a theory based exclusively on the idea that the unit of selection is the individual. Group or kin selection models are relevant components of the population genetics theory. Consider, for instance, the evolution of virulence in parasite organisms. Explanatory hypotheses have been advanced that are not only based on tradeoff at the individual level between virulence and transmission rate (for a review see Frank, 1996), but also on group selection (Miralles *et al.*, 1997).

There is an agreement between quasispecies theory and group selection models, because the cloud or distribution of closely related mutants fulfills the three conditions that the latter imposed to consider a group as the target of selection. First, we are dealing with replicative entities whose population structure promotes fast genetic divergence (Wade, 1976). Second, the members of the quasispecies are intimately related (Hull, 1980). Third, the whole distribution of mutants should be considered as an individual instead of a group (Hull, 1980). In summary, objections and explanatory power of both theories can be exchanged. Once stated the formal equivalence between quasispecies theory and group selection models, the major problem to be solved is to gain experimental evidence in favor of a supra individual unit of selection. However, few experimental or field observations have been reported, and the issue is still highly controversial.

We thank Prof. Amparo Latorre for critical reading and suggestions to improve the manuscript. This work has been supported by Grant PM97-0060-C02-02 from the Spanish Dirección General de Enseñanza Superior. A.B. and R.M. also have been recipients of fellowships from the Ministerio de Educación y Ciencia.

REFERENCES

Bracho, A., Moya, A. & Barrio, E. (1998) Contribution of Taq induced errors to the estimation of RNA virus diversity. *J. Gen. Virol.* 79, 2921–2928.

Cannings, C. (1974) Drift in haploid models. *Adv. Appl. Prob.* 7, 4–5.

Cannings, C. (1975) The latent roots of certain Markov chains arising in genetics: a new approach, II. Further haploid models. *Adv. Appl. Prob.* 7, 264–282.
Chao, L. (1990) Fitness of RNA virus decreased by Muller´s ratchet. *Nature (London)* 348, 454–455.
Clarke, D., Duarte, E., Moya, A., Elena, S. F., Domingo, E. & Holland, J. J. (1993) Genetic bottlenecks and population passages cause profound fitness differences in RNA viruses. *J. Virol.* 67, 222–228.
Clarke, D. K., Duarte, E. A., Elena, S. F., Moya, A., Domingo, E. & Holland, J. J. (1994) The Red Queen reigns in the kingdom of RNA viruses. *Proc. Natl. Acad. Sci. USA* 91, 4821–4824.
Conzelmann, K. K. (1998) Nonsegmented negative-strand RNA viruses: genetics and manipulation of viral genomes. *Annu. Rev. Genet.* 32, 123–162.
Cook, R. D. & Nassar, R. F. (1972) Dynamics of finite populations. I. The expected time to fixation or loss and the probability of fixation of an allele in a haploid population of variable size. *Biometrics* 28, 373–384.
De la Peña, M., Elena, S. F. & Moya, A. (2000) Effect of deleterious mutation-accumulation on the fitness of RNA bacteriophage MS2. *Evolution* 54, 686–691.
Domingo, E. & Holland, J. J. (1997) RNA virus mutations and fitness for survival. *Annu. Rev. Microbiol.* 51, 151–178.
Drake, J. W. & Holland, J. J. (1999) Mutation rates among RNA viruses. *Proc. Natl. Acad. Sci. USA* 96, 13910–13913.
Drobník, J. & Dlouhá, J. (1966) Statistical model of evolution of haploid organisms during simple vegetative reproduction. *J. Theor. Biol.* 11, 418–435.
Duarte, E. A., Clarke, D. K., Moya, A., Domingo, E. & Holland, J.J. (1992) Rapid fitness losses in mammalian RNA virus clones due to Muller´s ratchet. *Proc. Natl. Acad. Sci. USA* 89, 6015–6019.
Duarte, E. A., Clarke, D. K., Moya, A., Elena, S. F., Domingo, E. & Holland, J. J. (1993) Many-trillionfold amplification of single RNA virus particle fails to overcome the Muller´s ratchet effect. *J. Virol.* 67, 3620–3623.
Duarte, E. A., Novella, I. S., Ledesma, S., Clarke, D. K., Moya, A., Elena, S. F., Domingo, E. & Holland, J. J. (1994) Subclonal components of consensus fitness in an RNA virus clone. *J. Virol.* 68, 4295–4301.
Eigen, M. & Schuster, P. (1977) The hypercycle: a principle of natural self-organization. Part A: emergence of the hypercycle. *Naturwissenschaften* 64, 541–565.
Eigen, M. (1996) On the nature of virus quasispecies. *Trends Microbiol.* 4, 216–218.
Elena, S. F. (1999) Little evidence for synergism among deleterious mutations in a nonsegmented RNA virus. *J. Mol. Evol.* 49, 703–707.
Elena, S. F., Ekunwe, L., Hajela, N., Oden, S. A. & Lenski, R. E. (1998) Distribution of fitness effects caused by random insertion mutations in *Escherichia coli. Genetica* 102/103, 349–358.
Elena, S. F., Dávila, M., Novella, I. S., Holland, J. J., Domingo, E. & Moya, A. (1998) Evolutionary dynamics of fitness recovery from the debilitating effects of Muller's ratchet. *Evolution* 52, 309–314.
Elena, S. F., González-Candelas, F., Novella, I. S., Duarte, E. A., Clarke, D. K., Domingo, E., Holland, J. J. & Moya, A. (1996) Evolution of fitness in experimental populations of vesicular stomatitis virus. *Genetics* 142, 673–679.
Elena, S. F., Miralles, R. & Moya, A. (1997) Frequency-dependent selection in a mammalian RNA virus. *Evolution* 51, 984–987.
Elena, S. F. & Moya, A. (1999). Rate of deleterious mutation and the distribution of its effects on fitness in vesicular stomatitis virus. *J. Evol. Biol.* 12, 1078–1088.

Emigh, T. H. (1979a) The dynamics of finite haploid populations with overlapping generations. I. Moments, fixation probabilities and stationary distributions. *Genetics* 92, 323–337.

Emigh, T. H. (1979b) The dynamics of finite haploid populations with overlapping generations. II. The diffusion approximation. *Genetics* 92, 339–351.

Escarmís, C., Dávila, M., Charpentier, N., Bracho, A., Moya, A. & Domingo, E. (1996) Genetic lesions associated with Muller´s ratchet in an RNA virus. *J. Mol. Biol.* 264, 255–267.

Escarmís, C., Dávila, M. & Domingo, E. (1999) Multiple molecular pathways for fitness recovery of an RNA virus debilitated by operation of Muller's ratchet. *J. Mol. Biol.* 285, 495–505.

Felsenstein, J. (1971) The rate of loss of multiple alleles in finite haploid populations. *Theor. Pop. Biol.* 2, 391–403.

Frank, S. (1996) Models of parasite virulence. *Q. Rev. Biol.* 71, 37–78.

Gerrish, P. J. & Lenski, R. E. (1998) The fate of competing beneficial mutations in an asexual population. *Genetica* 102/103, 127–144.

Gillespie, J. H. (1973) Natural selection with varying selection coefficients—a haploid model. *Genet. Res.* 21, 115–120.

Gillespie, J. C. (1991) *The Causes of Molecular Evolution* (Oxford Univ. Press, Cambridge).

Gladstein, K. (1976) Loss of alleles in a haploid population with varying environment. *Theor. Pop. Biol.* 10, 383–394.

Haigh, J. (1978) The accumulation of deleterious genes in a population—Muller´s ratchet. *Theor. Pop. Biol.* 14, 251–267.

Hardin, G. (1960) The competitive exclusion principle. *Science* 131, 1292–1297.

Holland, J. J., de la Torre, J. C., Clarke, D. K. & Duarte, E. A. (1991) Quantitation of relative fitness and great adaptability of clonal populations of RNA viruses. *J. Virol.* 65, 2960–2967.

Hull, D. L. (1980). Individuality and selection. *Annu. Rev. Ecol. Syst.* 11, 311–332.

Karlin, S. & McGregor, J. (1971) On mutation selection balance for two-locus haploid and diploid populations. *Theor. Pop. Biol.* 2, 60–70.

Keightley, P. D. (1994) The distribution of mutation effects on viability in *Drosophila melanogaster*. *Genetics* 138, 1315–1322.

Kimura, M. (1983) *The Neutral Theory of Molecular Evolution* (Cambridge Univ. Press, Cambridge).

Lewontin, R. C. (1974) *The Genetic Basis of Evolutionary Change* (Columbia Univ. Press, New York).

Li, W.-H. (1997) *Molecular Evolution* (Sinauer, Sunderland, MA).

Miralles, R., Gerrish, P. J., Moya, A. & Elena, S. F. (1999) Clonal interference and the evolution of RNA virus. *Science* 285, 1745–1747.

Miralles, R., Moya, A. & Elena, S. F. (1997) Is group selection a factor modulating the virulence of RNA viruses? *Genet. Res.* 69, 165–172.

Miralles, R., Moya, A. & Elena, S. F. (1999) Effect of population patchiness and migration rates on the adaptation and divergence of vesicular stomatitis virus quasispecies populations. *J. Gen. Virol.* 80, 2051–2059.

Moran, P. A. P. (1957) A two locus haploid population with overlapping generations. *Camb. Philos.* 54, 463–467.

Mukai. T., Chigusa, S. I., Mettler, L. E. & Crow, J. F. (1972) Mutation rate and dominance of genes affecting viability in *Drosophila melanogaster*. *Genetics* 72, 335–355.

Muller, H. J. (1964) The relation of recombination to mutational advance. *Mutat. Res.* 1, 2–9.

Novella, I. S., Cilnis, M., Elena, S. F., Kohn, J., Moya, A., Domingo, E. & Holland, J. J. (1996) Large-population passages of vesicular stomatitis virus in interferon-treated cells select variants of only limited resistance. *J. Virol.* 70, 6414–6417.

Novella, I. S., Clarke, D. K., Quer, J., Duarte, E. A., Lee, C. H., Weaver, S. C., Elena, S. F., Moya, A., Domingo, E. & Holland, J. J. (1995) Extreme fitness differences in mammalian and insect hosts after continuous replication of vesicular stomatitis virus in sanfly cells. *J. Virol.* 69, 6805–6809.

Novella, I. S., Elena, S. F., Moya, A., Domingo, E. & Holland, J. J. (1995) Size of genetic bottlenecks leading to virus fitness loss is determined by mean initial population fitness. *J. Virol.* 69, 2869–2872.

Novella, I. S., Elena, S. F., Moya, A., Domingo, E. & Holland, J. J. (1996) Repeated transfer of small RNA virus populations leading to balanced fitness with infrequent stochastic drift. *Mol. Gen. Genet.* 252, 733–738.

Novella, I. S., Hershey, C. L., Escarmis, C., Domingo, E. & Holland, J. J. (1999) Lack of evolutionary stasis during alternating replication of an Arbovirus in insect and mammalian cells. *J. Mol. Biol.* 287, 459–465.

Nowak, M. (1992) What is a quasispecies? *Trends Ecol. Evol.* 7, 118–121.

Nowak, M. & Schuster, P. (1989) Error thresholds of replication in finite populations, mutation frequencies and the onset of Muller's ratchet. *J. Theor. Biol.* 137, 375–395.

Orr, H. A. (1998) The population genetics of adaptation: the distribution of factors fixed during adaptive evolution. *Evolution* 52, 935–949.

Robson, D. S. (1957) Some biometrical formulae for the analysis of quantitative inheritance systems involving two haploid or inbred diploid parents. *Genetics* 42, 487–498.

Strobeck, C. (1979). Haploid selection with n alleles in m niches. *Am. Nat.* 113, 439–444.

Trajstman, A. C. (1973) A two locus haploid population with overlapping generations. *Biometrics* 29, 701–711.

Van Valen, L. (1973) A new evolutionary law. *Evol. Theory* 1, 1–30.

VandePol, S. B., Lefrancois, L. & Holland, J. J. (1986) Sequences of the major antibody binding epitopes of the Indiana serotype of vesicular stomatitis virus. *Virology* 148, 312–325.

Wade, M. J. (1976) Group selection among laboratory populations of *Tribolium*. *Proc. Natl. Acad. Sci. USA* 73, 4604–4607.

Weaver, S. C., Brault, A. C., Kang, W. & Holland, J. J. (1999) Genetic and fitness changes accompanying adaptation of an arbovirus to vertebrate and invertebrate cells. *J. Virol.* 73, 4316–4326.

Yuste, E., Sánchez-Palomino, S., Casado, C., Domingo, E. & López-Galíndez, C. (1999) Drastic fitness loss in human immunodeficiency virus type 1 upon serial bottleneck events. *J. Virol.* 73, 2745–2751.

6

Effects of Passage History and Sampling Bias on Phylogenetic Reconstruction of Human Influenza A Evolution

ROBIN M. BUSH[†], CATHERINE B. SMITH[§], NANCY J. COX[§], AND WALTER M. FITCH[†]

In this paper we determine the extent to which host-mediated mutations and a known sampling bias affect evolutionary studies of human influenza A. Previous phylogenetic reconstruction of influenza A (H3N2) evolution using the hemagglutinin gene revealed an excess of nonsilent substitutions assigned to the terminal branches of the tree. We investigate two hypotheses to explain this observation. The first hypothesis is that the excess reflects mutations that were either not present or were at low frequency in the viral sample isolated from its human host, and that these mutations increased in frequency during passage of the virus in embryonated eggs. A set of 22 codons known to undergo such "host-mediated" mutations showed a significant excess of mutations assigned to branches attaching sequences from egg-cultured (as opposed to cell-cultured) isolates to the tree. Our second hypothesis is that the remaining excess results from sampling bias. Influenza surveillance is purposefully biased toward sequencing antigenically dissimilar strains in an effort to

[†]Department of Ecology and Evolutionary Biology, University of California, Irvine, CA 92697; and [§]Influenza Branch, Centers for Disease Control and Prevention, Atlanta, GA 30333

This paper was presented at the National Academy of Sciences colloquium "Variation and Evolution in Plants and Microorganisms: Toward a New Synthesis 50 Years After Stebbins," held January 27–29, 2000, at the Arnold and Mabel Beckman Center in Irvine, CA.

Abbreviations: HA, hemagglutinin; HI, HA inhibition; HM, host-mediated.

identify new variants that may signal the need to update the vaccine. This bias produces an excess of mutations assigned to terminal branches simply because an isolate with no close relatives is by definition attached to the tree by a relatively long branch. Simulations show that the magnitude of excess mutations we observed in the hemagglutinin tree is consistent with expectations based on our sampling protocol. Sampling bias does not affect inferences about evolution drawn from phylogenetic analyses. However, if possible, the excess caused by host-mediated mutations should be removed from studies of the evolution of influenza viruses as they replicate in their human hosts.

It is well known that some pathogenic microbes undergo adaptation in response to laboratory culture. Host-mediated (HM) mutations have been particularly well studied in the influenza A virus (Robertson, 1993). However, this phenomenon has been documented in many other viruses, such as HIV, Japanese encephalitis virus, hepatitis A, and Sendai virus as well (Graff *et al.*, 1994; Sawyer *et al.*, 1994; Cao *et al.*, 1995; Itoh *et al.*, 1997). Molecular evolution studies using such sequences thus risk drawing inferences about the adaptation of the pathogen to its natural host from data containing laboratory artifacts. Additional problems may result from analysis of data sets that do not represent random samples of natural pathogen populations, or for which the sampling design is unknown. Here we determine the extent to which HM mutations and a known sampling bias affect studies of influenza A evolution.

Recent phylogenetic reconstruction of the evolution of human influenza A hemagglutinin (HA) of the H3 subtype revealed a 40% excess of amino acid replacements assigned to the terminal branches of the tree (Bush *et al.*, 1999a). The 40% excess of coding changes on terminal branches was calculated by using expectations based on the relative number of internal and terminal branches of the tree in Fig. 1. This observation was made in the course of identifying codons at which mutation appeared to have been adaptive in evading the human immune system. Because we used phylogenetic trees to model HA evolution (Bush *et al.*, 1999b), it was critical for our analyses that the excess mutations not be caused by evolutionary processes other than the ongoing evolution of the virus during replication in the human host. We proposed a number of hypotheses to explain the excess, but did not explore them in detail. Instead we simply deleted all mutations assigned to terminal branches from our analyses. In this paper we have tested two hypotheses that help to explain our observation.

The first hypothesis is that the excess consists of mutations that were either not present or were at low frequency in the viral sample when isolated from its human host. Although such mutations may increase in

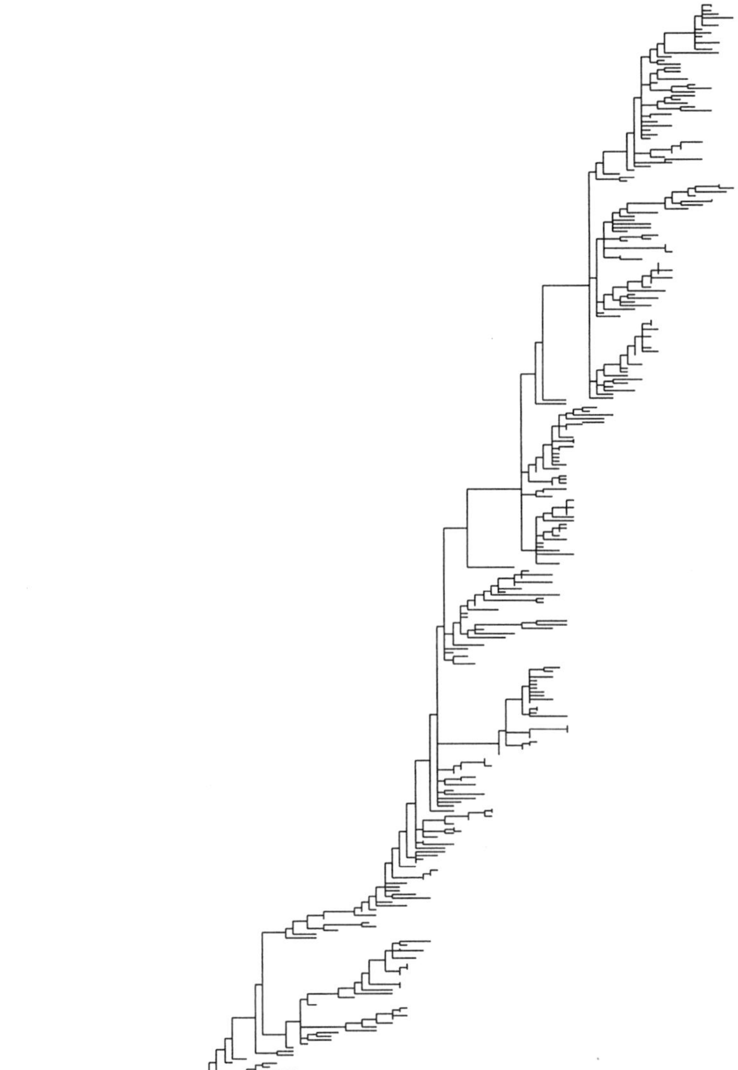

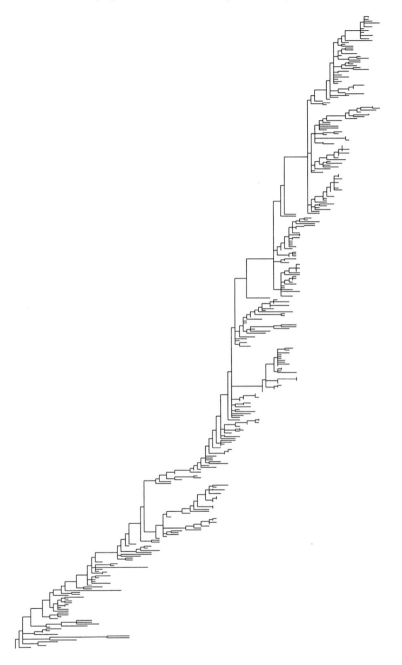

FIGURE 1. Maximum parsimony tree constructed from 357 HA1 genes of the human influenza virus type A subtype H3.

frequency in the laboratory because of genetic drift, for at least 22 HA1 codons an increase in frequency is thought to reflect a response to selective pressure for growth in embryonated chicken eggs (Bush *et al.*, 1999a). Such HM mutations most likely will appear on a phylogenetic tree as an additional mutation on a terminal branch, which is the branch attaching the sequence from a viral isolate to the tree. Phylogenetic reconstruction is based on similarity at all 329 codons. A HM mutation will alter only one of the 329 codons. Thus, a sequence of an isolate containing a HM mutation would in most cases still be most similar to the sequence from that isolate's closest relative. The effect of the HM mutation on the phylogenetic tree would be an increase in the length of the terminal branch joining the sequence from the egg-cultured isolate to the tree rather than a change in the point at which the branch is attached to the tree (Fig. 2).

The 22 suspected HM codons (Table 1) make up only 6.7% of the 329 codons in the HA1 domain, yet they account for 36.0% of the amino acid replacements across the HA tree in Fig. 1. Codons other than the set of 22 HM codons also may be found to undergo HM mutation with future study. There is thus great potential for error in inference if one assumes that HM mutations reflect evolution of influenza viruses within the human host. Here we test for the presence of HM mutations in our data set by examining the distribution of mutations in the HM and non-HM codons between branches attaching sequences from egg-cultured and cell-cultured isolates to the tree.

The second hypothesis to explain why we observed excess mutations assigned to the terminal branches of the HA tree is sampling bias. Our sequencing efforts are largely a contribution toward the World Health Organization influenza surveillance program. A priority in influenza surveillance is the identification of antigenically novel isolates from which previous infection with epidemic strains or prior immunization would not protect. The first level of screening for antigenic variants is the HA inhibition (HI) test, in which viral isolates are tested against postinfection ferret antiserum containing antibodies against HA from currently circulating strains of human influenza. We preferentially sequence the HA1 of isolates that appear, on the basis of the HI test, to be antigenically different from known circulating strains.

We illustrate how a bias against sequencing closely related viruses affects phylogenetic reconstruction in Fig. 2. In this hypothetical example, the tree on the left depicts the total population and each branch represents a single unique mutation. The tree on the right was constructed from a subset of eight relatively unrelated isolates. One of the 22 mutations used to construct the right-hand tree (on branch 4) reflects a HM change. Of the remaining 21 mutations, 15 are assigned to the eight terminal branches, and the remaining six mutations are assigned to the six internal branches.

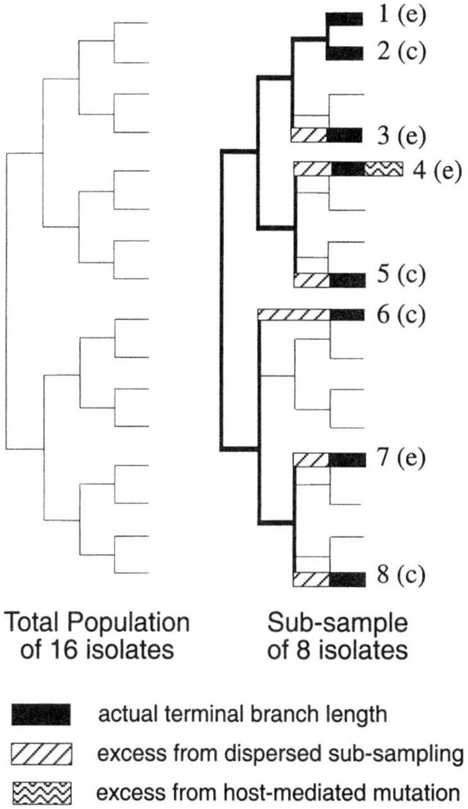

FIGURE 2. Partitioning mutations assigned to terminal branches of a phylogenetic tree. The tree on the left represents the evolution of a population of 16 viruses that each differ from their ancestor by one unique mutation. The tree on the right is a reconstruction after (i) sampling only eight of the viruses with a bias against sequencing closely related isolates and (ii) propagating the isolates in embryonated chicken eggs (e) or cell culture (c) in the laboratory. The tree constructed of sampled sequences is shown in black, with the terminal branches as thicker lines. The branch attaching sequence 4 (an egg-cultured isolate) to the tree is one mutation longer than it should be. The additional mutation is HM, that is, a mutation not present or at low frequency in the isolate before laboratory propagation. The branches attaching sequences from isolates 3–8 to the tree are longer than they would have been if our sample had included their nearest relatives. The increased length of branch 4 is in part caused by a process other than the ongoing evolution of the virus during replication in the human host. The remaining excess is simply a reflection of sampling bias, and thus does not affect evolutionary inferences made from the tree.

TABLE 1. Codons known to undergo HM mutations during propagation in egg culture

Codon	Rbs	AB	PosSel
111	0	0	0
126	0	1	0
137	1	1	0
138	1	1	1
144	0	1	0
145	0	1	1
155	1	1	0
156	0	1	1
158	0	1	1
159	0	1	0
185	0	0	0
186	0	1	1
193	0	1	1
194	1	1	1
199	0	0	0
219	0	0	0
226	1	0	1
229	0	0	0
246	0	0	0
248	0	0	0
276	0	0	0
290	0	0	0

Five of the 22 HM codons known to undergo HM mutations during propagation in egg culture are associated with the HA sialic acid receptor binding site (Rbs), 12 HM codons are in or near antibody combining sites A or B (AB). Eight HM codons have been identified as having been under positive selection (PosSel) to change the amino acid they encoded in the past (Bush *et al.*, 1999).

If mutations were assigned to terminal and internal branches in proportion to the relative number of each branch type, we would expect to have 12 mutations assigned to the terminal branches. However, we observed 15 mutations on the terminal branches, an excess of 25% over expectations. Thus intentionally sampling with a bias toward genetically divergent isolates results in those isolates being attached to the tree by longer branches than if their close relatives were also in the sample.

Unlike the excess mutations assigned to terminal branches that are caused by HM change, the excess caused by sampling bias is not of concern with respect to the evolutionary inference one might draw from the tree. Apportioning excess terminal branch lengths to the two different hypotheses is easily illustrated in a cartoon such as Fig. 2. In reality, we know that we have observed an excess of mutations on the terminal branches; however, we don't know precisely which branches or mutations are involved. In this paper we show how to determine the proportion, but not the actual identities of HM mutations present in such a data set. After partitioning the excess caused by HM mutations out of the data set, we can determine whether the remaining excess is consistent with what we would expect given our sampling scheme. This was done through comparison with trees produced from sampling a simulated data set.

DESCRIPTION OF DATA SET AND DEFINITION OF TERMS

Fig. 1 shows the phylogenetic tree for which we recently reported an excess of mutations assigned to the terminal branches (Bush *et al.*, 1999a, b). This tree was constructed by using the maximum parsimony routine of PAUP* 4.0b2 (Swofford, 1999) using 357 sequences, each 987 nt in length, produced from isolates collected between 1983 and September 30, 1997 (Bush *et al.*, 1999a, b). The terminal nodes of a tree are the sequences obtained from isolates in the laboratory. Internal nodes are the ancestors of the terminal nodes as reconstructed by the parsimony algorithm. Terminal branches attach terminal nodes, that is, the sequence from an isolate, to the tree. All other branches are internal branches.

We use the term egg isolates when referring to the 152 isolates that were propagated in embryonated chicken eggs in the laboratory. The egg isolates also may have been previously propagated in cell culture. We use the term cell isolates to refer to the 148 isolates propagated in cell culture but never in eggs. The remaining sequences were obtained from direct PCR (n = 3) or from isolates of partially unknown passage history (n = 54). The propagation histories of these isolates (GenBank accession nos. AF008656–AF008909 and AF180564–AF180666) can be found in the curated influenza database at Los Alamos National Laboratory (http://www.flu.lanl.gov/).

For this study we constructed additional trees by using two different samples of the original data set. Trees constructed using only the 152 sequences obtained from isolates propagated in eggs or using only the 148 sequences from viruses propagated in cells are referred to as the egg tree and the cell tree, respectively. Twenty two codons (Table 1) have been reported to undergo HM replacements in influenza isolates grown in eggs (Bush *et al.*, 1999a). We refer to these 22 codons as the HM codons and the other 307 codons as the non-HM codon set. Silent and nonsilent nucleotide substitutions were abbreviated by the letters S and NS, respectively.

Comparing the Phylogenetic Distribution of Nonsilent and Silent Substitutions

Analyses reported in this paper were performed by using all substitutions, only nonsilent substitutions, or only silent substitutions. Because the results in most cases were very similar, and because nonsilent and silent substitutions are distributed similarly across the internal and terminal branches of the tree in Fig. 1 (Table 2), all analyses reported below used only the nonsilent substitutions, unless stated otherwise.

Reconstructing Ancestral Character States

In the last step of the process by which parsimony algorithms assign mutations to the branches of a tree, mutations are assigned along each lineage starting at the root and moving along the lineage toward the terminal nodes. In some lineages on the tree in Fig. 1 there was flexibility as to which branches the mutations could be assigned. In our previous work, when there was a choice, we set our algorithm to delay assigning mutations as long as possible (Bush *et al.*, 1999a, b). That is, mutations were assigned to branches that were as far from the root as possible. We did this to minimize the extent to which HM mutations were assigned to the internal branches of the tree. In the present work, however, one goal is to identify and quantify HM mutations. HM mutations are most likely to be assigned to the terminal branches, thus we did not want to assign mutations to the terminal branches unless it was necessary to do so. Resetting our algorithm to assign mutations as close to the root as possible caused a net change of 14 replacements to be shifted from the terminal to internal branches. This reduced the excess of nonsilent substitutions on the terminal branches from 40.0% to 36.5% (Table 3). Thus our assignment procedure was not responsible for the majority of the observed excess. We retain the assignment procedure that minimizes the number of mutations assigned to the terminal branches for all analyses that follow.

HYPOTHESIS 1: HM MUTATIONS

We first determined whether there was evidence that HM mutations were contributing to the excess nonsilent substitutions on the terminal branches of the HA tree. We examined two sources of HM mutations.

TABLE 2. The distribution of nonsilent (NS) and silent (S) substitutions

Branch type	Number of branches	Exp NS	Obs NS	χ^2	Exp S	Obs S	χ^2
Terminal	357	503.1	510	0.12	405.9	399	0.12
Internal	355	242.9	235	0.24	196.1	204	0.25
Sum	712	745.0	745	0.36	603.0	603	0.37

Results of a 2 × 2 contingency test show that nonsilent and silent substitutions are similarly distributed across the terminal and internal branches of the tree in Fig. 1. Total $\chi^2 = 0.73$, df = 1, $P > 0.4$. Obs, observed; Exp, expected.

TABLE 3. The distribution of nonsilent (NS) substitutions across internal and terminal branches

Branch type	Number of branches	Exp NS	Obs NS	χ^2
Terminal	357	373.6	510	49.8
Internal	355	371.4	235	50.1
Sum	712	745.0	745	99.9

The tree in Fig. 1 has significantly more nonsilent substitutions assigned to its terminal branches than expected based on the relative numbers of internal and terminal branches, ($P < 0.05$, df = 1). Exp, expected; Obs, observed.

First, we looked for evidence that mutations were occurring at the 22 known HM codons. Second, we determined whether there were any additional codons, besides the 22 in the HM set, that showed evidence for undergoing HM mutations.

HM Mutations in the Egg and Cell Branches

If HM mutations were occurring in the 22 HM codons, then we should see excess mutations in the HM codons on the egg branches, or the terminal branches attaching sequences from egg-cultured isolates to the tree (Fig. 1). The expectations for this test are based on the distribution of mutations in the non-HM codons across the egg and cell branches. As would be expected if HM mutations were occurring, the set of 22 HM codons underwent a significantly greater number of nonsilent substitutions on the egg branches than expected based on the distribution of mutations at the non-HM codons (Table 4). The number of excess nonsilent substitutions caused by HM change can be estimated as follows. We first assume that the distribution of nonsilent substitutions in the non-HM codons to the egg and cell branches (49.6% and 50.4%, respectively) is

TABLE 4. The distribution of nonsilent (NS) substitutions in HM and non-HM codons across egg and cell branches

Branch type	Obs NS non-HM	Exp NS HM	Obs NS HM	χ^2
Egg branches	138	75.45	105	11.57
Cell branches	140	76.55	47	11.40
Sum	278	152.00	152	22.98

The HM codons had significantly more nonsilent substitutions on the terminal branches attaching sequences from egg-cultured isolates to the tree in Fig. 1 than on branches attaching sequences from cell-cultured isolates ($P < 0.05$, df = 1). Expectations are based on the distribution of nonsilent substitutions in non-HM codons. Obs, observed; Exp, expected.

unaffected by HM mutation. (We verify this assumption below.) We also assume that none of the 47 nonsilent substitutions in the HM codons on the cell branches were HM. Based on these assumptions the number of nonsilent substitutions we would expect on egg branches is 46.3, which is 58.7 fewer than the 105 observed (Table 4). If the 58.7 "excess" nonsilent substitutions were indeed HM, then approximately 8% of the 745 amino acid replacements in our data set did not occur within a human host.

HM Mutations in the Non-HM Codon Set

The 22 HM codons may not be the only codons undergoing HM mutation in the HA1 domain. There could be other codons that undergo HM mutation during propagation in eggs but have not as yet been identified as doing so. To explore this possibility we excluded the 22 HM codons from our data set and then contrasted the structure of two trees: one constructed by using only the 152 isolates known to have been grown in egg culture, and the other constructed by using the 148 isolates that had undergone passage in cell but not egg culture (not shown). The egg and cell data sets are similarly sized and contain isolates collected over the same range of time with the same sampling bias. If only the HM codons undergo HM mutation, the egg and cell trees should show similar distributions of replacements across the terminal and internal branches. However, if additional (non-HM) codons are accruing HM mutations, the egg tree should have a larger excess of mutations assigned to the terminal branches than the cell tree. We found a 30% excess of nonsilent substitutions on the terminal branches of both the egg tree and the cell tree (Table 5). Based on this analysis we find no evidence to support the hypothesis that codons in addition to the 22 in the HM codon set are undergoing HM mutations during laboratory passage unless they are doing so at the same rate at which they undergo mutations in response to passage in cell culture.

HYPOTHESIS 2: SAMPLING BIAS

We have shown that HM mutations in the 22 HM codons appear to be responsible for some of the excess mutations on the terminal branches of the HA1 tree in Fig. 1. We also have demonstrated that the non-HM codons do not appear to be undergoing HM mutations. We are now left to explain why there is still, after partitioning out the HM mutations, a 30% excess of mutations on the terminal branches of the egg and cell trees.

As illustrated in Fig. 2, a sampling scheme biased against sequencing closely related viruses will cause an excess of mutations to be assigned to the terminal branches of a tree. We preferentially sequence isolates that

TABLE 5. The phylogenetic distribution of nonsilent (NS) substitutions on trees constructed using only sequences from egg-cultured or cell-cultured isolates and using only non-HM codons

	Number of branches	Exp NS	Obs NS	χ^2
Branches on egg tree				
Terminal	152	119.3	155	10.7
Internal	150	117.7	82	10.8
Total	302	237.0	237	21.5
Branches on cell tree				
Terminal	148	121.3	158	11.1
Internal	146	119.7	83	11.2
Total	294	241.0	241	22.3

Trees constructed without the HM codons using sequences from isolates propagated in egg culture or in cell culture both showed significant excesses of nonsilent substitutions on their terminal branches ($P < 0.05$, df = 1 for both tests). The percent excess of nonsilent substitutions on the terminal branches was 30% for both the egg and cell trees.

we do not believe, based on HI tests, to be closely related to isolates already sequenced. For instance, in the 1996–1997 influenza season we sequenced only 7% of the isolates on which we performed HI assays. Because the isolates sent to the Centers for Disease Control and Prevention from the World Health Organization collaborating laboratories may themselves already be biased against commonly occurring isolates, the bias against sequencing closely related viruses is even greater than 7%. Based on this bias we expect the terminal branches on our trees to be longer on average than they would have been had we sampled randomly. Because we do not know the genetic structure of the influenza population circulating in nature, we cannot know how we actually sampled it. Thus, we cannot calculate the exact distribution of mutations we should expect on the terminal and internal branches of the tree constructed by using our sample. We can, however, determine whether the excess we have observed is consistent with what we would expect based on our sampling protocol.

We sampled a simulated viral population using various sampling schemes to determine the extent to which our observation is consistent with this hypothesis. We constructed a hypothetical population of 16 viral isolates and sampled it as illustrated in Fig. 3. The samples consisted of eight relatively unrelated isolates (the dispersed sample), eight closely related isolates (the clumped sample), and two collections of eight isolates sampled in an intermediate manner. To ensure that samples all included the total range of variation in the population, each included the uppermost and bottom-most isolate on the 16-isolate tree. The percent excess of

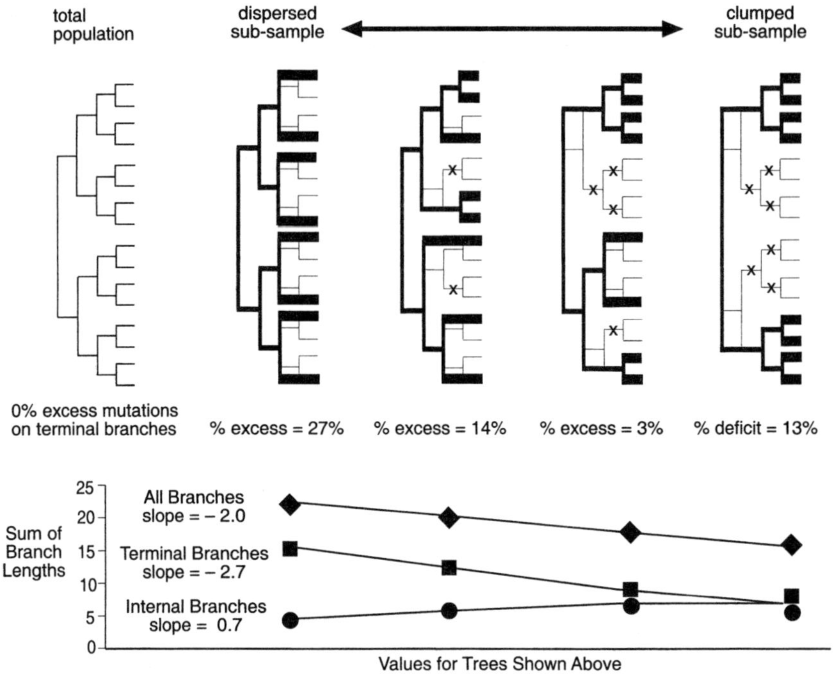

FIGURE 3. The effects of sampling bias on phylogenetic reconstruction. The tree on the left shows a hypothetical population of 16 isolates that each differ from their ancestor by one unique mutation. The four trees to the right show the original tree overlaid with the tree that would result from sampling only half of the total population. The tree constructed of sampled sequences is shown in black, with the terminal branches as thicker lines. Clumped sampling causes a decrease in the total genetic variation sampled. The mutations not captured in the sample would have been assigned only to internal branches, as shown by the symbol X. As a result, the proportion of mutations assigned to the internal and terminal branches changes with sampling dispersion, but not at the same rate (shown in the line plot at the bottom). Without knowledge of where a sample lies on such a continuum, there is no way to derive the expected proportion of mutations that should be assigned to the terminal and internal branches of a phylogenetic tree.

mutations assigned to the terminal branches of the eight-isolate trees was greatly influenced by the degree to which the sampled isolates were dispersed or clustered. The dispersed sample shows a 27.3% excess of mutations assigned to the terminal branches, the clumped sample has a 13% deficit.

The magnitude of the excess or deficit depends not only on the degree of dispersion, but also on the proportion of the total population sampled.

In the example in Fig. 3, 50% of the total population was sampled. If we were to increase the size of the total population from 16 to 64 and again sample only eight isolates, we would be sampling 12.5% of the total population. This is close to the percent of isolates (7%) that we sequenced based on results from HI tests. Sampling eight dispersed isolates of 64 results in a 47.5% excess of mutations on the terminal branches, a much greater excess than the 27.3% shown in Fig. 3. Thus, even though we do not know the actual distribution of genetic variation present in nature during the time span included in our study, and therefore do not know exactly how we sampled that variation, the magnitude of excess mutations assigned to the terminal branches of the tree in Fig. 1 is consistent with our sampling bias: we have sampled only a fraction of circulating viral strains and have done so in a consciously dispersed manner.

DISCUSSION

We found evidence suggesting that approximately 59 nonsilent substitutions assigned to the terminal branches of the HA tree in Fig. 1 were caused by HM mutations occurring in the set of 22 codons known to undergo HM mutation in chicken eggs in the laboratory. We have no way of identifying which 59 particular substitutions were HM except that they are among the 105 nonsilent substitutions assigned to branches attaching sequences of egg-cultured isolates to the tree. We found no evidence to suggest that HM mutations are occurring at the other 307 codons in the HA1. The majority of the excess mutations that were assigned to terminal branches of the HA tree are most likely simply the result of sampling bias. Detailed antigenic and genetic analysis of viruses collected during influenza surveillance is purposefully biased toward sequencing antigenically dissimilar strains in an effort to identify new antigenic and genetic variants that may signal the need to update the vaccine. Thus, viral isolates that are antigenically very similar to the predominant antigenic variant that circulates during a particular influenza season are sequenced less often than are antigenically variant strains.

The 59 apparently HM mutations represent 7.9% of the 745 nonsilent substitutions that occurred over the time period sampled. Thus, there is good reason for concern about HM mutations if one wants to draw inferences about evolution from this or any similarly affected data set. Culture in live cells is necessary for the propagation not only of viruses, but for many bacteria, such as the obligately intracellular rickettsial and chlamydial bacteria, as well. Laboratories involved in influenza surveillance have long been attuned to the presence of HM mutations. However, people obtaining influenza sequences from public databases might not suspect that the sequences could contain laboratory artifacts. In our previous analyses of these data (Bush *et al.*, 1999a, b) we dealt with this problem by removing all mutations assigned to terminal branches (70% of the total) from our analyses. Our results indicate that HM mutations are confined to the 22 HM codons, thus, we could take a less drastic approach in the future. For instance, we could assign missing data codes to the HM codons when sequences are obtained from egg-cultured isolates.

We have shown that the excess mutations that remain on the terminal branches after accounting for HM mutations is of a magnitude consistent with expectations given our sampling protocol. Despite our bias toward dispersed sampling, examination of Fig. 1 shows that our data set does contain a number of closely related isolates. To get an idea of how sensitive the calculation of percent excess mutations on the terminal branches is to the degree to which we sampled in a dispersed manner as opposed to clumped manner, we removed 10 of 357, or 2.8%, of the most genetically divergent isolates from our original data set, and constructed a new tree (not shown). The excess of replacements on the terminal branches was reduced from 40% to 32%. Removing 38, or 10.6%, of the most genetically divergent isolates reduced the excess to 28%. Thus the presence of even small numbers of genetically divergent isolates accounts for much of the excess of mutations assigned to the terminal branches of the HA tree.

Unlike the excess mutations on terminal branches caused by sampling bias, the excess caused by HM change could cause problems in studies of how influenza viruses evolve as they replicate in their human hosts. For instance, in our previous work identifying codons under positive selection, we examined the ratio of nonsilent to silent substitutions (Bush *et al.*, 1999a). If an isolate was sequenced shortly after a new amino acid replacement became fixed in a laboratory culture, sequencing viruses from that culture might fail to show silent substitutions that also had occurred during passage but that had been lost in the selective sweep. After fixation of the HM nonsilent substitution, silent substitutions once again would begin to accumulate. Because we do not know the exact circumstances under which HM mutations occurred in our data set relative to the time at which particular isolates were sequenced, we cannot make any predictions about the relative frequencies of nonsilent or silent substitutions in the HM codons as compared with the non-HM codons. However, we can examine the frequencies of nonsilent and silent substitutions in the two codon sets to learn more about how HM codons differ from the non-HM codon set. The HM codons showed significantly greater numbers of nonsilent substitutions than expected (Table 6). As shown in Table 1, eight of the 22 HM codons are among those we previously identified as being under positive selection to change the amino acid they encode. One interpretation of this result is that some of the HM codons are under selection to change the amino acid they encode to adapt to

TABLE 6. The distribution of silent and nonsilent substitutions in the HM vs. non-HM codons

All branches	Obs non-HM	Exp HM	Obs HM	χ^2
Nonsilent sub	478	142.76	268	109.88
Silent sub	560	167.24	42	93.79
Sum	1038	310.00	310	203.67

The HM codons showed a significant excess of nonsilent substitutions as opposed to silent substitutions compared to expectations based on the non-HM codon set ($P < 0.05$, df = 1).

growth in egg culture in addition to being under selection to evade the human immune response.

The observation of excess mutations assigned to the terminal branches of the HA tree is consistent with expectations based on two very different hypotheses. HM mutations appear to account for part of the excess. The majority of the excess is of a magnitude consistent with expectations based on our sampling protocol, which is biased against sequencing closely related viruses. Unlike the excess caused by sampling bias, excess mutations attributable to HM change reflect processes other than the ongoing evolution of the virus during replication in the human host, and thus should be identified and extracted before making evolutionary inference based on phylogenetic reconstruction of influenza evolution.

We gratefully acknowledge the technical expertise of Huang Jing and critical reviews by C. Bergstrom, B. Levin, A. Moya, and K. Subbarao. This work was supported by National Institutes of Health Grant 1R01AI44474–01 and by funds provided by the University of California for the conduct of discretionary research by Los Alamos National Laboratory, conducted under the auspices of the U.S. Department of Energy.

REFERENCES

Bush, R. M., Fitch, W. M., Bender, C. A. & Cox, N. J. (1999a) Positive selection on the H3 hemagglutinin gene of human influenza virus A. *Mol. Biol. Evol.* 16, 1457–1465.

Bush, R. M., Bender, C. A., Subbaro, K., Cox, N. J. & Fitch, W. M. (1999b) Predicting the evolution of human influenza A. *Science* 286, 1921–1925.

Cao, J. X. , Ni, H., Wills, M. R., Campbell, G. A., Sil, B. K., Ryman, K. D., Kitchen, I. & Barrett, A. D. (1995) Passage of Japanese encephalitis virus in HeLa cells results in attenuation of virulence in mice. *J. Gen. Virol.* 76, 2757–2764.

Graff, J., Normann, A., Feinstone, S. M. & Flehmig, B. (1994) Nucleotide sequence of wild-type hepatitis A virus GBM in comparison with two cell culture-adapted variants. *J. Virol.* 68, 548–554.

Itoh, M., Isegawa, Y., Hotta, H. & Homma, M. (1997) Isolation of an avirulent mutant of Sendai virus with two amino acid mutations from a highly virulent field strain through adaptation to LLC-MK2 cells. *J. Gen. Virol.* 78, 3207–3215.

Robertson, J. S. (1993) Clinical influenza virus and the embryonated hen's egg. *Rev. Med. Virol.* 3, 97–106.

Sawyer, L. S. W., Wrin, M. T., Crawford-Miksza, L., Potts, B., Wu, Y., Weber, P. A., Alfonso, R. D. & Hanson, C. V. (1994) Neutralization sensitivity of human immunodeficiency virus type 1 is determined in part by the cell in which the virus is propagated. *J. Virol.* 68, 1342–1349.

Swofford, D. L. (1999) PAUP*. Phylogenetic Analysis Using Parsimony (*and Other Methods). Version 4. Sinauer Associates, Sunderland, Massachusetts.

7
Bacteria are Different: Observations, Interpretations, Speculations, and Opinions About the Mechanisms of Adaptive Evolution in Prokaryotes

BRUCE R. LEVIN AND CARL T. BERGSTROM

To some extent, the genetic theory of adaptive evolution in bacteria is a simple extension of that developed for sexually reproducing eukaryotes. In other, fundamental ways, the process of adaptive evolution in bacteria is quantitatively and qualitatively different from that of organisms for which recombination is an integral part of the reproduction process. In this speculative and opinionated discussion, we explore these differences. In particular, we consider (i) how, as a consequence of the low rates of recombination, "ordinary" chromosomal gene evolution in bacteria is different from that in organisms where recombination is frequent and (ii) the fundamental role of the horizontal transmission of genes and accessory genetic elements as sources of variation in bacteria. We conclude with speculations about the evolution of accessory elements and their role in the adaptive evolution of bacteria.

For the most part, the genetic theory of adaptive evolution was developed by sexually reproducing eukaryotes, for sexually reproducing eukaryotes. Although there have been some wonderful theo-

Department of Biology, Emory University, Atlanta, GA 30322

This paper was presented at the National Academy of Sciences colloquium "Variation and Evolution in Plants and Microorganisms: Toward a New Synthesis 50 Years After Stebbins," held January 27–29, 2000, at the Arnold and Mabel Beckman Center in Irvine, CA.

retical studies of population genetics and evolution of bacteria—many of which are cited herein—the formal (mathematical) and informal (verbal) theory of the mechanisms of adaptive evolution in bacteria is modest in its volume, breadth, and level of integration.

Bacteria are haploid, reproduce clonally, and rarely subject their genomes to the confusion of recombination. Thus, one might believe (and both of us once did believe) that the genetic basis of adaptive evolution in these prokaryotes would be simpler than that of so-called higher organisms and that the theory to account for it would be straightforward extensions of that already developed for sexually reproducing eukaryotes. As we shall try to convince the reader, in this personal (read, opinionated) discussion, this situation is by no means simple. Bacteria are different. The mechanisms of adaptive evolution in the prokaryotic world raise a number of delicious theoretical and empirical questions that have only begun to be addressed.

OBSERVATIONS

Adaptation by the Acquisition of Genes and Accessory Genetic Elements from Without

The most striking feature of retrospective studies of genetic variation and molecular evolution in bacteria is the extent to which these organisms are chimeras. Much of the DNA of bacteria classified as *Escherichia coli* has been acquired relatively recently: more than 17% of the open reading frames of the *E. coli* K-12 genome was acquired in the last 100 or so million years from organisms with G + C ratios and codon usage patterns distinguishable from those of other strains of *E. coli* and closely related Enterobacteriacae (Lawrence and Ochman, 1998). Moreover, a substantial amount of the variation in bacteria is not in their chromosomal genes. Bacteria commonly carry arrays of active and retired accessory genetic elements (plasmids, prophages, transposons, and integrons), the composition of which also varies widely among members of the same bacterial species. Although, at any given time, some of these elements, such as insertion sequences and cryptic plasmids, may not carry genes that code for specific host-expressed phenotypes, others are responsible for the more interesting adaptations of bacteria to their environment. For example, many of the genes coding for the adhesins, toxins, and other characters responsible for the pathogenicity of bacteria, "virulence factors," either are present in clusters known as "pathogenicity islands" (Finlay and Falkow, 1997; Groisman and Ochman, 1996; Hacker *et al.*, 1997; Lee, 1996), which almost certainly had former lives as accessory elements or as parts thereof, or are borne on functional accessory elements, such as plasmids and prophages of temperate viruses. Some of these islands and virulence-encoding accessory elements may have had a long history with one or a few specific pathogenic species of bacteria (Groisman and Ochman, 1997). On the other hand, many are distributed among bacteria that, based on their less mobile chromosomal loci, are phylogenetically quite different (Czeczulin *et al.*, 1999). Plasmids and their running dogs, transposons, also bear many of the genes responsible for antibiotic resistance and, all too commonly, multiple antibiotic resistance (Falkow, 1975), and many of the other characters that make the lives of prokaryotic organisms as interesting and exciting as they are. Included among these plasmid-borne genes are those that code for the fermentation of exotic carbon sources, the detoxification of heavy metals, and the production of allelopathic agents, such as bacteriocins (Summers, 1996). Bacteria also commonly carry integrons, elements that acquire, accumulate, and control the expression of genes acquired from external sources (Hall, 1997, 1998; Mazel *et al.*, 1998; Row-Magnus and Mazel, 1999).

All of these islands and accessory genetic elements and even some seemingly ordinary chromosomal genes, like those for the resistant forms of the penicillin-binding proteins of *Streptococcus pneumoniae* (Dowson *et al.*, 1989), were acquired from other organisms (primarily, but possibly not exclusively, other bacteria). They were picked up in a number of ways: as free DNA (transformation), through bacteriophage (transduction), or by intimate contact with other bacteria (conjugation). Although genetic exchange may occur less frequently in bacteria than in sexual eukaryotes (for which recombination is an integral part of the reproductive process), the phylogenetic range across which genetic exchange can occur in bacteria is far broader than that in extant eukaryotes. From a prokaryotic perspective, sexual eukaryotes like ourselves are incestuous nymphomaniacs: we do "it" too far often and almost exclusively with partners that, from a phylogenetic perspective, are essentially identical to ourselves. To be sure, genes acquired by horizontal transfer can be a source of variation for adaptive evolution in "higher" eukaryotes, up to and including those high enough to publish (Courvalin *et al.*, 1995; Grillot-Courvalin *et al.*, 1998). However, for at least contemporary eukaryotes, it is sufficient (and sufficiently problematic) to develop a comprehensive genetic theory of adaptive evolution based on variation generated from within by mutation (broadly defined to include transposition and chromosomal rearrangement) and recombination among members of the same species. By contrast, in contemporary as well as ancient bacteria, the horizontal transfer of genes (and accessory genetic elements) from other species is a major source of variation and is fundamental to the genetic theory of adaptive evolution in these prokaryotes.

The extent to which genes and accessory genetic elements are exchanged among phylogenetically distant populations raises a number of issues beyond those of the genetic mechanisms of adaptive evolution addressed herein. Perhaps the most fundamental of these other issues are the classical ones of the definition of species and the remarkable discreetness of the genomic clusters of organisms we classify as members of the same species of bacteria. For interesting and appealing considerations of these issues, see Dykhuizen (1998), Dykhuizen and Green (1991), Majewski and Cohan (1999), and Palys *et al.* (1997).

INTERPRETATIONS, SPECULATIONS, AND OPINIONS

Adaptive Evolution by Mutation and Selection of Chromosomal Genes

Although genes and accessory elements acquired from external sources are responsible for many of the interesting adaptations of bacteria to their environments and their seemingly saltational evolution, the mundane processes of mutation and selection at chromosomal loci are by no means inconsequential for the long-term as well as day-to-day evolution of bacteria. On the other hand, at least quantitatively and possibly qualitatively, the chromosomal population genetics of adaptive evolution in bacteria are different from those of sexually reproducing eukaryotes.

Adaptive Evolution When There Is Little or No Chromosomal Gene Recombination

The primary differences between the chromosomal population genetics of bacteria and those of sexually reproducing eukaryotes arise as a consequence of the low rates of chromosomal gene recombination in bacterial populations. To be sure, if one looks hard enough, some natural mechanism of chromosomal gene recombination can probably be demonstrated to occur in most species of bacteria. And collectively, the chromosomal genomes of some species of bacteria may even be at or near linkage equilibrium (Maynard Smith *et al.*, 1993; Maynard Smith and Smith, 1998). Nonetheless, because bacteria reproduce clonally, by binary fission, any given population will be composed of relatively few genetically distinct lineages, with recombination between them occurring only on rare occasions. We postulate that for most natural populations of bacteria, the probability of a gene in one lineage (clone) being replaced by homologous recombination with a gene from another lineage of its own or another species is on the order of the mutation rate or lower (Levin, 1981, 1988; Levin and Lenski, 1983; see also Hartl and Dykhuizen, 1984). The low rates of recombination in bacterial populations have at least four important (and interesting) ramifications for at least the local (short-term) population genetics of adaptive evolution in bacterial populations.

Sequential evolution

Within each bacterial lineage, adaptive evolution will proceed by the sequential accumulation of favorable mutations, rather than by recombinational generation of gene combinations; in this respect, bacterial evolution will be similar to that depicted in the top portion of Muller's famous diagram of evolution in asexual and sexual populations (Crow and Kimura, 1965). The evolution of a better genotype *ABC* from its less fit ancestor *abc* will proceed in stages, one gene at a time, with the order and rate of evolution depending primarily on the fitness of the intermediates (*aBc*, *aBC*, etc.). If, individually and collectively, the intermediates have a selective advantage over the *abc* ancestor, this evolution can proceed quite rapidly (Evans, 1986). If, on the other hand, the intermediates are not favored, it may take a great deal of time before the best genotype, *ABC*, is assembled. Moreover, the best genotype will most likely arise by mutation from an intermediate form, rather than through the recombinational merger of different intermediate forms; for a more extensive and formal (mathematical) consideration of the process of mutation and selection in bacteria, see Gerrish and Lenski (1998).

Compensatory evolution

In bacteria, adaptation to the effects of deleterious genes that have become fixed (because of their being favored in another environment or for other reasons) is likely to be through amelioration of those effects by the ascent of compensatory mutations at other loci, rather than by the evolution of more fit variants at the deleterious loci (Bjorkman *et al.*, 1998; Schrag and Perrot, 1996). In sexual eukaryotes as well as bacteria, these compensatory mutations are likely to be the first to arise when, as is probable, there are more ways to improve fitness of the organism than mutation (reversion) at the deleterious locus. If recombination is common, however, these compensatory mutations and the genes whose deleterious effect they ameliorate will rapidly become separated, especially if these loci are not closely linked. Because of the low rate of recombination in bacterial populations and the fact that, when recombination does occur, only a small fraction of the genome is replaced, the deleterious and compensatory genes will remain in linkage disequilibrium for extensive periods, no matter how far apart they lie on the chromosome. As the frequency of the compensatory mutants increases in the population, the intensity of selection for better alleles at the deleterious locus will drop off. Moreover, if the compensatory mutations become fixed, adaptive valleys may be established and effectively preclude the ascent of even more fit mutations at the deleterious locus itself (Schrag *et al.*, 1997).

Low effective population size

Although the total sizes of bacterial populations may be enormous, their genetically effective sizes can be quite low as a consequence of periodic selection, i.e., selective sweeps of better-adapted mutants (Atwood *et al.*, 1951; Koch, 1974; Levin, 1981) or as a result of bottlenecks, contractions in population size associated with transmission to new hosts or microhabitats. These low effective population sizes and selective sweeps have at least two major ramifications for adaptive evolution. One is to purge variation accumulating in the population by selection as well as by genetic drift. Another is to make a given population less fit than it would be if it did not have to deal with the stochastic trials and tribulations of periodic selection and bottlenecks. Although, on Equilibrium Day, the most fit genotypes will ascend to their rightful places, on the days before, those best types may well be lost as the population passes through bottlenecks (Gerrish, 1998; Levin *et al.*, 2000).

The evolution of genes that augment the rate at which variation is generated

Because low rates of recombination allow modifier loci and the genes whose effects they modify to remain together for extensive periods, natural selection can favor modifying traits that are advantageous in the long run despite their short-term disadvantages (Eshel, 1973a, b). One example of such a situation is the evolution of mutators, genes that increase the rate at which variation is generated by mutation. Because they generate deleterious mutations and possibly for other reasons as well, mutator genes, which are commonly defective mismatch repair loci, are anticipated to be at a selective disadvantage. However, continuous changes or heterogeneities in the selective environment can cause bacteria bearing these mutators to ascend by hitchhiking with the beneficial mutants they generate (Taddei *et al.*, 1997; Tenaillon *et al.*, 1999). As anticipated by this theory, in long-term experimental (de Visser *et al.*, 1999; Sniegowski *et al.*, 1997) and natural populations of bacteria (Oliver, 1999) adapting to new and/or changing environments, one can find relatively high frequencies of mutator genes.

It's Not the Rate That Counts

Although in any given bacterial population, the absolute rate of chromosomal gene change by recombination may be as low or lower than that of mutation, the consequences of recombination for adaptive evolution in bacteria can be far more profound than those of mutation. By legitimate (homologous) or illegitimate (nonhomologous) recombination, bacteria can acquire new genes that have evolved in other often phylogenetically and ecologically distant populations. These genes can code for phenotypes that can expand or alter the ecological niche of their host bacterium (Cohan, 1996), and if they are favored in the recipient population, they will ascend. Given sufficient time and sufficiently intense selection, bacteria will acquire whatever genes they need either directly or indirectly through intermediate species. And, as we have learned from the evolution of antibiotic resistance, "sufficient time" need not be very long. For this reason, bacterial ecosystems have been characterized as a "global gene pool" (Maiden *et al.*, 1996). From the perspective of adaptive evolution, it is generally not the rate of recombination that is important but rather the existence of mechanisms for gene exchange, the range of "species" with which a populations of bacteria can (and do) exchange genes, and the intensity of selection for those genes.

Adaptive Evolution by the Transmission of Accessory Genetic Elements

From one perspective, the accessory genetic elements of bacteria are parasites and symbionts, and their population and evolutionary biology can be—and has been—treated in that context with little or no reference to their role as sources of variation for their host bacteria. However, as we have discussed above, much of the real "action" in adaptive evolution in bacteria is through genes borne on, transmitted by, and sequestered from these elements. And, from this perspective, the population and evolutionary dynamics of these elements form an integral part of the process of adaptive evolution in bacteria.

Variations on a Single Theme

Traditionally, we classify the accessory genetic elements of bacteria into functional, rather than phylogenetic groups: primarily as plasmids, transposons, and temperate phages. Furthermore, we typically draw a distinction between these peregrine elements and those with less mobile and autonomous life styles such as islands and integrons. However, the distinction between plasmids, phages, and transposons is somewhat artificial even on functional grounds. For example, when they are in bacteria, the prophage of the phage P1 is a plasmid and that of Mu is a transposon. Moreover, although some plasmids and a few transposons code for the machinery needed for their own infectious transfer by conjugation, the mobility of many plasmids and of most transposons, integrons, islands, and even ordinary genes is by hitchhiking on conjugative plasmids or phages or by being picked up as free DNA by hosts with transformation mechanisms. Although the modes of replication and transmission are critical for considerations of the population dynamics, existence conditions, and ecology of these different classes of accessory elements (Campbell, 1961; Chao and Levin, 1981; Condit *et al.*, 1988; Levin and Stewart, 1980; Stewart and Levin, 1977, 1984), they are of only secondary import for a general consideration of adaptive evolution in bacteria. From this perspective, the most important factor is the extent to which these elements are mobile and the range of bacterial hosts they infect. In this sense, the accessory bacterial genetic elements can be seen as arrayed along a continuum from phages, plasmids, and transposons to pathogenicity (and nicer) islands, integrons, and even stay-at-home chromosomal genes.

Why Be a Vagabond When You Can Stay at Home?

Accessory elements at the most mobile end of the continuum may be maintained as "genetic parasites," spreading by infectious transfer alone without bearing genes that augment the fitness of their host bacteria. A parasitic existence is almost certainly the case for purely lytic (virulent) phages (Levin *et al.*, 1997) and possibly for many temperate phages as well (Stewart and Levin, 1984). Although a formal possibility, we believe that it is unlikely that plasmids and transposons are purely parasitic (Levin, 1993) and even less likely that islands and integrons are maintained without at least occasionally paying for their dinner. The rates of infectious transfer of these elements are almost certainly not great enough to overcome the fitness burden their carriage imposes on their host bacteria (Bouma and Lenski, 1988; Levin, 1980; Modi and Adams, 1991) and their losses by vegetative segregation. If this assumption is correct, then these elements must bear genes that are at least sometimes beneficial to their host bacteria. After all, these accessory elements are vertically transmitted (in the course of cell division); thus, it would be to their advantage to carry genes that augment the fitness of their hosts. But can this "niceness" account for the maintenance of accessory genetic elements? If accessory element-borne genes provide a selective advantage to bacteria and the accessory elements themselves are either costly or unstable, why are those genes not sequestered by the host chromosome?

In a recent investigation, we, along with Marc Lipsitch (Bergstrom *et al.*, 2000), addressed this question from a slightly different angle, that of the existence conditions for plasmids. We demonstrated that, under broad conditions, if host-expressed genes have higher fitness when carried on the chromosome than on an infectiously transmitted plasmid, then those genes will be sequestered eventually by the chromosome. These usurpings of genes from the plasmid by the chromosome will be the case even when selection for those genes is intermittent, as is the case for antibiotic resistance and most other plasmid-encoded characters. A broader interpretation of this result is that, *if the mobility of host-adaptive genes has a cost, that mobility will be lost eventually*. Plasmids, transposons, and temperate phages, or the genes they carry will give up their vagabond lifestyle and become islands.

On the other side, using primarily simulation methods, we demonstrated two seemingly realistic situations under which those genes can be maintained for extended periods on infectiously transmitted accessory elements. The first of these is the continuous entry into that population of lineages that are more fit than existing ones. By being infectiously transmitted, those favored (or occasionally favored) genes will be able to make their way to the rising stars rather than being lost along with their has-been hosts. The second situation involves movement among two or more distinct bacterial ecotypes in an ecologically heterogeneous environment. Although the accessory element may be lost from any particular ecotype at any particular time, it can return via horizontal transfer, and in the long term, the host-beneficial genes can persist on accessory elements.

In this interpretation, genes originally carried on accessory elements (and even the elements themselves) are in a continuous state of flux with respect to their mobility and within-host stability. As the habitat of a bacterial population becomes more stable and/or the opportunities for its accessory elements to move to uninfected populations decline, selection will favor the incorporation of those elements or the favored genes they carry into the chromosome. The opposite will occur in more interesting times and places. Selection will favor the mobility of accessory elements, and broadly favored but narrowly available genes and former elements will be seduced back into a vagabond lifestyle. A corollary of this interpretation, if interpretations are allowed to have corollaries, is that genes that become widely popular for some environmental reason—such as the genes for antibiotic resistance after the human use of antibiotics—will be borne initially by more mobile accessory elements. Whether the mobile elements bearing these genes will be phages, conjugative or nonconjugative plasmids, or transposons or whether those genes would be acquired by transformation will depend on a variety of historical, genetic, and ecological factors that are specific to the bacteria and habitats involved.

There are other conditions that may favor the infectious transfer of accessory elements as well. For example, a number of characters expressed by bacteria of one lineage augment the fitness of bacteria of other lineages in their vicinity whether the beneficiaries carry those genes that code for that character or not. Examples include the production of secreted agents that kill competing bacterial species (e.g., antibiotics and bacteriocins) and somatic cells (e.g., toxins) and those that detoxify or condition the local environment for bacterial growth (e.g., β-lactamases and other enzymes that denature antibiotics exogenously; Lenski and Hattingh, 1986). One consequence is that bacteria not carrying these genes can free-ride on the efforts of those that do. In such cases, infectious gene transfer would provide a way to convert these "cheaters" into good citizens that share the burden as well as the advantages of carrying and expressing these genes (Smith, 2000).

Evolving to Evolve: The Evolution of Infectious Gene Transfer

Thus far, we have considered the mechanisms that can *maintain* infectiously transmitted genetic elements over evolutionary time. But how did these mechanisms and the capacity for acquisition of genes from without evolve in the first place? We believe that, for accessory genetic elements, the most parsimonious (and possibly even correct) answers to these questions are those that treat the individual genetic elements, rather than their hosts, as the objects of natural selection.

Infectious Gene Transfer as a Product of Coincidental Evolution?

A number of years ago, Richard Lenski and B.R.L. (Levin, 1988; Levin and Lenski, 1983) considered the mechanisms responsible for conjugative plasmids and phages to serve as vehicles (vectors) for the infectious transfer of host genes: plasmid-mediated conjugation and phage-mediated transduction. They postulated that these forms of bacterial sex are coincidental to the infectious transfer of the elements themselves and to the presence of recombination repair enzymes in their host bacteria. Although we are now in a new and doubtless more enlightened millennium, this coincidental evolution hypothesis remains plausible and parsimonious, albeit still not formally tested. Clearly, generalized transduction and plasmid-mediated recombination (such as that by the F+ plasmid of *E. coli* K-12) are to the disadvantage of the phage and plasmid, respectively. Coincidental evolution also seems a reasonable hypothesis for the propensity of conjugative plasmids and bacteriophages to pick up hitchhiking accessory elements, such as mobilizable nonconjugative plasmids. However, the ability to hitchhike (mobilizability) may well be an evolved character of the hitchhiking element.

Can coincidental evolution also explain transformation and transformability (competence)? The mechanisms that naturally transforming bacteria have for picking up free DNA from the environment, protecting it from destruction by restriction enzymes, and incorporating it into their genomes are complex and highly evolved. A number of hypotheses have been presented for the selective pressures responsible for the evolution of transformation. One of these is consistent with a coincidental evolution hypothesis. In accord with this hypothesis, transformation evolved as a mechanism for acquiring food (nucleotides) from the external environment, and recombination is a coincidental side effect of DNA entering the cell (Redfield, 1993). Alternatively, it has been proposed that transformation evolved specifically as a mechanism to acquire genes from without as templates to repair double-stranded breaks (Michod *et al.*, 1988; Wojciechowski *et al.*, 1989, but see also Mongold, 1992; Redfield, 1993). To these hypotheses, we would like to add the following, not mutually exclusive alternative. We postulate that the ability to acquire genes from other organisms as a source of variation for adaptive evolution is the selective pressure responsible for the evolution and maintenance of transformation. In this interpretation, transformation evolved and is maintained through processes similar to the one proposed for the evolution of mutators (Taddei *et al.*, 1997; Tenaillon *et al.*, 1999), but in this case, the source of variation is external rather than internal. An analogous mechanism may also favor the evolution and maintenance of transposons, integrons, and other elements or processes that, once acquired, augment the rate at which variability is generated. (J. Smith, personal communication).

Conclusion

We have argued that the mechanisms of adaptive evolution are quantitatively and qualitatively different in bacteria than they are in sexual eukaryotes. This difference is primarily a consequence of the frequency of homologous gene recombination being low in bacteria and high in sexual eukaryotes and of the phylogenetic range of gene exchange being broad in bacteria and narrow in contemporary eukaryotes. Also contributing to this difference is the prominent role of viruses, plasmids, and other infectiously transmitted accessory genetic elements as bearers and vectors of genes responsible for adaptive evolution and their seemingly negligible role in this capacity in contemporary eukaryotes.

Clearly, both of these regimes of recombination and horizontal gene transfer are associated with groups of organisms that have been successful. To be sure, there may well be bacteria in which horizontal gene transfer plays little role in adaptive evolution (*Mycobacterium tuberculosis*, perhaps) and higher eukaryotes that rarely, if ever, engage in homologous gene recombination (Smith, 1992). There may also be species of bacteria in which recombination occurs at high rates and species of contemporary eukaryotes in which infectiously transmitted viruses and other accessory elements play a prominent role as sources of variation for adaptive evolution. We propose, however, that these are the exceptions. We postulate that evolutionary success of bacteria as a group is a consequence of their capacity to acquire and express genes from a vast and phylogenetically diverse array of species and that the success of eukaryotes as a group relies on the high rates at which genes are shuffled by recombination.

Coda

In this report, we have focused on how the mechanisms of adaptive evolution in bacteria are different from those of sexually reproducing eukaryotes. We have not considered how they are similar, nor have we covered the wonderful studies in which bacteria have been used as a model system, a tool, for studying evolution experimentally and actually testing, rather than just championing, general hypotheses about the ecological, genetic, biochemical, and molecular basis of evolution. For an excursion into that literature, we refer the reader to the work of Julian Adams, Jim Bull, Allan Campbell, Lin Chao, Patricia Clark, Fred Cohan, Tony Dean, Dan Dykhuizen, David Gordon, Barry Hall, Dan Hartl, Ryzard Korona, Richard Lenski, E. C. Lin, Rick Michod, Judy Mongold, Robert Mortlock, Rosie Redfield, Peg Riley, Paul Sneigowski, Francois Taddei, Michael Travisano, Paul Turner, Holly Wichman, Miro Radman, Lone Simonsen, and others, who will, doubtless, reprimand us for our unintentional transgression of not including their names in this list.

This article is dedicated to the memory of Ralph V. Evans (1958–1990). Ralph had been doing theoretical and experimental studies of adaptive evolution in bacteria. If not for his premature death by cancer, his name and contributions would have had a prominent role in this discussion.

We wish to thank Jeff Smith, Jim Bull, Andy Demma, and John Logsdon for useful comments and suggestions, some of which we have actually listened to and incorporated into this discussion. This endeavor has been supported by National Institutes of Health Grants GM33872 and AI40662 and National Institutes of Health Training Grant T32-AI0742.

REFERENCES

Atwood, K. C., Schneider, L. K. & Ryan, F. J. (1951) Periodic selection in *Escherichia coli. Proc. Nat. Acad. Sci. USA* 37, 146–155.

Bergstrom, C. T., Lipsitch, M. & Levin, B. (2000) Natural selection, infectious transfer and the existence conditions for bacterial plasmids. *Genetics*, 155, 1505–1519.

Bjorkman, J., Hughes, D. & Andersson, D. I. (1998) Virulence of antibiotic-resistant Salmonella typhimurium. *Proc. Natl. Acad. Sci. USA* 95, 3949–3953.

Bouma, J. E. & Lenski, R. E. (1988) Evolution of a bacteria/plasmid association. *Nature* 335, 351–352.

Campbell, A. (1961) Conditions for the existence of bacteriophage. *Evolution* 15, 153–165.

Chao, L. & Levin, B. R. (1981) Structured habitats and the evolution of anticompetitor toxins in bacteria. *Proc. Natl. Acad. Sci. USA* 78, 6324–6328.

Cohan, F. M. (1996) The role of genetic exchange in bacterial evolution. *ASM News* 62, 631–636.

Condit, R., Stewart, F. M. & Levin, B. R. (1988) The population biology of bacterial transposons: A priori conditions for maintenance as parasitic DNA. *American Naturalist* 132, 129–147.

Courvalin, P., Goussard, S. & Grillot-Courvalin, C. (1995) Gene transfer from bacteria to mammalian cells. *Comptes Rendus de l Academie des Sciences—Serie Iii, Sciences de la Vie* 318, 1207–1212.

Crow, J. F. & Kimura, M. (1965) Evolution in sexual and asexual populations. *American Naturalist* 99, 439–450.

Czeczulin, J. R., Whittam, T. S., Henderson, I. R., Navarro-Garcia, F. & Nataro, J. P. (1999) Phylogenetic analysis of enteroaggregative and diffusely adherent Escherichia coli. *Infection & Immunity* 67, 2692–2699.

de Visser, J. A. G. M., Zeyl, C. W., Gerrish, P. J., Blanchard, J. L., Lenski, R. E. (1999) Diminishing returns from mutation supply rate in asexual populations. *Science* 283, 404–406.

Dowson, C. G., Hutchison, A., Brannigan, J. A., George, R. C., Hansman, D., Linares, J., Tomasz, A., Smith, J. M., Spratt, B. G. (1989) Horizontal transfer of penicillin-binding protein genes in penicillin-resistant clinical isolates of Streptococcus pneumoniae. *Proc. Natl. Acad. Sci. USA* 86, 8842–8846.

Dykhuizen, D. E. (1998) Santa Rosalia revisited: why are there so many species of bacteria? *Antonie van Leeuwenhoek* 73, 25–33.

Dykhuizen, D. E. & Green, L. (1991) Recombination in Escherichia coli and the definition of biological species. *Journal of Bacteriology* 173, 7257–7268.

Eshel, I. (1973a) Clone-selection and optimal rates of mutation. *Journal of Applied Probability* 10, 728–738.

Eshel, I. (1973b) Clone selection and the evolution of modifying features. *Theoretical Population Biology* 4, 196–208.

Evans, R. (1986) Niche expansion in bacteria: can infectious gene exchange affect the rate of evolution? *Genetics* 113, 775–795.

Falkow, S. (1975) *Infectious Multiple Drug Resistance* (Pion Press, London).

Finlay, B. B. & Falkow, S. (1997) Common themes in microbial pathogenicity revisited. *Microbiology and Molecular Biology Reviews* 61, 136–169.

Gerrish, P. E. (1998) *Dynamics of Mutation and Selection in Asexual Populations* (Ph.D. Dissertation, Michigan State University).

Gerrish, P. J. & Lenski, R. E. (1998) The fate of competing beneficial mutations in an asexual population. *Genetica* 102–103, 127–144.

Grillot-Courvalin, C., Goussard, S., Huetz, F., Ojcius, D. M. & Courvalin, P. (1998) Functional gene transfer from intracellular bacteria to mammalian cells. *Nature Biotechnology* 16, 862–866.

Groisman, E. A. & Ochman, H. (1996) Pathogenicity islands: bacterial evolution in quantum leaps. *Cell* 87, 791–794.

Groisman, E. A. & Ochman, H. (1997) How Salmonella became a pathogen. *Trends in Microbiology* 5, 343–349.

Hacker, J., Blum-Oehler, G., Muhldorfer, I. & Tschape, H. (1997) Pathogenicity islands of virulent bacteria: structure, function and impact on microbial evolution. *Molecular Microbiology* 23, 1089–1097.

Hall, R. M. (1997) Mobile gene cassettes and integrons: moving antibiotic resistance genes in gram-negative bacteria. *Ciba Foundation Symposium* 207, 192–202; discussion 202–205.

Hall, R. M. (1998) The role of gene cassettes and integrons in the horizontal transfer of genes in Gram-negative bacteria. In *Horizontal Gene Transfer*, eds. Syvanen, M. & Kado, C. I. (Chapman & Hall, New York), pp. 53–62.

Hartl, D. L. & Dykhuizen, D. E. (1984) The population genetics of Escherichia coli. *Annual Review of Genetics* 18, 31–68.

Koch, A. L. (1974) The pertinence of the periodic selection phenomenon to prokaryote evolution. *Genetics* 77, 127–142.

Lawrence, J. G. & Ochman, H. (1998) Molecular archaeology of the Escherichia coli genome. *Proc. Natl. Acad. Sci. USA* 95, 9413–9417.

Lee, C. A. (1996) Pathogenicity islands and the evolution of bacterial pathogens. *Infectious Agents & Disease* 5, 1–7.

Lenski, R. E. & Hattingh, S. E. (1986) Coexistence of two competitors on one resource and one inhibitor: a chemostat model based on bacteria and antibiotics. *Journal of Theoretical Biology* 122, 83–93.

Levin, B. R. (1980) Conditions for the existence of R-plasmids in bacterial populations. In *Fourth International Symposium on Antibiotic Resistance*, eds. Mitshashi, S., Rosival, L. & Krcmery, V. (Springer, Smolenice, Czechoslovakia), pp. 197–202.

Levin, B. R. (1981) Periodic selection, infectious gene exchange and the genetic structure of E. coli populations. *Genetics* 99, 1–23.

Levin, B. R. (1988) The evolution of sex in bacteria. In *The Evolution of Sex: A Critical Review of Current Ideas*, eds. Michod, R. & Levin, B. R. (Sinauer, Sunderland, MA), pp. 194–211.

Levin, B. R. (1993) The accessory genetic elements of bacteria: existence conditions and (co)evolution. *Current Opinion in Genetics & Development* 3, 849–854.

Levin, B. R. & Lenski, R. E. (1983) Coevolution of bacteria and their viruses and plasmids. In *Coevolution*, eds. Futuyama, D. J. & Slatkin, M. (Sinauer, Sunderland, MA), pp. 99–127.

Levin, B. R., Perrot, V. & Walker, N. (2000) Compensatory Mutations, Antibiotic Resistance and the Population Genetics of Adaptive Evolution in Bacteria. *Genetics.* In Press.

Levin, B. R. & Stewart, F. M. (1980) The population biology of bacterial plasmids: a priori conditions for the existence of mobilizable nonconjugative factors. *Genetics* 94, 425–443.

Levin, B. R., Stewart, F. M. & Chao, L. (1977) Resource-limited growth, competition, and predation: a model and experimental studies with bacteria and bacteriophage. *American Naturalist* 977, 3–24.

Maiden, M. C. J., Malorny, B. & Achtman, M. (1996) A global gene pool in the neisseriae. *Molecular Microbiology* 21, 1297–1298.

Majewski, J. & Cohan, F. M. (1999) Adapt globally, act locally: the effect of selective sweeps on bacterial sequence diversity. *Genetics* 152, 1459–1474.

Maynard Smith, J., Smith, N. H., Orourke, M. & Spratt, B. G. (1993) How clonal are bacteria. *Proc. Natl. Acad. Sci. USA* 90, 4384–4388.

Maynard-Smith, J. & Smith, N., Eds. (1998) *The Genetic Population Structure of Pathogenic Bacteria* (Oxford University Press, Oxford).

Mazel, D., Dychinco, B., Webb, V. A. & Davies, J. (1998) A distinctive class of integron in the Vibrio cholerae genome. *Science* 280, 605–608.

Michod, R. E., Wojciechowski, M. F. & Hoelzer, M. A. (1988) NA repair and the evolution of transformation in the bacterium Bacillus subtilis. *Genetics* 118, 31–39.

Modi, R. I. & Adams, J. (1991) Coevolution in bacteria-plasmid populations. *Evolution* 45, 656–667.

Mongold, J. A. (1992) DNA repair and the evolution of transformation in Haemophilus influenzae. *Genetics* 132, 893–898.

Oliver, A., Canton, R., Campo, P., Baquero, F. & Blazquez, J. (2000) High frequency of hypermutable Pseudomonas aeruginosa in cystic fibrosis lung infection. *Science* 288, 1251–1253.

Palys, T., Nakamura, L. K. & Cohan, F. M. (1997) Discovery and classification of ecological diversity in the bacterial world: the role of DNA sequence data. *International Journal of Systematic Bacteriology* 47, 1145–1156.

Redfield, R. J. (1993) Genes for breakfast: the have-your-cake-and-eat-it-too of bacterial transformation. *Journal of Heredity* 84, 400–404.

Row-Magnus, D. A. & Mazel, D. (1999) Recent gene capture. *Current Opinion in Microbiology* 2, 483–488.

Schrag, S. & Perrot, V. (1996) Reducing antibiotic resistance. *Nature* 28, 120–121.

Schrag, S., Perrot, V. & Levin, B. (1997). Adaptation to the fitness cost of antibiotic resistance in *Escherichia coli*. *Proceedings of the Royal Society London* 264, 1287–1291.

Smith, J. (2000) The social evolution of bacterial pathogenesis. In Prep.

Smith, J. M. (1992) Clonal histories-age and the unisexual lineage. *Nature* 356, 661–662.

Sniegowski, P. D., Gerrish, P. J. & Lenski, R. E. (1997) Evolution of high mutation rates in experimental populations of E. coli. *Nature* 387, 703–705.

Stewart, F. M. & Levin, B. R. (1977) The population biology of bacterial plasmids: a priori conditions for the existence of conjugationally transmitted factors. *Genetics* 87, 209–228.

Stewart, F. M. & Levin, B. R. (1984) The population biology of bacterial viruses: Why be temperate. *Theoretical Population Biology* 26, 93–117.

Summers, D. K. (1996) *The Biology of Plasmids* (Blackwell, Oxford).

Taddei, F., Radman, M., Maynard-Smith, J., Toupance, B., Gouyon, P. H. & Godelle, B. (1997) Role of mutator alleles in adaptive evolution. *Nature* 387, 700–702.

Tenaillon, O., Toupance, B., Le Nagard, H., Taddei, F. & Godelle, B. (1999) Mutators, population size, adaptive landscape and the adaptation of asexual populations of bacteria. *Genetics* 152, 485–493.

Wojciechowski, M. F., Hoelzer, M. A. & Michod, R. E. (1989) DNA repair and the evolution of transformation in Bacillus subtilis. II. Role of inducible repair. *Genetics* 121, 411–422.

Part III

PROTOCTIST MODELS

The mitochondrial genome of kinetoplasts is a highly derived genome in which frameshift errors in reading frames are corrected at at the messenger RNA level. "RNA editing" refers to these post-transcriptional modifications, of which two types are known. One consists of the precise insertion or deletion of U residues, so as to produce open reading frames in the messenger RNAs encoded in the organelle DNA known as the maxicircle. The other editing system is a modification of 34 C's into 34 U's in the anticodon of transfer RNA molecules that thus can decode the UGA stop codon as tryptophan. Larry Simpson and colleagues ("Evolution of RNA Editing in Trypanosome Mitochondria," Chapter 8) seek to unravel the evolution of these two peculiar genetic systems. With support from computer simulations, the authors elaborate an evolutionary scenario that proposes an ancient but unique evolutionary origin for both systems, which may have arisen shortly after the divergence of the trypanosomes and their relatives from the euglenoids.

Stephen M. Rich and Francisco J. Ayala ("Population Structure and Recent Evolution of *Plasmodium Falciparum*," Chapter 9) summarize data showing absence of synonymous nucleotide polymorphisms in diverse genes from *Plasmodium falciparum*, the agent of malignant malaria. The inference is that the extant world populations of *P. falciparum* originated from a single ancestral cell in recent times, estimated to be less than 50,000 years. This inference seems at first incompatible with the existence of numerous amino acid and other polymorphisms in the antigenic genes of the parasite. Rich and Ayala analyze allelic sequences of antigenic genes and conclude that they are consistent with a recent origin of the world populations of *P. falciparum*. The antigenic polymorphisms come about rapidly by mass natural selection acting on sequence variations originated at high rates by intragenic recombination of short DNA repeats.

8

Evolution of RNA Editing in Trypanosome Mitochondria

LARRY SIMPSON*†, OTAVIO H. THIEMANN§, NICHOLAS J. SAVILL¶‡, JUAN D. ALFONZO*, AND D. A. MASLOV**

Two different RNA editing systems have been described in the kinetoplast-mitochondrion of trypanosomatid protists. The first involves the precise insertion and deletion of U residues mostly within the coding regions of maxicircle-encoded mRNAs to produce open reading frames. This editing is mediated by short overlapping complementary guide RNAs encoded in both the maxicircle and the minicircle molecules and involves a series of enzymatic cleavage-ligation steps. The second editing system is a C_{34} to U_{34} modification in the anticodon of the imported $tRNA^{Trp}$, thereby permitting the decoding of the UGA stop codon as tryptophan. U-insertion editing probably originated in an ancestor of the kinetoplastid lineage and appears to have evolved in some cases by the replacement of the original pan-edited cryptogene

*Howard Hughes Medical Institute and †Department of Microbiology, Immunology, and Molecular Genetics, University of California, Los Angeles, CA 90095; ¶School of Biological Sciences, Manchester University, Manchester, United Kingdom M13 9PT; §Laboratory of Protein Crystallography and Structural Biology, Physics Institute of Sao Carlos, University of Sao Paulo, Av. Dr. Carlos Botelho 1465, PO Box 369, Sao Carlos, SP, Brazil 13560-970; and **Department of Biology, University of California, 3401 Watkins Drive, Riverside, CA 92521

This paper was presented at the National Academy of Sciences colloquium "Variation and Evolution in Plants and Microorganisms: Toward a New Synthesis 50 Years After Stebbins," held January 27–29, 2000, at the Arnold and Mabel Beckman Center in Irvine, CA.

Abbreviations: gRNA, guide RNA; kDNA, kinetoplast DNA.

‡Present address: Department of Mathematics, Heriot-Watt University, Edinburgh, EH14 4AS, United Kingdom

with a partially edited cDNA. The driving force for the evolutionary fixation of these retroposition events was postulated to be the stochastic loss of entire minicircle sequence classes and their encoded guide RNAs upon segregation of the single kinetoplast DNA network into daughter cells at cell division. A large plasticity in the relative abundance of minicircle sequence classes has been observed during cell culture in the laboratory. Computer simulations provide theoretical evidence for this plasticity if a random distribution and segregation model of minicircles is assumed. The possible evolutionary relationship of the C to U and U-insertion editing systems is discussed.

The term RNA editing describes several types of posttranscriptional modifications of RNAs that involve either specific insertion/deletion or modifications of nucleotides (Smith *et al.*, 1997). The uridine (U)-insertion/deletion type of editing has so far only been found to occur in the mitochondria of kinetoplastid protists (Alfonzo *et al.*, 1997; Stuart *et al.*, 1998). We recently showed that C to U nucleotide modification editing also occurs in the mitochondria of these cells (Alfonzo *et al.*, 1999). The origin and evolution of these two genetic systems is the subject of this paper.

KINETOPLASTID PROTISTS CONSIST OF TWO MAJOR GROUPS: THE TRYPANOSOMATIDS AND THE BODONIDS

Kinetoplastid protists belonging to the Euglenozoa phylum, according to rRNA phylogenetic trees, represent one of the earliest mitochondrial-containing extant branches of the eukaryotic lineage (Cavalier-Smith, 1997). This view may change in the future, as protein-based phylogenies favor a later divergence of Euglenozoa (Budin and Philippe, 1998; Germot and Philippe, 1999; Philippe and Forterre, 1999). However, even in such a case, this phylum still demonstrates a long and independent evolutionary history and is well separated from other eukaryotic groups. Taxonomists previously have proposed the existence of two suborders in the Kinetoplastida, the Trypanosomatina and Bodonina. All of the pathogenic trypanosomatids belong to the suborder, Trypanosomatina, and to the single family, Trypanosomatidae. Phylogenetic reconstructions using nuclear SSU rRNA sequences have confirmed the separation of the trypanosomatids as a derived late-emerging group. The trypanosomes, which initially were thought to be paraphyletic, with *Trypanosoma brucei* as an early-diverging branch (Fernandes *et al.*, 1993; Landweber and Gilbert, 1994; Maslov *et al.*, 1994), are now thought more likely to be monophyletic (Alvarez *et al.*, 1996; Lukes *et al.*, 1997) (Fig. 1). There are two major clades of trypanosomatids, the trypanosomes and the clade of *Leishmania*,

Crithidia, Leptomonas, Phytomonas, Herpetomonas, and *Blastocrithidia.* An early divergence within the trypanosome lineage led to separate salivarian (e.g., *T. brucei*) and nonsalivarian trypanosomes (Haag *et al.*, 1998). One study further splits the nonsalivarian trypanosomes into two clades, consisting of bird trypanosomes, such as *Trypanosoma avium,* and stercorarian trypanosomes such as *Trypanosoma cruzi* (Haag *et al.*, 1998).

The bodonid group is poorly studied and is probably paraphyletic (Wright *et al.*, 1999). The rRNA tree in Fig. 1 includes multiple species from this lineage. The deepest branches of the bodonid lineage include the poorly studied free-living organisms *Bodo designis, Rhynchobodo,* and *Dimastigella.* This is followed by a mixed clade of free-living *Bodo caudatus, Cryptobia helicis,* and the parasitic *Cryptobia salmositica* and *Trypanoplasma borreli.* Another free-living organism, *Bodo saltans,* may represent the clos-

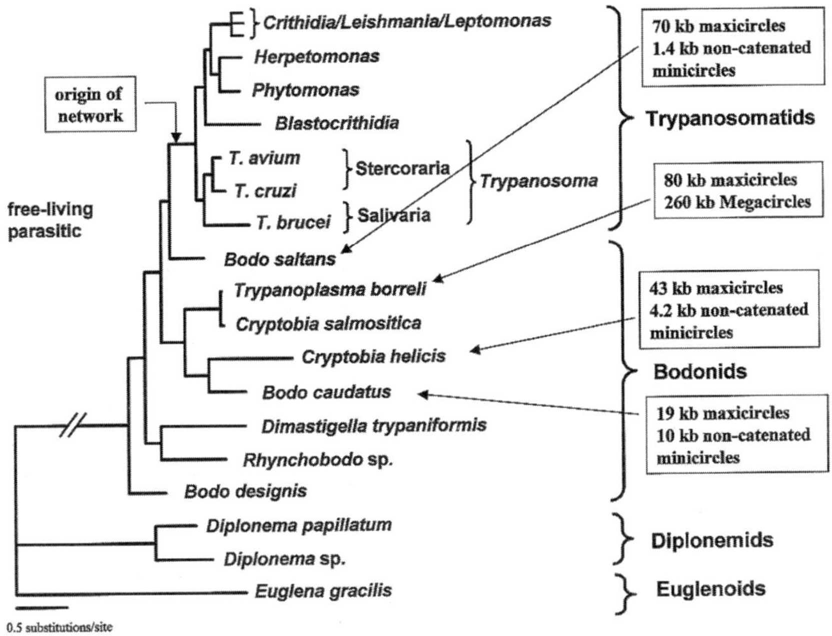

FIGURE 1. Phylogenetic tree of Kinetoplastida based on SSU rRNA sequences. Only representative species for each trypanosomatid lineage are shown. The sequences of B. designis, C. helicis, and B. saltans are from unpublished data of D. Doleîzel, M. Jirkû, D.A.M., and J. Lukeš. The tree was constructed by the method of maximum likelihood. The horizontal bar corresponds to 0.05 substitutions per site. This tree represents a tentative result based on a more extensive reconstruction using additional species (D. Doleîžel, M. Jirkû, D.A.M., and J. Lukeš, unpublished results).

est relative of trypanosomatids. *B. caudatus, T. borreli, C. helicis, and B. saltans* are the only bodonids for which mitochondrial molecular data are available. *Diplonema papillatum* represents either a sister group to the kinetoplastids or a sister group to the euglenoids (Maslov *et al.*, 1999). It is of some interest that the UGA stop codon is used to encode tryptophan in the mitochondrial genome of all kinetoplastid species including *Diplonema,* but not in the Euglenoids (Yasuhira and Simpson, 1997).

KINETOPLASTIDS CONTAIN A SINGLE EXTENDED TUBULAR MITOCHONDRION WITH AN UNUSUAL MITOCHONDRIAL DNA

The trypanosomatids contain a single tubular mitochondrion (Paulin, 1975; Simpson and Kretzer, 1997) that has a single mass of mitochondrial DNA situated within the matrix adjacent to the basal body of the flagellum (Simpson, 1986). The trypanosomatid mitochondrial or kinetoplast DNA (kDNA) consists of a single highly structured disk-shaped network composed of thousands of catenated minicircles ranging in size from as small as 465 bp in *Trypanosoma vivax* (Borst *et al.*, 1985) to as large as 10,000 bp in *T. avium* (Yurchenko *et al.*, 1999), and a smaller number of catenated maxicircles ranging in size from 23 to 36 kb. The maxicircles are the homologues of the informational mtDNA molecules found in other eukaryotes and contain 18 tightly clustered protein-coding genes and two rRNA genes; the gene order is conserved in all trypanosomatid species examined. No tRNA genes are encoded in the maxicircle. The mitochondrial tRNAs are encoded in the nucleus and are imported into the mitochondrion (Simpson *et al.*, 1989; Hancock and Hajduk, 1990).

The bodonids also have a single mitochondrion but the mtDNA is less structured and the molecules are not catenated. The DNA appears in thin sections as large oval fibrous structures. In *C. helicis* there are multiple nodules of DNA in the mitochondrion, an organization that has been termed pan-kinetoplastic (Vickerman, 1977; Lukes *et al.*, 1998). The kDNA in *C. helicis* consists of 43-kb maxicircles and several thousand 4.2-kb noncatenated minicircles (Lukes *et al.*, 1998), in *B. saltans* 70-kb maxicircles and multiple noncatenated 1.4-kb minicircles [which encode two guide RNAs (gRNAs) each] (Blom *et al.*, 1998), and in *B. caudatus* 19-kb maxicircles and 10- to 12-kb minicircles (Hajduk *et al.*, 1986). The kDNA in *T. borreli,* however, contains 80-kb maxicircles and 180-kb minicircle homologues (megacircles) (Lukes *et al.*, 1994; Maslov and Simpson, 1994; Yasuhira and Simpson, 1996). Sequence information is available only for fragments of the maxicircle equivalents from *B. saltans* and *T. borreli* and for five gRNAs from *T. borreli* and 14 gRNAs from *B. saltans*. It is of interest that the *Cyb, COI,* and *COIII* gene order in the *T. borreli* maxicircle and also the *COII* and *ND5* gene order in *B. saltans* differ from that in the trypanosomatids, which is consistent with the large evolutionary distance separating bodonids and trypanosomatids.

U-INSERTION/DELETION RNA EDITING

Mechanism

The transcripts of 12 (the precise number varies with the species) of the maxicircle protein-coding genes are edited posttranscriptionally by the insertion and occasional deletion of uridine (U) residues mostly within coding regions, thereby correcting frameshifts and producing translatable mRNAs. The minicircles encode gRNAs, which are complementary to mature edited sequences if G:U as well as canonical base pairs are allowed (Blum *et al.*, 1990). The gRNAs have 3′ nonencoded oligo(U) tails that may be involved in stabilizing the initial interaction of the gRNA and the mRNA by either RNA–RNA or RNA–protein interactions (Blum and Simpson, 1990; Kapushoc and Simpson, 1999). Fifteen gRNAs are encoded in intergenic regions of the maxicircle DNA of *Leishmania tarentolae*. This division of the mitochondrial genome into two physically separate genomes, with the RNA transcripts of one interacting with the incomplete mRNA transcripts of the other to produce translatable mRNAs is unprecedented and is suggestive of an unique evolutionary origin.

The mechanism of U-insertion/deletion editing involves a series of enzymatic cleavage-ligation steps, with the precise cleavages determined by base pairing with the cognate gRNAs (Alfonzo *et al.*, 1997). A single gRNA mediates the editing of a "block" of approximately 1–10 sites. Multiple overlapping gRNAs mediate the editing of a "domain" (Maslov and Simpson, 1992). The overall 3′ to 5′ polarity of editing site selection within a domain is determined by the creation of upstream mRNA "anchor" sequences by downstream editing (Fig. 2). Editing usually also proceeds 3′ to 5′ within a single block. A variable extent of "misediting" at the junction regions between fully edited and unedited sequences also has been observed, and this varies from gene to gene and from species to species (Decker and Sollner-Webb, 1990; Sturm and Simpson, 1990a; Sturm *et al.*, 1992). Misediting, which appears to be a consequence both of correct guiding by an incorrect gRNA and by stochastic errors in the editing process, is not deleterious, as misedited sequences appear to be re-edited correctly, in a 3′ to 5′ polarity (Sturm and Simpson, 1990a; Sturm *et al.*, 1992; Syrne *et al.*, 1996).

Comparative Analysis of Editing in Different Kinetoplastid Species

The U-insertion/deletion type of RNA editing has been detected in multiple trypanosomatid species. The extent of editing for several genes

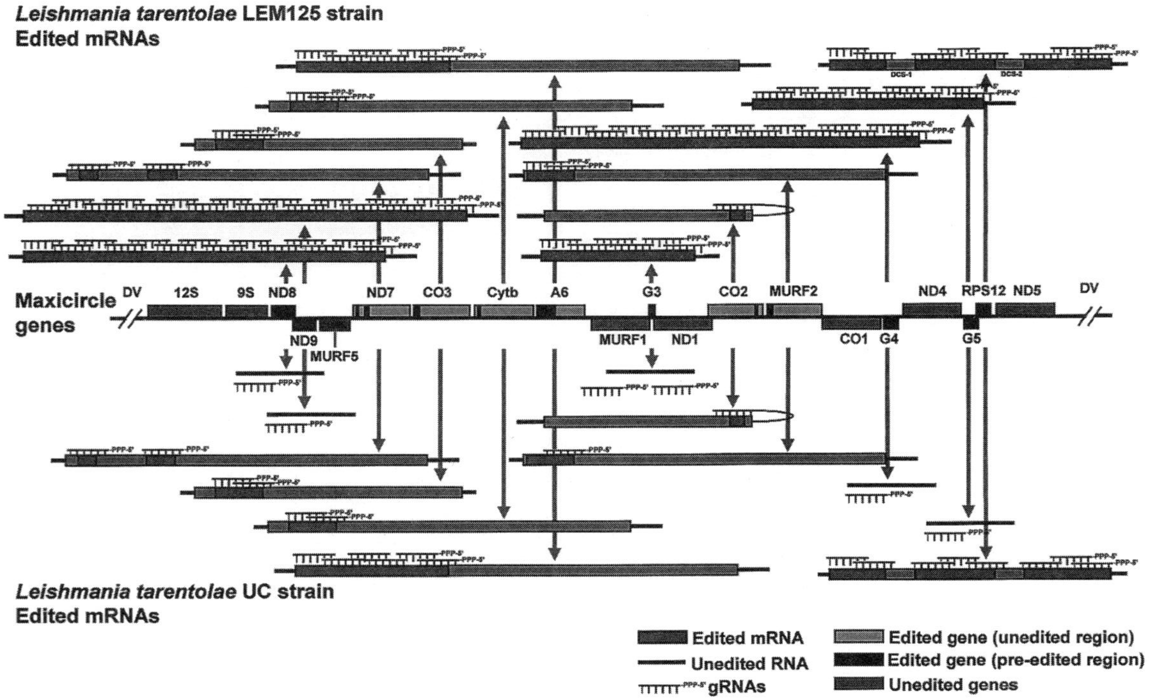

FIGURE 2. Diagram of the extent of gRNA-mediated editing of maxicircle cryptogenes in the old laboratory UC strain of L. tarentolae and the recently isolated LEM125 strain. The overlapping gRNAs that give rise to the overall 3′ to 5′ editing within a domain are indicated. In the LEM125 strain, all of the approximately 80 predicted gRNAs are indicated although only 47 gRNAs have been identified.

varies in different species. For example, the ND7 gene in the trypanosomes, *T. brucei* and *T. cruzi, is* pan-edited in two domains, whereas in the *Leishmania-Crithidia* clade this gene is edited only at the 5′ end of each domain. The A6 gene in the trypanosomes is pan-edited, whereas in the *Leishmania-Crithidia* clade, the editing of the A6 gene shows a gradient of restriction to the 5′ end of the single domain, from *L. tarentolae,* to *Herpetomonas muscarum,* to *Phytomonas serpens,* and to *Blastocrithidia culicis* (Maslov *et al.,* 1994).

To date, the deepest lineage in which U-insertion editing has been detected is the bodonid group. Because minicircles, which presumably encode gRNAs, are observed in *B. caudatus, B. saltans,* and *C. helicis,* this suggests that the free noncatenated state is a primitive feature. Catenation of minicircles to form the kDNA network probably arose in an ancestor of the trypanosomatids as a molecular mechanism designed to avoid minicircle losses by missegregation. Concatenation of minicircles in the 180-kb megacircle as observed in *T. borreli* might have independently arisen as another solution to the same problem. However, additional analyses of the kDNA structure in bodonids are required to shed more light on kDNA evolution.

The only mitochondrial gene isolated from the deeper branching *Diplonema* (Maslov *et al.,* 1999) and *Euglena gracilis* (Tessier *et al.,* 1997; Yasuhira and Simpson, 1997) is the *COI* gene, which is unedited. In addition, no evidence was obtained for small gRNA-like molecules in *E. gracilis* mitochondria by 5′ capping experiments (Yasuhira and Simpson, 1997). This preliminary evidence does not, of course, eliminate the possibility of editing in these cells, but the simplest hypothesis is that this type of editing evolved in the mitochondrion of an ancestral bodonid after the split from the euglenoid lineage.

Minicircle-Encoded gRNAs in Two Strains of *L. tarentolae*

The only species for which the entire complement of gRNAs is known (Maslov and Simpson, 1992) is the UC strain of *L. tarentolae,* which has been maintained as the promastigote form in culture in various laboratories for more than 60 years. There are 15 maxicircle-encoded gRNAs and 17 minicircle-encoded gRNAs in this strain. Five pan-edited genes (G1–G5) show a complete absence of productive editing in this strain, as evidenced by an inability to PCR-amplify mature edited transcripts by standard methods. These genes are productively edited in *T. brucei.* Two of the minicircle-encoded gRNAs in the *L. tarentolae* UC strain, gLt19 (=gG4-III) and gB4 (=gND3-IX), represent nonessential gRNAs for these nonfunctional editing cascades. This was determined by analyzing the minicircle-encoded gRNA complement of LEM125, a recently isolated strain of *L. tarentolae* (Thiemann *et al.*, 1994). LEM125 has the same 15 maxicircle gRNA genes but also has an estimated 80 total minicircle-encoded gRNAs, of which 30 have been cloned and sequenced and the remainder inferred to be present because of the existence of completely edited mRNA transcripts in this strain. These additional gRNAs mediate the editing of three components of complex I of the respiratory chain, *ND3, ND8,* and *ND9,* and also two unidentified genes, which were termed G-rich regions 3 and 4 (G3 and G4). It was proposed that multiple gRNA-encoding minicircle sequence classes had been lost from the UC strain probably because of a lack of a requirement for complex I activity in culture (Simpson and Maslov, 1994; Thiemann *et al.*, 1994). The presence of productively edited *ND8, ND9,* G3 (=CR3), G4 (=CR4), and *ND3* mRNAs in *T. brucei* and the presence of productively edited G3 mRNA in *P. serpens* (D.A.M., unpublished results) implies that the corresponding minicircle-encoded gRNAs also exist in these species, and this provides phylogenetic evidence for our hypothesis that the ancestral cell had a complete complement of minicircle classes. In addition, the presence of two minicircle-encoded gRNAs, gG4-III and gND3-IX, in the UC strain, which are remnants of the complete editing cascade of gRNAs for these two genes in LEM125, corroborates this evidence. To propose a loss of multiple minicircle classes from the UC strain is also more parsimonious than to propose a gain of multiple classes in the LEM125 strain. And finally, the existence of a 5′ terminal block of misedited sequence in the LEM125 *ND3* mRNA (Thiemann *et al.*, 1994) is indicative that this gene originally was completely edited and has lost the 5′ terminal gRNA.

Minicircles from *L. tarentolae* and other members of the *Leishmania-Crithidia* clade contain a single gRNA gene situated at a constant distance from the origin of replication (Sturm and Simpson, 1990b; Yasuhira and Simpson, 1995). Minicircles from *T. brucei,* however, also have a single origin of replication but contain three gRNA genes situated between 18-mer inverted repeats (Pollard *et al.*, 1990), and minicircles from *T. cruzi* contain four gRNA genes situated within each of the four variable regions between four origins of replication (Avila and Simpson, 1995). The total number of different minicircle sequence classes in *T. brucei* is estimated to be 200–300 (Stuart, 1979), which would yield a total of 600–900 gRNAs. Although only 72 gRNAs have been identified so far in *T. brucei* (Souza *et al.*, 1997), it is clear that there are extensive redundant gRNAs, which are gRNAs of different sequence but possessing the identical editing information because of the allowed G:U base pairing (Corell *et al.*, 1993). In fact, 28 of the 72 identified gRNAs are redundant over the entire length of the gRNA. Only a single redundant gRNA pair has been observed in *L. tarentolae* (Thiemann *et al.*, 1994). *T. brucei* also contains gRNAs with several mismatches in the anchor or guiding regions, which may be nonfunctional, but there is no evidence for or against this suggestion.

Retroposition Model for Loss of Editing in Evolution

Based on these observations and on the known 3′–5′ polarity of editing, a retroposition model was proposed to explain both the gradual restriction of editing to the 5′ end of domains and the complete loss of editing in some cases (Landweber, 1992; Simpson and Maslov, 1994). We proposed that partially edited mRNAs were being frequently converted to cDNAs by a postulated mitochondrial reverse transcriptase activity, and those cells that had replaced the original pan-edited cryptogene with a partially edited gene would survive a loss of an entire minicircle sequence class encoding a specific gRNA involved in that editing cascade. The retention of editing at the 5′ end of a domain may allow regulation of translation by creation of a methionine initiation codon and a possible ribosome-binding site. This model is based on the assumption that minicircles are distributed randomly to daughter cells upon cell division.

Replication and Segregation of Minicircles

One possible mechanism involved in the random distribution of minicircles is the mode of replication and segregation. The mitochondrial S phase is fairly synchronous with the nuclear S phase, although the kinetoplast network physically divides just before the nucleus (Simpson and Braly, 1970). Closed minicircles are apparently randomly removed from the side of the network facing the basal body by a topoisomerase II activity and migrate by an unknown mechanism to one of two replisomes (Ferguson *et al.*, 1992) that are located at the two antipodes of the kDNA nucleoid body (Simpson and Simpson, 1976; Ferguson *et al.*, 1992; Shapiro and Englund, 1995). After replication, the daughter molecules remain nicked or gapped, which may be a mechanism to ensure replication of each minicircle. The daughter minicircles then are recatenated into the periphery of the network. There is microscopic evidence that the networks in *Leishmania* and *Crithidia* (and also *T. cruzi)* are actually rotating, and this movement produces a complete peripheral distribution of newly replicated minicircles (Ferguson *et al.*, 1994; Robinson and Gull, 1994; Guilbride and Englund, 1998). The networks in the middle of S phase consist of an expanding ring of nicked circles and a central core of closed circles, and at the end of S phase the networks consist entirely of nicked circles. The minicircles then become closed and then the network segregates into two daughter networks as the single mitochondrion divides (Pérez-Morga and Englund, 1993). This mechanism of replication appears to introduce a certain amount of randomness into minicircle segregation. In other words, sister minicircles may not necessarily end up in different daughter cells. A pulse–chase experiment performed with *C. fasciculata* cells at the light microscope level previously showed that newly replicated minicircles are spread throughout the network after one cell cycle is completed (Simpson *et al.*, 1974).

In the case of *T. brucei*, the network apparently does not rotate and two dumbbell-shaped masses of nicked replicated minicircles accumulate at either end of the nucleoid body, which then divides in half into the daughter cells (Ferguson *et al.*, 1994; Robinson and Gull, 1994). In this case there does not appear to be a mechanism for randomization throughout the network, other than the possible random selection and migration to the antipodal replisomes.

Plasticity of Minicircle Sequence Class Copy Number in *L. tarentolae* in Culture

The number of minicircles per network in *L. tarentolae* was assayed by counting 4′,6-diamidino-2-phenylindole (DAPI)-stained networks in a cell counting chamber using a fluorescent microscope and measuring the DNA concentration spectrophotometrically. Quantitative dot blot hybridization using an oligonucleotide probe complementary to the conserved CSB-3 12-mer sequence yielded values of 12,600 ± 300 and 12,700 ± 800 for the UC and LEM125 strains, respectively. Similar dot blot hybridization analysis showed that the copy number of maxicircle DNA molecules was very similar in the UC and LEM125 strains (32 ± 2 and 25 ± 2 copies per network, respectively).

Quantitation of the copy numbers of 17 specific minicircle sequence classes in the UC strain was previously performed by Southern blot analysis using specific oligonucleotide probes for specific gRNAs. We have repeated these analyses with both UC strain kDNA and LEM125 strain kDNA, by dot blot hybridization of *Msp*I-digested kDNA (all minicircles have at least one *Msp*I site), and a known amount of specific cloned minicircles using primers specific to each gRNA. A primer to the conserved 12-mer sequence was used as a loading control. The results in Table 1 show that homologous minicircle sequence class frequencies are extremely variable, both between strains and between different kDNA isolates from the same strain taken after several years of culture. In general, the LEM125 strain kDNA exhibited lower copy numbers for the sequence classes in common between the strains, which is consistent with LEM125 possessing a more complex minicircle repertoire.

In the UC strain kDNA as mentioned above, two gRNAs, pLtl9 (= G4-III) and pB4 (= gND3-IX), are nonfunctional, in that all of the other

TABLE 1. Minicircle copy numbers in kDNA networks from UC and LEM 1 25 strains of *L. tarentolae*

	UC		LEM 125
	% Minicircles per network		% Minicircles per network
Minicircle class	1992*	1994	
RPS1 2-I	3.7	4.7	0.2
RPS12-II	1.4	0.2	0.8
RPS12-III	10.5	1.5	2.5
RPS12-IV	2.1	0.4	0.8
RPS1 2-V	5.0	0.2	0.4
RPS12-VII	0.9	15.0	2.8
RPS12-VIII	0.3	0.9	0.1
A6-I	2.2	0.4	6.0
A6-II	1.1	0.4	1.8
A6-III	3.8	1.0	0.5
A6-IV	3.3	0.1	0.1
A6-V	2.1	2.0	0.2
A6-VI	3.2	0.8	1.0
COIII-I	1.9	0.5	0.8
COIII-II	3.7	1.7	3.6
ND8-I	—	ND	4.8
ND8-II	—	ND	9.2,
ND8-IV	—	ND	3.9
G4-III(Lt19)	25	66.8	1.7
ND3-II	—	ND	0.2
ND3-III	—	ND	6.4
ND3-V	—	ND	2.8
ND3-VI	—	ND	5.4
ND3-IX (B4)	29.8	3.4	3.1

ND, not detected above background.
*Data taken from Maslov and Simpson, 1992.

minicircle-encoded gRNAs in those editing cascades are missing from this strain. It is of interest that these nonfunctional minicircles showed the greatest plasticity in frequency.

As was found previously (Maslov and Simpson, 1992), there was no correlation of minicircle copy number and gRNA relative abundance (data not shown).

Computer Simulations of Minicircle Sequence Class Plasticity

Using a population dynamics model of minicircle segregation, Savill and Higgs (1999) recently have shown that random segregation can indeed account for much of the above experimental observations on minicircle plasticity. The copy number of every minicircle class in every cell in a population is tracked over many generations. In every generation each cell replicates its minicircles, hence doubling the copy number of all its classes. Then the cells divide and the daughter cells receive a certain number of copies of each class. The actual number of copies is randomly chosen according to a binomial distribution that models a purely random segregation process. All daughter cells that receive the full complement of minicircle classes and have fewer than 12,000 minicircles in total are randomly chosen to populate the next generation up to a maximum population size. These two conditions model the reasonable assumptions that (*i*) if a cell does not receive any copies of a particular class it is therefore missing a gRNA and hence its mRNA cannot be correctly edited, which is assumed to be lethal, and (*ii*) the network is restricted in its maximum size because of physical constraints.

A typical simulation of a hypothetical species with 17 minicircle classes is shown in Fig. 3. It clearly demonstrates that random segregation causes fluctuations in the average minicircle class copy number from one generation to the next. Moreover, it also leads to the experimentally observed distribution of many classes having very low copy number and a few having very high copy number. No two runs are ever the same, thus explaining why homologous minicircle classes in different strains have different copy numbers.

The loss of minicircle classes during the long culture history of the UC strain also was modeled by starting with 70 classes, of which 15 are required and 55 are not. Fig. 4 shows the number of generations for each unnecessary class to be lost, from the time when the UC strain was first cultured. Many classes are lost fairly rapidly within the first few hundred generations, but it takes successively longer for the remaining classes to be lost, and the last few classes may take tens of thousands of generations to be lost. Moreover, by averaging over many simulations we found that the last remaining unnecessary class was also the most abundant class in 27% of cases. Therefore, random segregation can explain the observed long persistence time of unnecessary classes and their high abundance. However, as shown in Fig. 3, the highest frequency achieved by the most abundant class for a hypothetical species with 17 classes (similar to the UC strain) only reaches about 30% and never as large as the 67% observed in the 1,994 UC cells. This large abundance of one class is similar to the situation in the CFC1 strain of *C. fasciculata*, in which one minicircle se-

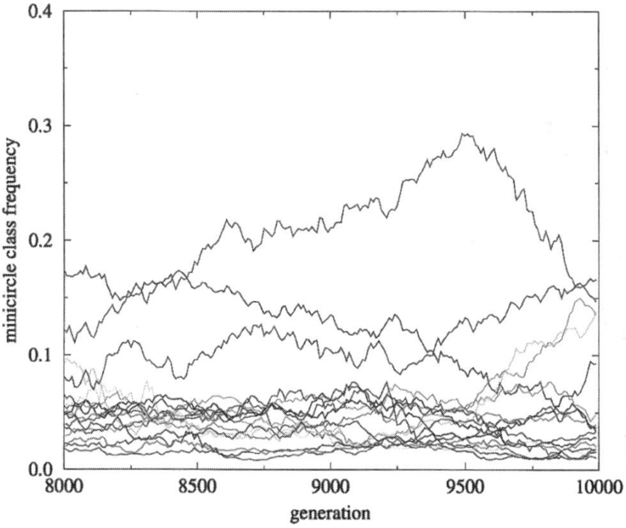

FIGURE 3. The average frequencies of 17 minicircle classes undergoing random segregation over 2,000 generations. Random segregation causes fluctuations in the frequencies, giving rise to the experimentally observed distribution of many classes having low frequency and few classes having high frequency. Initially at generation zero, every class in every cell has 170 copies. There are 1,000 cells that have maximum network sizes of 12,000 minicircles.

quence class shows over 90–95% abundance. It appears that random segregation alone cannot explain the large abundance of these classes, and therefore other selective forces must be present.

If the additional following assumptions are made, simulations can explain the experimental results: (*i*) The network has a minimum allowable size. If the network is too small, it may not abut the replisomes. (*ii*) The number of copies of each necessary minicircle class is regulated by an unknown mechanism. (*iii*) The number of copies of each unnecessary minicircle class is unregulated, i.e., once a minicircle becomes unnecessary—by loss of other gRNAs in a cascade, its copy number is not regulated and can vary freely. The model is modified so that if the total number of minicircles in a daughter cell falls below a predetermined threshold or if the copy number of each necessary class exceeds a predetermined threshold, the cell does not survive into the next generation. For simplicity, in the model this threshold is set to the same value for all necessary classes, but in reality it may vary between classes. The lower threshold for each necessary class is one copy, as in the original model. Again, in reality this may not be true. Fig. 5 shows a simulation where the minimum

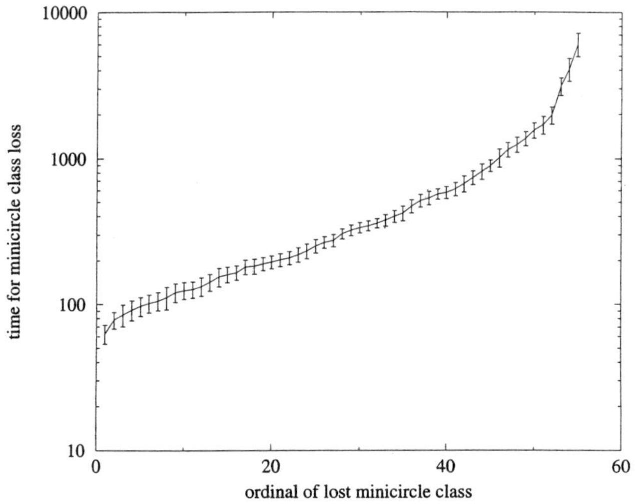

FIGURE 4. The number of generations for consecutive unnecessary minicircle classes to be lost. The last few classes take many thousands of generations to be lost. The simulations are initially run for 2,000 generations with all classes being necessary. This is to lose the artificial initial conditions. Then, 55 classes become unnecessary, i.e., if their frequencies reach zero in a cell, the cell is still viable. The loss time was averaged over 10 simulations; error bars show ± 2 SEM. Previously published in Savil and Higgs (1999).

kinetoplast threshold size is set to 10,000 minicircles and the upper threshold for necessary classes is set to 200. Fifteen classes are required and 55 are not. Initially all classes have the same copy number of 170, giving a total of 11,900 minicircles per cell, which lies between the assumed lower and upper thresholds for the kinetoplast size (i.e., 10,000 and 12,000 minicircles, respectively). The figure shows the cumulative proportion of minicircles of all necessary classes, all unnecessary classes, and the proportion of the most abundant class. Initially the proportions are 21% (170 × 15/11,900), 79% (170 × 55/11,900), and 1.4% (170/11,900), respectively. Because of assumption *ii*, the proportion of minicircles of necessary classes cannot exceed 30% (15 × 200/10,000). Therefore, the unnecessary classes must make up the difference for the kinetoplasts to maintain their minimum sizes. However, as unnecessary classes are lost because of random segregation over time, there are fewer classes that can make up this difference. Finally, there will be only one unnecessary class left to make up at least 70% of the minicircles in the kinetoplast. This class is now necessary only to maintain kinetoplast size, and the function of encoding gRNAs

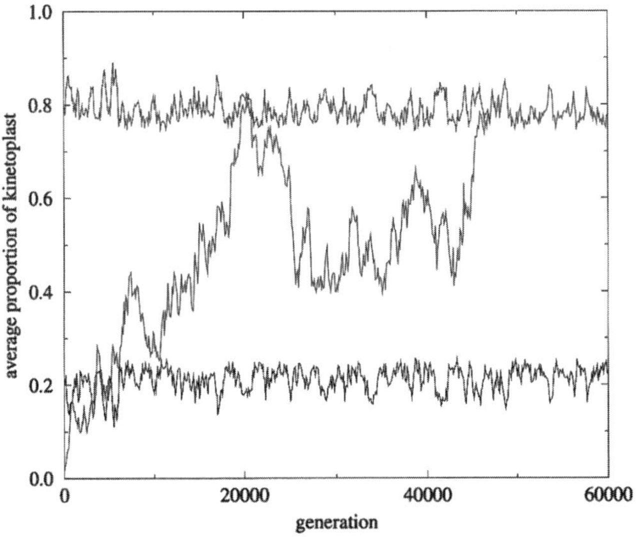

FIGURE 5. As unnecessary classes are lost, one unnecessary class must become highly abundant to maintain the size of the kinetoplast, i.e., 10,000 minicircles. This is because the 15 necessary classes are restricted to a maximum of only 200 copies. Initially all classes have a copy number of 170, and 55 classes are unnecessary and 15 are necessary. The top line shows the cumulative frequency of all unnecessary classes, the bottom line the cumulative frequency of all necessary classes, and the middle line the frequency of the most abundant class at any given time.

has now been replaced by a buffering function. By adjusting the parameters, it is even possible to obtain an unnecessary class with over 90% abundance, as in the CFC1 strain (Pérez-Morga and Englund, 1993). This successful simulation of large frequencies for unnecessary minicircle classes actually provides support for assumption *ii*.

The model of random segregation also makes several interesting predictions. At every generation some daughter cells become unviable and do not survive into the next generation because they do not receive the full complement of minicircle classes. Hence some fraction of the total population of daughter cells is viable; we term this the daughter cell viability. We find that cell viability increases with increasing kinetoplast size and decreasing number of minicircle sequence classes (Fig. 6). If the cells have some mechanism that more evenly segregates sister minicircles between daughter cells, cell viability increases. This implies that there could be some selection pressure for trypanosomatids to segregate their minicircles more evenly, which may have led to the development of the rotating network in the *Leishmania-Crithidia* clade.

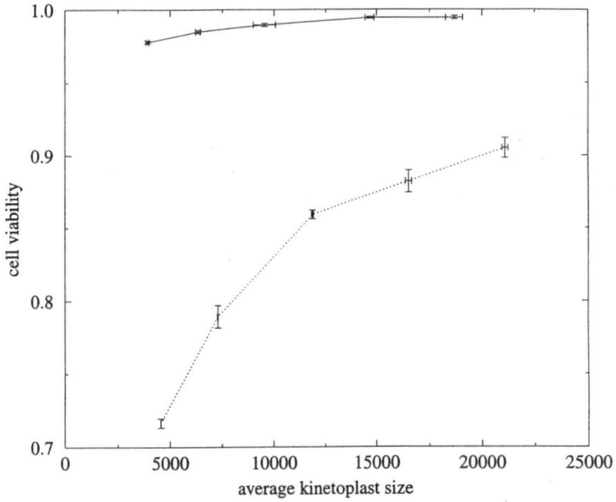

FIGURE 6. Cell viability increases as the average kinetoplast size increases and as the number of minicircle classes decreases (dotted line 70 classes, solid line 17 classes). This is because classes have more copies and hence there is more chance that a daughter cell receives the full complement of classes. Averages were taken over 10 simulations; error bars show ± 2 SEM. Previously published in Savil and Higgs (1999).

In the case of *T. brucei,* random segregation of the 250+ sequence classes would lead to a predicted cell viability in this model of less than 0.5, and hence population extinction. However, incorporating the information that each minicircle in this species encodes multiple gRNAs and that genetic exchange occurs, it has been shown that the model can produce the observed situation of evolutionary viability and multiple redundant and nonfunctional gRNAs (N.J.S. and P. G. Higgs, unpublished results). Mutation of the gRNA genes and drift in the minicircle copy numbers lead to an ever-increasing number of necessary classes encoding ever fewer functional gRNAs per minicircle.

C TO U EDITING AND THE ORIGIN OF URIDINE-INSERTION EDITING IN TRYPANOSOMES

UGA Codon Reassignment

Kinetoplastids use a nonuniversal genetic code in which the UGA stop codon is read as tryptophan (de la Cruz *et al.,* 1984). The codon capture hypothesis (Inagaki *et al.,* 1998; Osawa *et al.,* 1992) proposes that evolutionary reassignment of a stop codon involves first the disappearance of the stop codon and replacement with synonymous codons, then mutations in the peptide chain release factor so as not to recognize the stop codon, and finally duplication and mutation of a tRNA gene to allow decoding of the old codon with a new meaning. The occurrence of a nonuniversal genetic code in mitochondrial genomes is thought to be a derived character that arose independently in different organisms. In the Euglenozoa phylum, the use of a nonuniversal code is limited to the kinetoplastids (and diplonemids) (Yasuhira and Simpson, 1997; Maslov *et al.*, 1999). However, the appearance of a new gene for a tRNA decoding UGA for tryptophan did not occur in these species, perhaps because of the early loss of all mitochondrial-encoded tRNA genes.

C to U Editing of $tRNA^{Trp}$

Alfonzo *et al.* (1999) recently reported that, at least in the case of *L. tarentolae*, the problem of decoding UGA was solved by evolving an editing activity that changes the first position of the anticodon of the mitochondrial imported $tRNA^{Trp}$ from C to U (CCA to UCA in the anticodon), thereby allowing the decoding of UGA codon as tryptophan (Fig. 7). The evidence for this editing involved the observation of a loss of a *Hin*fI restriction site in a cDNA copy of the mitochondrial $tRNA^{Trp}$, which was confirmed by sequencing the reverse transcription–PCR-amplified product, and by direct analysis of the mitochondrial $tRNA^{Trp}$ by poisoned

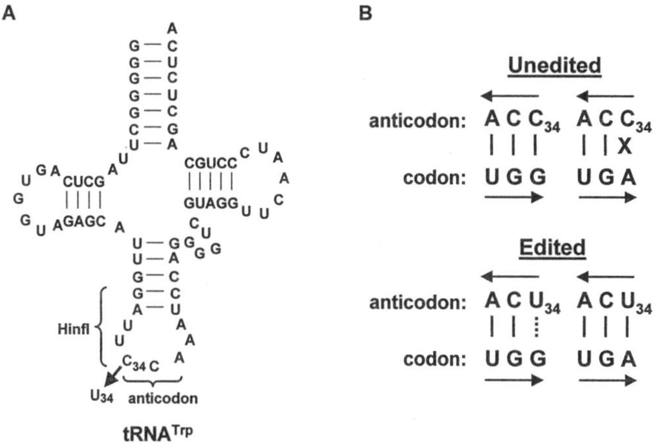

FIGURE 7. C to U editing of the anticodon of the mitochondrial-imported $tRNA^{Trp}$. (A) $tRNA^{Trp}$ showing the editing of C34. The HinfI site that is destroyed by the C to U editing event is also indicated. (B) The C34 to U34 editing allows the decoding of the UGA codon as tryptophan.

primer extension experiments. More than 40% of the mitochondrial $tRNA^{Trp}$ is edited at C34. A C to U editing event also has been described for the cytosolic 7SL RNA in *Leptomonas* (Ben Shlomo *et al.*, 1999). C to U editing is found in many phylogenetically diverse organisms, both in organellar and nuclear genomes, suggesting that this site-specific modification represents an ancient evolutionary activity (Covello and Gray, 1989; Morl *et al.*, 1995; Navaratnam *et al.*, 1995).

The following hypothetical scenario could explain the origin of this tRNA editing and the tryptophan codon change in trypanosomes. We propose that tRNA importation into the mitochondrion was developed at a very early stage of evolution and that tRNA genes in the kDNA subsequently were lost because of redundancy. The original state included the encoding of tryptophan by UGG and the CCA anticodon in the tRNA. We also assume that a pre-existing activity performing some other function in the cell produced a promiscuous C to U modification in the anticodon of the imported $tRNA^{Trp}$ at a low frequency. G to A transition mutations, perhaps driven by AT mutational pressure, led to the replacement of UGA with UAA stop codons and this was followed by mutations that affected the interaction of release factor with UGA. Similar mutational pressure led to the replacement of TGG tryptophan codons with TGA in essential mitochondrial genes, and this would have made the C to U tRNA editing indispensable for cell survival. This scenario combines the model of Covello and Gray (1993) for the evolution of RNA editing systems in general and a modified codon capture hypothesis (Inagaki *et al.*, 1998). The editing of an imported tRNA offers a new mechanism for codon capture that does not require a gene duplication event. The alternative hypothesis of duplication and mutation of an existing nuclear-encoded tRNA gene [$tRNA^{Trp}$ (CCA)] did not occur perhaps because of the problems involved in maintaining a suppressor tRNA [$tRNA^{Trp}$(UCA)] in the cytosol.

The Relationship of C to U Editing and U-Insertion Editing

It is interesting that 7% of the UGA tryptophan codons are created by U-insertion editing (Table 2). This observation places some time constraints on hypotheses for the appearance of U-insertion editing. We previously have proposed a scenario for the origin of U-insertion editing (Simpson and Maslov, 1999) based on the models of Covello and Gray (1993) and Cavalier-Smith (1997), which involved the pre-existence of editing enzymatic activities that were used for other biochemical functions, genetic drift in a mitochondrial gene, appearance of complementary gRNAs by partial gene duplication and antisense transcription, and finally utilization of editing for gene regulation. Stoltzfus (1999) has

TABLE 2. Tryptophan and stop codons in *L. tarentola mitochondria*

	Trp			Created by editing			Stop	
Gene	UGA	UGG	Total	UGA	UGG	Total	UAA	UAG
ND8	1	0	1	0	0	0		X*
ND9	2	1	3	1	0	2		X*
MURF5	1	1	2	0	0	0	X	
ND7	3	0	3	0	0	0	X	
COIII	8	0	8	0	0	0	X	
Cyb	15	1	16	0	0	0	X	
MURF4	2	0	2	2	0	2	X	
MURF1	4	1	5	0	0	0	X	
G3	0	1	1	0	1	0	X	
ND1	3	0	3	0	0	0	X	
COII	7	0	7	0	0	0		X
MURF2	3	0	3	0	0	0	X	
COI	13	1	14	0	0	0	X	
G4	1	2	3	1	2	3	X	
ND4	8	1	9	0	0	0		X
G5	2	2	4	2	2	4	X	
RPS12	1	0	1	1	0	1	X	
ND5	15	1	16	0	0	0	X	
Total	89	12	101	7	5	12	14	4 (2*)

*Created by editing.

pointed out that DNA polymerases have a bias toward single nucleotide deletions and that, for this reason, an increase in the number of edited sites is more likely than a loss of an editing site. The same author also proposed that if gRNAs arose by duplication and antisense transcription, this must have occurred before the genetic drift that gave rise to the pre-edited sequence, because gRNAs are complementary to the edited sequence and not to the pre-edited sequence. If one accepts this proposal, then the observed guiding of U-insertions to produce UGA tryptophan codons suggests that this codon reassignment also occurred before the appearance of U-insertion editing.

It should be noted that the presence of guiding G residues in the gRNAs that base pair with inserted Us presents a potential problem for the gene duplication scenario for the origin of gRNAs. One possible mechanism could have been deamination of the guiding A to produce inosine by an adenosine deaminase acting on RNA-like activity (Polson *et al.*, 1996; Yang *et al.*, 1997; Gerber *et al.*, 1998; Keller *et al.*, 1999) and retroposition of the mutated gRNA back into the genome to replace the original gRNA gene. Another mechanism could have been a replication-associated deletion of a C in a mitochondrial gene that was corrected by U-specific insertional editing activity at the transcript level guided by the original G residue in the gRNA.

An alternative scenario for the origin of gRNAs can be derived from an analysis of computational algorithms for searching for possible gRNA genes that are complementary to candidate cryptogenes (Von Haeseler *et al.*, 1992) in which it was shown that known gRNA sequences are in or very close to the statistical noise. Based on these results, one could speculate that the primordial gRNA was derived from some other mitochondrial RNA fragment that by chance base-paired with the mRNA downstream of the U-deletion site and contained a guiding A or G residue that could base-pair with an inserted U and thereby overcome this frameshift mutation and allow translation of the mRNA.

CONCLUSIONS

The mitochondrial genome of the kinetoplastid protists is a highly derived genome in which frameshift errors in the reading frames of 12 of the 18 genes are corrected at the RNA level by U-insertion/deletion editing, which probably arose in the early bodonid-kinetoplast lineage after divergence of the euglenoids. The sequence information for these corrections is partially located in a physically separate guide RNA genome. The most primitive type of organization of this genome may have been similar to that seen in the bodonids, *B. saltans, B. caudatus,* and *C. helicis,* in which the gRNA genes are present on multiple plasmid-like molecules. The next steps in evolution may have either been a concatenation of the plasmids into megacircles such as in *T. borreli* or a catenation of the plasmids into a network such as in the trypanosomatids.

The fact that daughter cells must receive a complete complement of all of the minicircle sequence classes encoding the gRNAs required for editing has led to the evolution of mechanisms for the random distribution of minicircles within the single network. The highly structured organization of the catenated minicircles within the network must have placed additional constraints on the evolution of this system. Two different types of mechanisms evolved, both based on a decatenation of minicircles from the network and replication at two antipodal nodes before recatenation of the daughter molecules. In the *Leishmania-Crithidia* clade (and *T. cruzi),* random selection of closed molecules and rotation of the network during the recatenation process in S phase produced a high degree of randomization. We have shown that computer simulations provide evidence that random segregation of minicircles during replication can account for many of the phenomena observed in *L. tarentolae* and possibly for the observed restriction of editing that has occurred in evolution. However, to explain the high abundance of unnecessary minicircle classes in the

UC strain of *L. tarentolae* and the CFC1 strain of *C. fasciculata,* further assumptions need to be made concerning the regulation of minicircle copy number.

In the *T. brucei* clade in which the network does not rotate, the randomization is mainly a function of random selection of closed molecules. However, the homogenizing effect of genetic exchange that occurs in the tsetse vector (Bogliolo *et al.*, 1996; Hope *et al.*, 1999), but which does not appear to occur at a detectable level in *Leishmania*, is another factor that may affect random distribution of minicircles to daughter cells. Relevant to this is the fact that two closely related trypanosome species, *T. equiperdum* and *T. evansi,* which have lost the sexual cycle in the fly and are transmitted by sexual intercourse and by mechanism transmission by tabanid flies, respectively, have networks consisting of one of several single minicircle sequence classes and mutated or deleted maxicircle DNA (Frasch *et al.*, 1980; Barrois *et al.*, 1982; Borst *et al.*, 1987; Songa *et al.*, 1990; Lun *et al.*, 1992; Shu and Stuart, 1994).

Another derived feature of the kinetoplastid mitochondrial genome is the complete lack of tRNA genes and the importation of all mitochondrial tRNAs from the cytosol (Simpson *et al.*, 1989). To decode UGA as tryptophan, the imported $tRNA^{Trp}$ is edited by a C to U modification within the anticodon (Alfonzo *et al.*, 1999).

RNA editing appears to have arisen in evolution multiple times in different organisms as a way to correct errors and modulate genetic sequences at the RNA level. In kinetoplastid protists, two types of editing that apparently arose early in the evolution of the kinetoplast-mitochondrion are intimately tied in with the unusual mitochondrial genome unique to these organisms. This provides yet another example of the evolutionary diversity of lower eukaryotes.

REFERENCES

Alfonzo, J.D., Blanc,V., Estevez, A. M., Rubio, M. A. & Simpson, L. (1999). C to U editing of the anticodon of imported mitochondrial tRNATrp allows decoding of the UGA stop codon in Leishmania tarentolae. *EMBO J.* 18, 7056–7062.

Alfonzo, J. D., Thiemann, O. & Simpson, L. (1997). The mechanism of U insertion/deletion RNA editing in kinetoplastid mitochondria. *Nucl. Acids Res.* 25, 3751–3759.

Alvarez, F., Cortinas, M. N. & Musto, H. (1996). The analysis of protein coding genes suggests monophyly of Trypanosoma. *Mol. Phylogenet. Evol.* 5, 333–343.

Avila, H. & Simpson, L. (1995). Organization and complexity of minicircle-encoded guide RNAs from *Trypanosoma cruzi. RNA* 1, 939–947.

Barrois, M., Riou, G. & Galibert, F. (1982). Complete nucleotide sequence of minicircle kinetoplast DNA from Trypanosoma equiperdum. *Proc. Natl. Acad. Sci. USA* 78, 3323–3327.

Ben Shlomo, H., Levitan, A., Shay, N. E., Goncharov, I. & Michaeli, S. (1999). RNA editing associated with the generation of two distinct conformations of the trypanosomatid Leptomonas collosoma 7SL RNA. *J. Biol. Chem.* 274, 25642–25650.

Blom, D., De Haan, A., Van den Berg, M., Sloof, P., Jirku, M., Lukes, J. & Benne, R. (1998). RNA editing in the free-living bodonid *Bodo saltans*. *Nucl. Acids Res.* 26, 1205–1213.

Blum, B., Bakalara, N. & Simpson, L. (1990). A model for RNA editing in kinetoplastid mitochondria: "Guide" RNA molecules transcribed from maxicircle DNA provide the edited information. *Cell* 60, 189–198.

Blum, B. & Simpson, L. (1990). Guide RNAs in kinetoplastid mitochondria have a nonencoded 3' oligo-(U) tail involved in recognition of the pre-edited region. *Cell* 62, 391–397.

Bogliolo, A. R., Lauria-Pires, L. & Gibson,W. C. (1996). Polymorphisms in *Trypanosoma cruzi*: Evidence of genetic recombination. *Acta Tropica* 61, 31–40.

Borst, P., Fase-Fowler, F. & Gibson, W. (1987). Kinetoplast DNA of Trypanosoma evansi. *Mol. Biochem. Parasitol.* 23, 31–38.

Borst, P., Fase-Fowler, F., Weijers, P., Barry, J., Tetley, L. & Vickerman, K. (1985). Kinetoplast DNA from Trypanosoma vivax and T. congolense. *Mol. Biochem. Parasitol.* 15, 129–142.

Budin, K. & Philippe, H. (1998). New insights into the phylogeny of eukaryotes based on ciliate Hsp70 sequences. *Mol. Biol. Evol.* 15, 943–956.

Byrne, E. M., Connell, G. J. & Simpson, L. (1996). Guide RNA-directed uridine insertion RNA editing in vitro. *EMBO J.* 15, 6758–6765.

Cavalier-Smith, T. (1997). Cell and genome coevolution: Facultative anaerobiosis, glycosomes and kinetoplastan RNA editing. *Trends Genet.* 13, 6–9.

Corell, R. A., Feagin, J. E., Riley, G. R., Strickland, T., Guderian, J. A., Myler, P. J. & Stuart, K. (1993). *Trypanosoma brucei* minicircles encode multiple guide RNAs which can direct editing of extensively overlapping sequences. *Nucl. Acids Res.* 21, 4313–4320.

Covello, P. S. & Gray, M. W. (1989). RNA editing in plant mitochondria. *Nature* 341, 662–666.

Covello, P. S. & Gray, M. W. (1993). On the evolution of RNA editing. *Trends Genet.* 9, 265–268.

de la Cruz, V., Neckelmann, N. & Simpson, L. (1984). Sequences of six structural genes and several open reading frames in the kinetoplast maxicircle DNA of Leishmania tarentolae. *J. Biol. Chem.* 259, 15136–15147.

Decker, C. J. & Sollner-Webb, B. (1990). RNA editing involves indiscriminate U changes throughout precisely defined editing domains. *Cell* 61, 1001–1011.

Ferguson, M., Torri, A. F., Ward, D. C. & Englund, P. T. (1992). In situ hybridization to the Crithidia fasciculata kinetoplast reveals two antipodal sites involved in kinetoplast DNA replication. *Cell* 70, 621–629.

Ferguson, M. L., Torri, A. F., Pérez-Morga, D., Ward, D. C. & Englund, P. T. (1994). Kinetoplast DNA replication: Mechanistic differences between *Trypanosoma brucei* and *Crithidia fasciculata*. *J. Cell Biol.* 126, 631–639.

Fernandes, A. P., Nelson, K. & Beverley, S. M. (1993). Evolution of nuclear ribosomal RNAs in kinetoplastid protozoa: Perspectives on the age and origins of parasitism. *Proc. Natl. Acad. Sci. USA* 90, 11608–11612.

Frasch, A., Hajduk, S., Hoeijmakers, J., Borst, P., Brunel, F. & Davison, J. (1980). The kDNA of Trypanosoma equiperdum. *Biochim. Biophys. Acta* 607, 397–410.

Gerber, A., Grosjean, H., Melcher, T. & Keller, W. (1998). Tad1p, a yeast tRNA-specific adenosine deaminase, is related to the mammalian pre-mRNA editing enzymes ADAR1 and ADAR2. *EMBO J.* 17, 4780–4789.

Germot, A. & Philippe, H. (1999). Critical analysis of eukaryotic phylogeny: a case study based on the HSP70 family. *J. Eukaryot. Microbiol.* 46, 116–124.

Guilbride, D. L. & Englund, P. T. (1998). The replication mechanism of kinetoplast DNA networks in several trypanosomatid species. *J. Cell Sci.* 111, 675–679.

Haag, J., O'hUigin, C. & Overath, P. (1998). The molecular phylogeny of trypanosomes: evidence for an early divergence of the Salivaria. *Mol. Biochem. Parasitol.* 91, 37–49.

Hajduk, S., Siqueira, A. & Vickerman, K. (1986). Kinetoplast DNA of Bodo caudatus : a noncatenated structure. *Mol. Cell Biol.* 6, 4372–4378.

Hancock, K. & Hajduk, S. L. (1990). The mitochondrial tRNAs of *Trypanosoma brucei* are nuclear encoded. *J. Biol. Chem.* 265, 19208–19215.

Hope, M., MacLeod, A., Leech, V., Melville, S., Sasse, J., Tait, A. & Turner, C. M. (1999). Analysis of ploidy (in megabase chromosomes) in Trypanosoma brucei after genetic exchange [In Process Citation]. *Mol. Biochem. Parasitol.* 104, 1–9.

Inagaki, Y., Ehara, M., Watanabe, K. I., Hasashi-Ishimaru, Y. & Ohama, T. (1998). Directionally evolving genetic code: the UGA codon from stop to tryptophan in mitochondria. *J. Mol. Evol.* 47, 378–384.

Kapushoc, S. T. & Simpson, L. (1999). In vitro uridine insertion RNA editing mediated by cis-acting guide RNAs. *RNA.* 5, 656–669.

Keller, W., Wolf, J. & Gerber, A. (1999). Editing of messenger RNA precursors and of tRNAs by adenosine to inosine conversion. *FEBS Lett.* 452, 71–76.

Landweber, L. F. (1992). The evolution of RNA editing in kinetoplastid protozoa. *BioSystems* 28, 41–45.

Landweber, L. F. & Gilbert, W. (1994). Phylogenetic analysis of RNA editing: A primitive genetic phenomenon. *Proc. Natl. Acad. Sci. USA* 91, 918–921.

Lukes, J., Arts, G. J., Van den Burg, J., De Haan, A., Opperdoes, F., Sloof, P. & Benne, R. (1994). Novel pattern of editing oregions in mitochondrial transcripts of the cryptobiid *Trypanoplasma borreli. EMBO J.* 13, 5086–5098.

Lukes, J., Jirku, M., Avliyakulov, N. & Benada, O. (1998). Pankinetoplast DNA structure in a primitive bodonid flagellate, *Cryptobia helicis. EMBO J.* 17, 838–846.

Lukes, J., Jirku, M., Dolezel, D., Kral'ová, I., Hollar, L. & Maslov, D. A. (1997). Analysis of ribosomal RNA genes suggests that trypanosomes are monophyletic. *J. Mol. Evol.* 44, 521–527.

Lun, Z.-R., Brun, R. & Gibson, W. (1992). Kinetoplast DNA and molecular karyotypes of *Trypanosoma evansi* and *Trypanosoma equiperdum* from China. *Mol. Biochem. Parasitol.* 50, 189–196.

Maslov, D. A., Avila, H. A., Lake, J. A. & Simpson, L. (1994). Evolution of RNA editing in kinetoplastid protozoa. *Nature* 365, 345–348.

Maslov, D. A. & Simpson, L. (1992). The polarity of editing within a multiple gRNA-mediated domain is due to formation of anchors for upstream gRNAs by downstream editing. *Cell* 70, 459–467.

Maslov, D. A. & Simpson, L. (1994). RNA editing and mitochondrial genomic organization in the cryptobiid kinetoplastid protozoan, Trypanoplasma borreli. *Mol. Cell. Biol.* 14, 8174–8182.

Maslov, D. A., Yasuhira, S. & Simpson, L. (1999). Phylogenetic affinities of Diplonema within the Euglenozoa as inferred from the SSU rRNA gene and partial COI protein sequences. *Protist* 150, 33–42.

Morl, M., Dorner, M. & Paabo, S. (1995). C to U editing and modifications during the maturation of the mitochondrial tRNA(Asp) in marsupials. *Nucl. Acids Res.* 23, 3380–3384.

Navaratnam, N., Bhattacharya, S., Fujino, T., Patel, D., Jarmuz, A. L. & Scott, J. (1995). Evolutionary origins of *apoB* mRNA editing: Catalysis by a cytidine deaminase that has acquired a novel RNA-binding motif at its active site. *Cell* 81, 187–195.

Osawa, S., Jukes, T. H., Watanabe, K. & Muto, A. (1992). Recent evidence for evolution of the genetic code. *Microbiol. Rev.* 56, 229–264.

Paulin, J. J. (1975). The chondriome of selected trypanosomatids. A three-dimensional study based on serial thick sections and high voltage electron microscopy. *J. Cell Biol.* 66, 404–413.

Pérez-Morga, D. & Englund, P. T. (1993). The structure of replicating kinetoplast DNA networks. *J. Cell Biol.* 123, 1069–1079.

Philippe, H. & Forterre, P. (1999). The rooting of the universal tree of life is not reliable. *J. Mol. Evol.* 49, 509–523.

Pollard, V. W., Rohrer, S. P., Michelotti, E. F., Hancock, K. & Hajduk, S. L. (1990). Organization of minicircle genes for guide RNAs in Trypanosoma brucei. *Cell* 63, 783–790.

Polson, A. G., Bass, B. L. & Casey, J. L. (1996). RNA editing of hepatitis delta virus antigenome by dsRNA-adenosine deaminase [see comments] [published erratum appears in *Nature* 1996 May 23; 381(6580):346]. *Nature* 380, 454–456.

Robinson, D. R. & Gull, K. (1994). The configuration of DNA replication sites within the *Trypanosoma brucei* kinetoplast. *J. Cell Biol.* 126, 641–648.

Savill, N. J. & Higgs, P. G. (1999). A theoretical study of random segregation of minicircles in trypanosomatids. *Proc. R. Soc. Lond B Biol. Sci.* 266, 611–620.

Shapiro, T. A. & Englund, P. T. (1995). The structure and replication of kinetoplast DNA. *Annu. Rev. Microbiol.* 49, 117–143.

Shu, H.-H. & Stuart, K. (1994). Mitochondrial transcripts are processed but are not edited normally in *Trypanosoma equiperdum* (ATCC 30019) which has kDNA sequence deletion and duplication. *Nucl. Acids Res.* 22, 1696–1700.

Simpson, A. & Simpson, L. (1976). Pulse-labeling of kinetoplast DNA:Localization of two sites of synthesis within the netowrks and kinetics of labeling of closed minicircles. *J. Protozool.* 23, 583–587.

Simpson, A. M., Suyama, Y., Dewes, H., Campbell, D. & Simpson, L. (1989). Kinetoplastid mitochondria contain functional tRNAs which are encoded in nuclear DNA and also small minicircle and maxicircle transcripts of unknown function. *Nucl. Acids Res.* 17, 5427–5445.

Simpson, L. (1986). Kinetoplast DNA in trypanosomid flagellates. *Int. Rev. Cytol.* 99, 119–179.

Simpson, L. & Braly, P. (1970). Synchronization of Leishmania tarentolae by hydroxyurea. *J. Protozool.* 17, 511–517.

Simpson, L. & Kretzer, F. (1997). The mitochondrion in dividing Leishmania tarentolae cells is symmetric and becomes a single asymmetric tubule in non-dividing cells due to division of the kinetoplast portion. *Mol. Biochem. Parasitol.* 87, 71–78.

Simpson, L. & Maslov, D. A. (1994). RNA editing and the evolution of parasites. *Science* 264, 1870–1871.

Simpson, L. & Maslov, D. A. (1999). Evolution of the U-insertion/deletion RNA editing in mitochondria of kinetoplastid protozoa. *Ann. N. Y. Acad. Sci.* 870, 190–205.

Simpson, L., Simpson, A. & Wesley, R. (1974). Replication of the kinetoplast DNA of Leishmania tarentolae and Crithidia fasciculata. *Biochim. Biophys. Acta* 349, 161–172.

Smith, H. C., Gott, J. M. & Hanson, M. R. (1997). A guide to RNA editing. *RNA* 3, 1105–1123.

Songa, E. B., Paindavoine, P., Wittouck, E., Viseshakul, N., Muldermans, S., Steinert, M. & Hamers, R. (1990). Evidence for kinetoplast and nuclear DNA homogeneity in *Trypanosoma evansi* isolates. *Mol. Biochem. Parasitol.* 43, 167–180.

Souza, A. E., Hermann, T. & Göringer, H. U. (1997). The guide RNA database. *Nucl. Acids Res.* 25, 104–106.

Stoltzfus, A. (1999). On the possibility of constructive neutral evolution. *J. Mol. Evol.* 49, 169–181.

Stuart, K. (1979). Kinetoplast DNA of Trypanosoma brucei: Physical map of the maxicircle. *Plasmid* 2, 520–528.

Stuart, K., Kable, M. L., Allen, T. E. & Lawson, S. (1998). Investigating the mechanism and machinery of RNA editing. *Methods* 15, 3–14.
Sturm, N. R., Maslov, D. A., Blum, B. & Simpson, L. (1992). Generation of unexpected editing patterns in Leishmania tarentolae mitochondrial mRNAs: misediting produced by misguiding. *Cell* 70, 469–476.
Sturm, N. R. & Simpson, L. (1990a). Partially edited mRNAs for cytochrome b and subunit III of cytochrome oxidase from Leishmania tarentolae mitochondria: RNA editing intermediates. *Cell* 61, 871–878.
Sturm, N. R. & Simpson, L. (1990b). Kinetoplast DNA minicircles encode guide RNAs for editing of cytochrome oxidase subunit III mRNA. *Cell* 61, 879–884.
Tessier, L. H., Van der Speck, H., Gualberto, J. M. & Grienenberger, J. M. (1997). The *cox1* gene from *Euglena gracilis*: A protist mitochondrial gene without introns and genetic code modifications. *Curr. Genet.* 31, 208–213.
Thiemann, O. H., Maslov, D. A. & Simpson, L. (1994). Disruption of RNA editing in *Leishmania tarentolae* by the loss of minicircle-encoded guide RNA genes. *EMBO J.* 13, 5689–5700.
Vickerman, K. (1977). DNA throughout the single mitochondrion of a kinetoplast flagellate: Observations on the ultrastructure of Cryptobia vaginalis. *J. Protozool.* 24, 221–233.
Von Haeseler, A., Blum, B., Simpson, L., Sturm, N. & Waterman, M. S. (1992). Computer methods for locating kinetoplastid cryptogenes. *Nucl. Acids Res.* 20, 2717–2724.
Wright, A. D., Li, S., Feng, S., Martin, D. S. & Lynn, D. H. (1999). Phylogenetic position of the kinetoplastids, Cryptobia bullocki, Cryptobia catostomi, and Cryptobia salmositica and monophyly of the genus Trypanosoma inferred from small subunit ribosomal RNA sequences. *Mol. Biochem. Parasitol.* 99, 69–76.
Yang, J. H., Sklar, P., Axel, R. & Maniatis, T. (1997). Purification and characterization of a human RNA adenosine deaminase for glutamate receptor B pre-mRNA editing. *Proc. Natl. Acad. Sci. USA* 94, 4354–4359.
Yasuhira, S. & Simpson, L. (1995). Minicircle-encoded guide RNAs from *Crithidia fasciculata*. *RNA* 1, 634–643.
Yasuhira, S. & Simpson, L. (1996). Guide RNAs and guide RNA genes in the cryptobiid kinetoplastid protzoan, Trypanoplasma borreli. *RNA* 2, 1153–1160.
Yasuhira, S. & Simpson, L. (1997). Phylogenetic affinity of mitochondria of *Euglena gracilis* and kinetoplastids using cytochrome oxidase I and hsp60. *J. Mol. Evol.* 44, 341–347.
Yurchenko, V., Hobza, R., Benada, O. & Lukes, J. (1999). *Trypanosoma avium*: Large minicircles in the kinetoplast DNA. *Exp. Parasitol.* 92, 215–218.

9
Population Structure and Recent Evolution of Plasmodium falciparum

STEPHEN M. RICH* AND FRANCISCO J. AYALA†

Plasmodium falciparum is the agent of malignant malaria, one of mankind's most severe maladies. The parasite exhibits antigenic polymorphisms that have been postulated to be ancient. We have proposed that the extant world populations of P. falciparum have derived from one single parasite, a cenancestor, within the last 5,000–50,000 years. This inference derives from the virtual or complete absence of synonymous nucleotide polymorphisms at genes not involved in immune or drug responses. Seeking to conciliate this claim with extensive antigenic polymorphism, we first note that allele substitutions or polymorphisms can arise very rapidly, even in a single generation, in large populations subject to strong natural selection. Second, new alleles can arise not only by single-nucleotide mutations, but also by duplication/deletion of short simple-repeat DNA sequences, a process several orders of magnitude faster than single-nucleotide mutation. We analyze three antigenic genes known to be extremely polymorphic:

*Division of Infectious Diseases, Tufts University School of Veterinary Medicine, North Grafton, MA 01536; and †Department of Ecology and Evolutionary Biology, University of California, Irvine, CA 92697

This paper was presented at the National Academy of Sciences colloquium "Variation and Evolution in Plants and Microorganisms: Toward a New Synthesis 50 Years After Stebbins," held January 27–29, 2000, at the Arnold and Mabel Beckman Center in Irvine, CA.

Abbreviations: CSP, circumsporozoite protein; MSP-1 and MSP-2, merozoite surface proteins 1 and 2, respectively; CR, central region; NR, not repetitive; RAT, repeat allotype; RHR, repeat homology region.

Csp, Msp-1, and Msp-2. We identify regions consisting of tandem or proximally repetitive short DNA sequences, including some previously unnoticed. We conclude that the antigenic polymorphisms are consistent with the recent origin of the world populations of P. falciparum inferred from the analysis of nonantigenic genes.

THE MALARIA PLAGUE AND CONTROL EFFORTS

The World Health Organization estimates that there are 300–500 million clinical cases of malaria per year, more than 1 million children die in sub-Saharan Africa, and more than 2 billion people are at risk throughout the world (WHO, 1995). *Plasmodium falciparum* is the agent of malignant malaria, the most fatal version of the disease. Malaria has been an elusive target for medical intervention. Epidemiological control efforts were first directed against the *Anopheles* mosquito vectors, which soon evolved resistance to massively applied insecticides. Current efforts against the mosquito vectors seek to produce transgenic mosquitoes that are unable to transmit *Plasmodium,* followed by massive release of the transformed vectors in endemic regions.

Greater efforts yet are invested in the development of protective vaccines or remedial drugs directed against the parasite. These exertions are handicapped, however, by the parasite's rapid evolution of drug resistance and antigens. Underlying this evolution is a wealth of genetic variation that arises rapidly by rearrangement of modular repeating elements that generate ever newly protected phenotypes.

The merozoite form of the *Plasmodium* parasite found in the human bloodstream is haploid. A fraction of these haploids differentiate into gametocytes, which are taken up in the mosquito's blood meal. Gametes fuse in the mosquito midgut to form transient diploids, which then undergo meiosis to yield haploid infectious forms, called sporozoites. Protective immunity against *P. falciparum* was demonstrated in the 1970s by immunization of human patients with irradiated sporozoites (Clyde *et al.,* 1973). Parasite genes that code for antigenic determinants subsequently have been isolated and characterized. Notable among the genes intensively investigated and chosen for vaccine development are those encoding surface proteins of the sporozoite (*Csp,* coding for the circumsporozoite protein) and the merozoite (*Msp-1* and *Msp-2,* coding for the merozoite surface proteins 1 and 2). The success of efforts to develop an effective malaria vaccine is contingent on determining the extent of diversity of these genes and on identifying the mechanisms by which this variation is generated and persists in populations of *P. falciparum.*

Assessment of DNA sequence variation in *P. falciparum* has been based almost exclusively on examination of genes coding for antigenic determinants, where amino acid polymorphisms (nonsynonymous nucleotide polymorphisms) are common and likely to be affected by natural selection. Numerous studies have indicated that *Csp, Msp-1, Msp-2,* and other antigenic genes are polymorphic and that their multiple allelic forms differ in their ability to abrogate recognition by the host's immune response (Tanabe *et al.*, 1987; Smythe *et al.*, 1991; Ukhayakumar *et al.*, 1994; Zevering *et al.*, 1994; Babiker and Walliker, 1997). These observations have been interpreted as instantiation of widespread polymorphism throughout the genome. Yet, we have investigated allelic variation in a diverse set of gene loci and found a complete absence of silent site polymorphism (Rich *et al.*, 1998) and have proposed a recent derivation (within thousands of years) of the extant *P. falciparum* world populations from a single propagule.

It seems paradoxical that *P. falciparum* antigenic genes would be so highly polymorphic, because these genes must have shared the recent allelic homogenization caused by the population bottleneck we have inferred. Indeed, some authors have hypothesized that the polymorphisms of genes encoding *P. falciparum* surface proteins are very old, even older than the species itself.

We shall argue herein that the antigenic gene polymorphisms of *P. falciparum* are consistent with the conclusion drawn from the analysis of synonymous DNA sites, that the current world populations of the parasite are of recent origin, derived from a single strain within the last several thousand years. We will review our previous analysis of *Csp* (Rich *et al.*, 1998) and then we will examine the *Msp-1* and *Msp-2* polymorphisms.

EVOLUTIONARY ASSOCIATION OF *P. FALCIPARUM* WITH THE HOMINID LINEAGE

Fig. 1 is a phylogenetic tree of *Plasmodium* species derived from *Csp* gene sequences (Escalante *et al.*, 1995; for very similar trees based on other genes see Escalante and Ayala, 1994; Ayala *et al.*, 1998, 1999). Estimates of divergence times are shown in Table 1.

It is apparent that the three human parasites, *P. falciparum, Plasmodium malariae,* and *Plasmodium vivax* are very remotely related to each other, so that the evolutionary divergence of these three human parasites greatly predates the origin of the hominids. *Plasmodium ovale,* a fourth human parasite, is also remotely related to the other three (Qari *et al.*, 1996). These results are consistent with the diversity of physiological and epidemiological characteristics of these four *Plasmodium* species (Coatney *et al.*, 1971; López-Antuñano and Schumunis, 1993).

P. falciparum is more closely related to *Plasmodium reichenowi,* the chimpanzee parasite, than to any other *Plasmodium* species. The time of divergence between these two *Plasmodium* species is estimated at 8–12 million

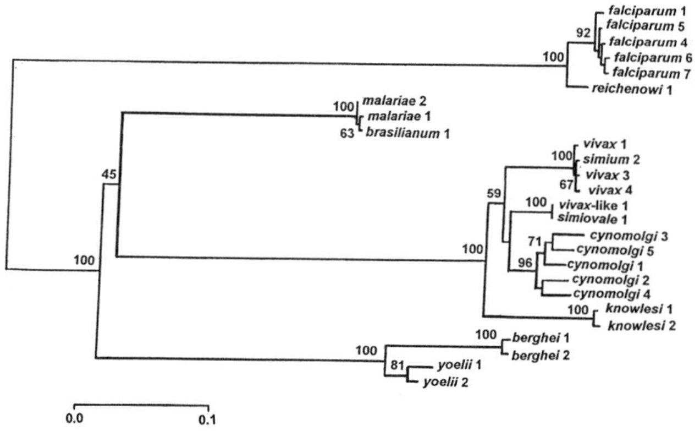

FIGURE 1. Phylogeny of 12 Plasmodium species inferred from Csp gene sequences. P. falciparum, malariae, and vivax are human parasites; berghei and yoelii are rodent, and all others are primate parasites. The numbers refer to different strains. Bootstrap values above branches assess the reliability of the branch clusters; values above 70 are considered statistically reliable. Reprinted with permission from Ayala *et al.* 1999.

years ago, which is roughly consistent with the time of divergence between the two host species, human and chimpanzee. A parsimonious interpretation of this state of affairs is that *P. falciparum* is an ancient human parasite, associated with our ancestors since the divergence of the hominids from the great apes. Fig. 1 shows that *P. malariae*, a human parasite, is genetically indistinguishable from *Plasmodium brasilianum*, a parasite of New World monkeys; similarly, human *P. vivax* is genetically indistinguishable from *Plasmodium simium*, also a parasite of New World monkeys. It follows that lateral transfer between hosts has occurred in recent times, at least in these two cases. Whether the transfer has been from humans to monkeys or vice versa is a moot question (for discussion, see Ayala *et al.*, 1999).

TABLE 1. Time (in million years) of divergence, between *Plasmodium* species, based on genetic distances at two gene loci (see Figs. 1 and 2; adapted from Escalante *et al.* [1995], and Escalante and Ayala [1994])

Plasmodium	*rRNA*	*CSP*
falciparum vs. *reichenowi*	11.2 ± 2.5	8.9 ± 0.4
vivax vs. *monkey**	20.9 ± 3.8	25.2 ± 2.1
vivax vs. *malariae*	75.7 ± 8.8	103.5 ± 0.6
falciparum vs. *vivax/malariae*	75.7 ± 8.8	165.4 ± 1.6

**brasilianum* not included.

MALARIA'S EVE: RECENT ORIGIN OF *P. FALCIPARUM* WORLD POPULATIONS

Silent (i.e., synonymous) nucleotide polymorphisms are appropriate for estimating the age of genes, because silent nucleotide polymorphisms are often adaptively neutral (or very nearly so). Thus, silent nucleotide polymorphisms reflect the mutation rate and the time elapsed since their divergence from a common ancestor. Table 2 summarizes data for 10 genes (Rich *et al.*, 1998). The gene sequences analyzed derive from isolates of *P. falciparum* geographically representative of the global malaria endemic regions (see table 1 in Rich *et al.*, 1998; and, for the *Csp* gene, Rich *et al.*, 1997). A scarcity of synonymous polymorphisms also has been observed in a separate study of 10 *P. falciparum* genes, most of them antigenic (Escalante *et al.*, 1998).

As we have expounded elsewhere (Rich *et al.*, 1998; Ayala *et al.*, 1998, 1999), five possible hypotheses may account for the absence of silent polymorphisms in *P. falciparum*: (*i*) persistent low effective population size, (*ii*) low rates of spontaneous mutation, (*iii*) strong selective constraints on silent variation, (*iv*) one or more recent selective sweeps affecting the genome as a whole, and (*v*) a demographic sweep, i.e., a recent population bottleneck, so that extant world populations of *P. falciparum* would have recently derived from a single ancestral strain. We have concluded that only the fifth hypothesis is consistent with the observations and have used coalescent theory to estimate the age of the ancestral strain or "cenancestor" (Table 3).

The issue arises of how to account for a recent demographic sweep in *P. falciparum*. One possible hypothesis is that *P. falciparum* has become a human parasite in recent times, by lateral transfer from some other host species (Waters *et al.*, 1991). This hypothesis is contrary to available evidence (Escalante and Ayala, 1994; Escalante *et al.*, 1995). An alternative explanation is that human parasitism by *P. falciparum* has long been highly restricted geographically and has dispersed throughout the Old World continents only within the last several thousand years, perhaps within the last 10,000 years, after the Neolithic revolution (Coluzzi, 1994, 1997, 1999). Three possible scenarios may have led to this historically recent dispersion: (*i*) changes in human societies, (*ii*) genetic changes in the host-parasite-vector association that have altered their compatibility, and (*iii*) climatic changes that entailed demographic changes (migration, density, etc.) in the human host, the mosquito vectors, and/or the parasite.

One factor that may have impacted the widespread distribution of *P. falciparum* in human populations from a limited original focus, probably in tropical Africa, may have been changes in human living patterns, particularly the development of agricultural societies and urban centers that increased human population density (Livingstone, 1958; Wiesenfeld, 1967;

TABLE 2. Polymorphisms in 10 loci of *P. falciparum*

Gene	Chromosome location	Length, bp	Sequences in the sample, n_i	Nonsynonymous polymorphisms, D_n	Synonymous polymorphisms, D_s	Synonymous sites analyzed	
						4-fold, $n_i l_i$	2-fold, $n_i m_i$
Dhfr	4	609	32	4	0	2,144	4,128
Ts	4	1,215	10	0	0	1,250	2,640
Dhps	8	1,269	12	5	0	1,536	2,724
Mdr1	5	4,758	3	1	0	1,350	2,088
Rap1	—	2,349	9	8	0	1,092	1,668
Calm	14	441	7	0	0	364	602
G6pd	14	2,205	3	9	0	726	1,404
Hsp86	7	2,241	2	0	0	532	910
Tpi	—	597	2	0	0	180	262
CsP 5′end	3	387	25	7	0	688	2,010
Csp 3′end	3	378	25	17	0	1,050	1,625
Total	—	—	—	51	0	10,912	20,061

Modified from Ayala *et al.* (1999).

TABLE 3. Estimated times to the cenancestor of the world populations of *P. falciparum*

Estimated mutation rate × 10^{-9}			
μ_a	μ_b	t_{95}	t_{50}
7.12	2.22	24,511	5,670
3.03	0.95	57,481	13,296

Adapted from Rich *et al.* (1998) and Ayala *et al.* (1998). t_{95} and t_{50} are the upper boundaries of the confidence intervals. Thus, in the first row the cenancestor lived less than 24,511 years ago with a 95% probability, and less than 5,670 years ago with a 50% probability. μ_a and μ_b are the estimated neutral mutation rates of 4-fold and 2-fold degenerate codons, respectively.

de Zulueta, 1973, 1994; Coluzzi, 1997, 1999; Sherman, 1998). Genetic changes that have increased the affinity within the parasite-vector-host system also seem to be a viable explanation for a recent expansion. Coluzzi (1997, 1999) has cogently argued that the worldwide distribution of *P. falciparum* is recent and has come about, in part, as a consequence of a recent dramatic rise in vectorial capacity caused by repeated speciation events in Africa of the most anthropophilic members of the species complexes of the *Anopheles gambiae* and *Anopheles funestus* mosquito vectors. The biological processes implied by this account may have, in turn, been associated with, and even depended on the onset of agricultural societies in Africa and climatic changes, specifically the gradual increase in ambient temperatures after the Würm glaciation, so that about 6,000 years ago climatic conditions in the Mediterranean region and the Middle East made possible the spread of *P. falciparum* and its vectors beyond tropical Africa (de Zulueta, 1973, 1994; Coluzzi, 1997, 1999).

Sherman (1998) has noticed the late introduction and low incidence of *falciparum* malaria in the Mediterranean region, which postdates historical times. Hippocrates (460–370 B.C.) describes quartan and tertian fevers, but there is no mention of severe malignant tertian fevers, which suggests that *P. falciparum* infections did not yet occur in classical Greece, as recently as 2,400 years ago. The late introduction of *falciparum* malaria into the Mediterranean region and the Middle East has been attributed to the low vectorial efficiency of the indigenous anopheline mosquitoes (Coluzzi, 1997, 1999). Once the demographic and climate conditions became suitable for the propagation of *P. falciparum*, natural selection would have facilitated the evolution of *Anopheles* species that were highly anthropophilic and effective *falciparum* vectors (de Zulueta, 1973; Coluzzi, 1997, 1999).

The selective sweep hypothesis (*iv*) is, in a way, a special case of the demographic sweep hypothesis (*v*); i.e., a particular strain may have spread throughout the world and replaced all other strains impelled by natural selection. Natural selection can account for the absence of synonymous variation at any one of the 10 loci shown in Table 2, if the particular gene itself (or a gene with which it is linked) has been subject to a recent worldwide selective sweep, without sufficient time for the accumulation of new synonymous mutations. However, the 10 genes are located on, at least, six different chromosomes (Table 2), and thus six independent selective sweeps would need to have occurred more or less concurrently, which seems *prima facie* unlikely. A selective sweep simultaneously affecting all chromosomes could happen if one accepts the hypothesis that the population structure of *P. falciparum* is predominantly clonal, rather than sexual (see Escalante and Ayala, 1994; Ayala *et al.*, 1999). This hypothesis is controversial, although we have argued that it may indeed be the case, the capacity for sexual reproduction of the parasite notwithstanding (Rich *et al.*, 1997; Ayala *et al.*, 1999).

THE RECENT ORIGIN OF *P. FALCIPARUM* POPULATIONS VIS-À-VIS ANTIGENIC POLYMORPHISMS

The absence of synonymous polymorphisms in most *P. falciparum* genes must be made congruous with the substantial levels of polymorphism observed in such antigenic genes as *Csp*, *Msp-1*, and *Msp-2*. We propose that nucleotide polymorphism arises in antigenic genes promoted by natural selection acting on two different "mutation" processes. First, the familiar process of single-site nucleotide mutation generates amino acid replacements that give rise to polymorphisms at antigenic sites subject to diversifying selection. Second, there is intragenic recombination that generates variation at a rapid rate in repetitive segments (often occurring in tandem) of antigenic genes. The variation generated by intragenic recombination is also subject to diversifying natural selection because it contributes to the parasite's ability to evade the immune response of the human host. We will show that some of the reported nucleotide variation between antigenic alleles is an artifact stemming from misalignment of gene sequences that are of different lengths as a consequence of unequal numbers of repetitions generated by intragenic recombination.

THE CSP GENE

The *Csp* gene is comprised of two terminal regions that are not repetitive (5′ NR and 3′ NR), which embrace a central region (CR) made up of a variable number (mostly, between 40 and 50) of tandemly arranged 12-nt-long repeats. As shown in Table 2, there are no silent polymorphisms in the 5′NR and 3′NR regions of the gene, which is part of the evidence supporting a recent origin of *P. falciparum* populations.

The repetitive amino acid sequences encoded within the CR are remarkably conserved (only two amino acid motifs are known in *P. falciparum,* NANP and NVDP; Table 4), but there is a fair deal of synonymous nucleotide polymorphism among the repeats (Table 5). We have introduced the concept of the repeat allotype (RAT) to refer to variant nucleotide sequences that encode a single amino acid motif (Rich *et al.,* 1997). Among the known *Csp* gene sequences of *P. falciparum,* there are 10 RATs that encode the NANP motif and four RATs that encode the NVDP motif, with an average of about 10 RATs per gene sequence (range 9–11; see Table 6). Table 4 displays the arrangement of the two amino acid motifs in 25 gene sequences of *P. falciparum* and one of *P. reichenowi.* The alignment of the RATs can be found in Rich *et al.* (1997; see also Ayala *et al.,* 1999). The only known sequence of *Csp* in *P. reichenowi* is somewhat shorter than those of *falciparum* (35 rather than about 45 repeats per sequence, on average), but has a similar number of distinct RATs (10, the

TABLE 4. Composition of the CR of the *Csp* gene

Sequence	Repeat motifs	Number of repeats		
		1	2	3
M15505	1212111111111111111211111111111111111111111111	43	3	0
M83173	1212111111111111111211111111111111111111111111	43	3	0
M83149	12121211111111111111111111111111111111111111	41	3	0
M83150	12121111111111111112111111111111111111111111111	44	3	0
M83156	121211111111111111111111111111111111111111111111111	49	2	0
M83158	1212121211111111111111111111111111111111111111	42	4	0
M83161	1212121111111111111111111211111111111111111.	39	4	0
M83163	12121111111111111112111111111111111111111111111	43	3	0
M83164	12121111111111111112111111111111111111111111111	46	3	0
M83165	1212121111111111111111111111111111111111111ili	43	3	0
M83166	1212121211111111111111111111111111111111111111	42	4	0
M83167	12121211111111111111111111111111111111111111111	46	3	0
M83168	1212121211111111111111111111111111111111111111	42	4	0
M83169	12121211111111111111111111111111111111111111	41	3	0
M83170	1212121211111111111111111111111111111111111111	42	4	0
M83174	1212121111111111211111111111111111111111111	39	4	0
M19752	12121211111111111111111111111111111111111111	41	3	0
M83172	121212111111111111111121111111111111111111	38	4	0
K02194	12121211111111111111121111111111111111111	37	4	0
M57499	12121212111111111111111111111111111111111111	40	4	0
U20969	1212121111111111111112111111111111111111	36	4	0
M83886	121212111111111111111112111111111111111111	38	4	0
M22982	12121211111111111111111112111111111111111111	40	4	0
X15363	12121211111111111111111112111111111111111111	40	4	0
M57498	12121211111111111111121111111111111111111	37	4	0
P. reichenowi	12121212131213131311111111111111111	26	5	4

The repeat motifs NANP, NVDP, and NVNP are represented by 1, 2, and 3, respectively. Adapted from Ayala *et al.,* (1999).

TABLE 5. Amino acid and nucleotide sequence of the RATs and their incidence

	Motif		*falciparum*		*reichenowi*	
RAT	Amino acid	Nucleotide	%	Number	%	Number
A	NANP	aatgcaaaccca	55.1	566	38.5	10
B	NANP	........t..t	16.1	165	30.8	8
C	NANP	........t...	7.6	78	—	
D	NANP	.....c..t..a	6.2	64	3.8	1
E	NANP	c	6.2	64	—	—
F	NANP	..c........c	5.1	52	—	—
G	NANP	.....c......	3.1	32	7.7	2
H	NANP	...........t	0.3	3	—	—
I	NANP	..c.........	0.2	2	3.8	1
J	NANP	......c....c	0.1	1	—	—
Z	NANP	........t..c	—	—	15.4	4
M	NVDP	....t.g.t...	52.3	46	20.0	1
N	NVDP	....t.g.t..c	31.8	28	40.0	2
O	NVDP	..c.t.g.t..t	14.8	13	—	—
P	NVDP	....t.g.t..t	1.1	1	20.0	2
X	NVNP	....t...t..c	—	—	100.0	4

Modified from Rich *et al.* (1997); see Ayala *et al.* (1999).

same as the *falciparum* average) and three rather than only two amino acid motifs, two of them identical to those of *falciparum*.

Nearly all of the synonymous site differences observed in the CR are between RATs that exist within any single allele. This is a strong indication that while RAT diversity may have an ancient origin, it has been maintained within individual alleles and therefore can withstand even the most constricted bottleneck. For example, all 25 *Csp* CR alleles contain at least one copy of each of the most common RATs (A, B, C, D, E, and F, which amount to 96% of all NANP repeats; and M and N, which amount to 84% of all NVDP repeats). If any one of these alleles were the sole survivor following a bottleneck, it alone would possess nearly all of the

TABLE 6. Number of RATs in the *Csp* gene sequences of *P. falciparum* and in *P. reichenowi*

	Number of sequences	Different RATs per sequence			Total RATs per sequence									
		9	10	11	35	40	41	42	43	44	46	47	49	51
P. falciparum	25	12	8	5		1	2	2	2	6	8	1	2	1
P. reichenowi	1		1		1									

diversity currently known for the species; intragenic recombination between the RATs originally present in one allele can generate size polymorphisms in the resulting alleles. The process of bottleneck reduction, ensued by generation of new variations through intragenic recombination, may have occurred numerous times in the evolution of the species, and may continue to do so, given the nature of the parasite lifestyle and its propensity for being confronted by population bottlenecks, for example, upon colonization of new geographic regions or during seasonal epidemic relapses.

We have proposed that most of the variation in antigenic genes is attributable to duplication and/or deletion of the repeated segments within the genes, which is simply an instance of the general slipped-strand process for generating length variation in repetitive DNA regions (Fig. 2). This process occurs by several mechanisms, each of which is well understood at the molecular level and may involve either intrahelical or

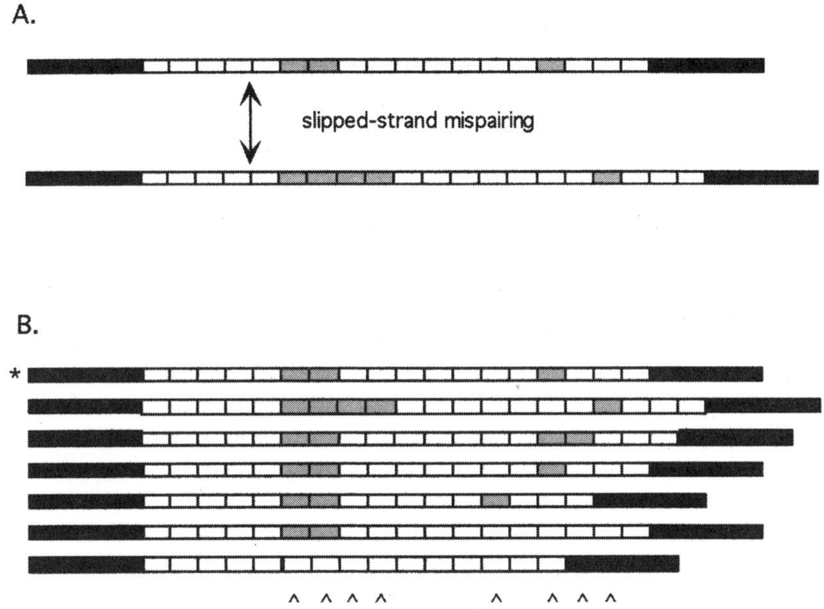

FIGURE 2. A model of RAT evolution. Black boxes represent flanking single-copy regions; gray and white boxes represent different RATs. (A) A single slippage event yields a duplication of two gray RATs. (B) Six new alleles, derived from a single ancestor (indicated by *) after several cell generations. Slippage produces deletions as well as duplications. Karats at the bottom mark artifactual substitutions appearing when the alleles are aligned. Reprinted with permission from Ayala *et al.* (1999).

interhelical exchange of DNA (Levinson and Gutman, 1987). Intragenic recombination often is associated with the evolution of minisatellite or microsatellite DNA loci, such as those recently described in *P. falciparum* (Su and Wellems, 1996; Anderson *et al.*, 1999). However, intragenic recombination also has been implicated in generating variability within coding regions in a variety of eukaryotes; including the *Drosophila* yolk protein gene and the human α_2-globin gene, to cite just two examples (Ho *et al.*, 1996; Oron-Karni *et al.*, 1997).

New RATs can arise by one of two processes: (*i*) replacement or silent substitutions in a codon, and (*ii*) the slippage mechanism that leads to RAT proliferation. The two amino acid motifs and the different RAT types have arisen by the first process. The variation in the number of RATs arises by the second process. The second process occurs with a frequency several orders of magnitude greater than the first process (Schug *et al.*, 1998).

How much of the variation now present in the *Csp* CR region of *falciparum* may have arisen by the second process? Notice that only two amino acid motifs are present in the whole set of 25 *Csp* sequences and that both motifs are present in every one of the sequences (Tables 4 and 5). Thus, there is no evidence that any replacement substitution has occurred in the recent evolution of *P. falciparum*.

CRYPTIC REPEATS IN THE *MSP-1* POLYMORPHISM

The *Msp-1* gene codes for MSP-1 (also referred to as MSA-1, P195, and otherwise), which is a large 185- to 215-kDa protein precursor that is proteolytically cleaved into several membrane protein constituents. The known alleles of *Msp-1* belong to one or the other of two allelic classes (group I and group II). There is considerable nucleotide substitution and length variation between the two classes but much less variation within each class (Tanabe *et al.*, 1987; Hughes, 1992). The two classes are commonly designated by the strains in which they were originally identified: K1 (group I) and MAD20 (group II).

Tanabe *et al.* (1987) partitioned MSP-1 into 17 blocks, based on the degree of amino acid polymorphism (Table 7). They classified seven blocks (blocks 2, 4, 6, 8, 10, 14, and 16) as highly variable; five blocks (blocks 7, 9, 11, 13, and 15) as semiconserved, and five blocks (blocks 1, 3, 5, 12, and 17, which include the two terminal segments) as conserved. The "highly variable" (as well as the "semiconserved") amino acid polymorphisms occur only when comparisons are made between the two allele groups, whereas amino acid, as well as synonymous, nucleotide polymorphisms are very low within each allele group. An exception is block 2, which encodes a set of repetitive tripeptides and thus is subject to the same intragenic recombination described above for *Csp*, as a mechanism

TABLE 7. Nucleotide diversity (π) within and between group I and II alleles of the *P. falciparum Msp-1* genes

Block	Length, codons	Synonymous			Nonsynonymous		
		Group I	Group II	Group I + group II	Group I	Group II	Group 1 + group II
1	55	0.019	0.021	0.017	0.017	0.010	0.013
2	55	0.106	0.185	0.150	0.449	0.497	0.553
3	202	0.038	0.006	0.042	0.018	0.000	0.023
4	31	0.031	0.000	0.020	0.307	0.000	0.215
5	35	0.000	0.000	0.070	0.000	0.000	0.026
6	227	0.000	0.000	0.282	0.004	0.001	0.300
7	73	0.000	0.000	0.361	0.003	0.000	0.072
8	95	0.000	0.000	0.338	0.000	0.003	0.711
9	107	0.000	0.023	0.409	0.005	0.043	0.126
10	126	0.008	0.000	0.448	0.011	0.000	0.394
11	35	0.000	0.000	0.128	0.000	0.000	0.068
12	79	0.000	0.000	0.000	0.000	0.000	0.000
13	84	0.000	0.042	0.040	0.005	0.007	0.052
14	60	0.000	0.018	0.212	0.002	0.005	0.371
15	89	0.000	0.000	0.216	0.001	0.003	0.089
16	217	0.002	0.032	0.277	0.027	0.027	0.185
17	99	0.002	0.019	0.007	0.010	0.027	0.016

Blocks are as defined by Tanabe *et al.* (1987). Some block lengths vary between group I and II alleles; the value given is the average length of group I and II alleles.

for generating polymorphism (in block 4 there is considerable nonsynonymous polymorphism among group I alleles). Table 7 gives the nucleotide diversity (π) for synonymous and nonsynonymous substitutions for each of the 17 blocks, both within and between groups (see Fig. 3). The most extensive amino acid polymorphism between the two allele groups occurs in block 8, which has been assumed to have no simple repeats, but that we will show below to be composed of tandem and proximal repeats (see Fig. 4).

The dimorphism observed among group I and II alleles within block 2 has been shown to result by processes analogous to those within the *Csp* central repeat region (Frontali and Pizzi, 1991; Frontali, 1994). The occurrence of repetitive DNA within other blocks has not been described to date. However, we have identified repeats within several of the most polymorphic *Msp-1* blocks; in particular, blocks 4, 8, and 14, which heretofore were assumed to be NR. We focus on the repeats detected within block 8, identified by Tanabe *et al.* (1987) as showing the lowest amino acid similarity between groups (10%; $\pi = 0.711$ in Table 7). We have identified three group-specific repeats within this block, two in group I alleles (R1a and R1b), and one in group II alleles (R2a). R2a is a 9-bp repeat

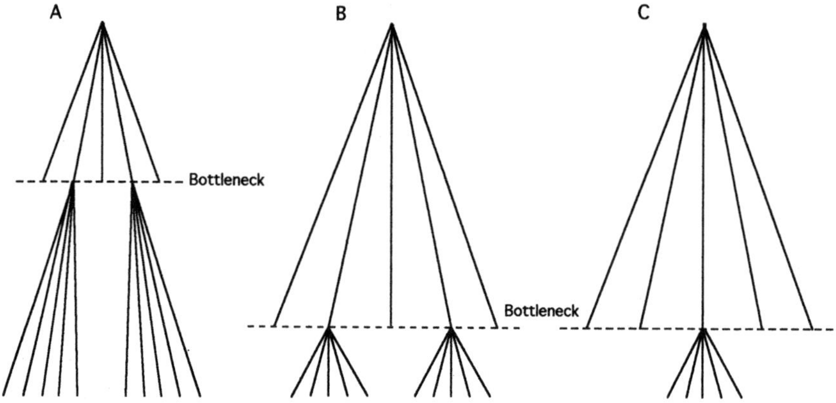

FIGURE 3. Three possible models of the evolution of Msp-1 group I and group II alleles. (A) After an ancient bottleneck, only two alleles survive; these two alleles each give rise to new alleles over time. We expect the two allele groups to be very heterogeneous within groups, and more so between groups, with respect to both synonymous and nonsynonymous substitutions. (B) After a recent bottleneck, only two alleles survive, each of which give rise to new alleles. Alleles within a group are fairly similar to each other but alleles from different groups are very heterogeneous throughout the length of the gene, with respect to synonymous and nonsynonymous substitutions. (C) After a recent bottleneck, only one allele survives that gives rise to new alleles over time. Alleles within and between groups are similar, except for occasional (mostly synonymous) substitutions and for differences generated by intragenic recombination, evidenced by the presence of repeats. A and B are inconsistent with the data in Table 6.

tandemly replicated five times in all group II alleles (the five uppermost alleles in Fig. 3). R1a is a 7-bp repeat replicated five times, and R1b is a 6-bp repeat replicated four times in all group I alleles. The occurrence of repeats within this very short stretch of DNA is a highly significant departure from chance (Ayala *et al.,* 1999). We have searched the recently completed genomic sequences of *P. falciparum* chromosomes 2 and 3. The nucleotide sequences of repeats R1a, R1b, and R2a appear 25, 116, and 11 times, respectively, within the 947 kb of chromosome 2. Within the 1,060 kb of chromosome 3, the R1a, R1b, and R2a are present 39, 52, and seven times, respectively. None of the three nucleotide repeats ever appears in tandem on either chromosome 2 or 3. The average distance between each occurrence on these chromosomes is >20 kb, corroborating that their repeated occurrence in the short 147-bp segment of *Msp-1* block 8 is a strong departure from random expectation. The *Msp-1* gene is located on chromosome 9, which has not yet been assembled as a complete nucleotide

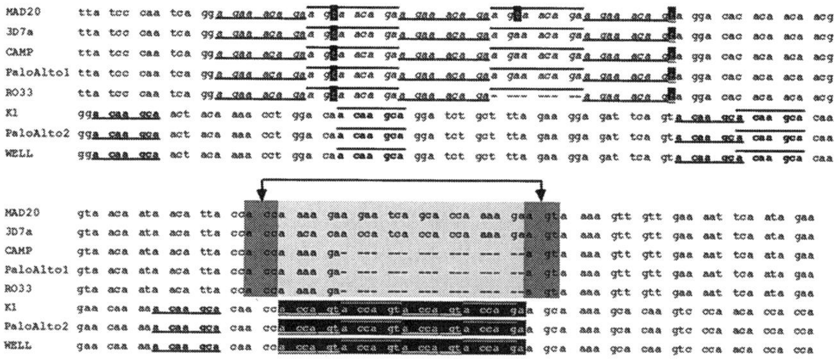

FIGURE 4. Partial alignment of MspI (block 8) group I and II alleles. Alternating odd and even occurrence of a repeat is indicated by underline and overbar, respectively. Region R2a consists of five tandem repeats of a 9-bp sequence (agaaacaga, in italics) highlighted in the five group II alleles (Upper); one copy is missing in the RO33 allele. Regions R1a and R1b consist of two repeats, measuring 7-bp (acaagca, in boldface; repeated five times) and 6-bp (accagt, shown in inverted text; repeated four times) found in group I alleles. The five 7-bp repeats (except for two) are separated by several codons, whereas the 6-bp repeats occur in tandem. There are no repeat sequences shared between groups I and II; however, the 6-bp repeat in group I alleles clearly derives from a deletion of the intervening lightly shaded portion of group II alleles, followed by duplication of the resulting accagt motif (junction indicated by arrows). In this regard, the Camp, Palo Alto-1, and RO33 alleles are intermediate between MAD20/3D7 and K1/Palo Alto-2/Wellcome alleles.

sequence; but the distribution of these nucleotide repeats is not likely to differ markedly between chromosomes by chance alone.

We have made two interesting observations while searching chromosome 2 and 3 sequences for the presence of these nucleotide repeats. First, five of the 11 R2a repeats found on chromosome 2 are located within a 558-bp region corresponding to a predicted secreted antigen that appears similar to the glutamic acid-rich protein gene. Second, 67 of the 116 R1a nucleotide repeats on chromosome 2 occur as the 3′ terminus of a 39-nt repeat within the *pfEMP* member of the *var* gene family, which is an important component of *P. falciparum* antigenic variation. The observation of highly significant repeats within regions of the *Msp-1* gene previously thought not to be repetitive makes it clear that the extensive between-group nucleotide diversity between the two allelic groups is attributable to the same kinds of repeat variation and rapid divergence known in other antigenic determinants.

MSP-2 POLYMORPHISM

The *Msp-2* gene codes for MSP-2 (or MSA-2), a glycoprotein anchored, like MSP-1, in the merozoite membrane, but 45 kDa in size, and thus much smaller than MSP-1. The *Msp-2* of *P. falciparum* shows much greater variability in length, amino acid content, and number of repeats than *Csp*, but the pattern of allele polymorphism in *Msp-2* is consistent with the hypothesis that it has rapidly arisen by intragenic recombination.

Similar to CSP, MSP-2 is characterized by conserved N and C termini, with 43 and 74 residues, respectively (Smythe *et al.*, 1991). Bracketed within these segments, is the highly variable repeat region. Two allelic families have been identified and named after the isolates in which they were first identified. The FC27 family is characterized by at least one copy of a 32-aa sequence and a variable number of repeats, 12 aa in length. The 3D7/Camp family contains tandem amino acid repeats measuring 4–10 aa in length (Felger *et al.*, 1994).

The 3D7/Camp alleles are more variable in length and sequence of repeat types than those of the FC27 family (Felger *et al.*, 1997). Fenton *et al.* (1991) have proposed a model to explain the origin of repeat diversity within the 3D7/Camp family of alleles. They divided the 3D7/Camp family into distinct allelic subclasses, which included types A1 and A3, distinguished by amino acid repeats of different length. For example, A1 alleles possess 4-aa motifs, whereas a repeating 8-aa motif occurs in A3. Fenton *et al.* (1991) have shown that the allelic subclasses within the 3D7/Camp family are derived from a common ancestral nucleotide sequence and that the diversity arises from duplication and deletion of repeat subunits.

Recently, Dubbeld *et al.* (1998) have cloned and sequenced the *Msp-2* gene of *P. reichenowi*, which is a "unique mosaic of *P. falciparum* allelic forms and species-specific elements." We have used the methods described by Fenton *et al.* (1991) to determine whether the *Msp-2* of *P. reichenowi* provides insight into the ancestry of the FC27 and 3D7/Camp families. Fig. 5 shows the amino acid sequence alignment of two *P. falciparum* MSP-2s with the *P. reichenowi* MSP-2. The *P. falciparum* alleles from the 3D7 and OKS isolates are representative of the 3D7/Camp and FC27 families, respectively. The two *P. falciparum* alleles are identical at nucleotide sites encoding the N and C termini, but exhibit little similarity, even at the amino acid level, in the intervening repeat region.

A closer look at the nucleotides within the central portion of the gene manifests the homology of three distinct regions, which we define as repeat homology regions (RHRs). RHR1 shows common ancestry between the *P. reichenowi Msp-2* and the 3D7 *Msp-2* alleles (Fig. 6, black shading). Diversity within this region results from proliferation of a ggtgct hexamer (Fenton *et al.*, 1991). This hexamer is ancestral to both the 3D7/Camp and

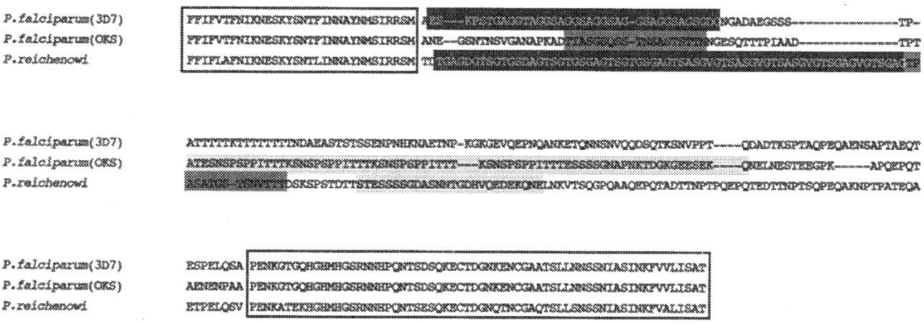

FIGURE 5. Amino acid alignment of P. falciparum (3D7 and OKS) and P. reichenowi Msp-2 alleles. Open boxes demarcate the conserved N and C termini. The inferred RHRs are shaded in black (RHR1), dark gray (RHR2), and light gray (RHR3). The nucleotide alignments for the inferred repeats of these regions are shown in Figs. 6 and 7.

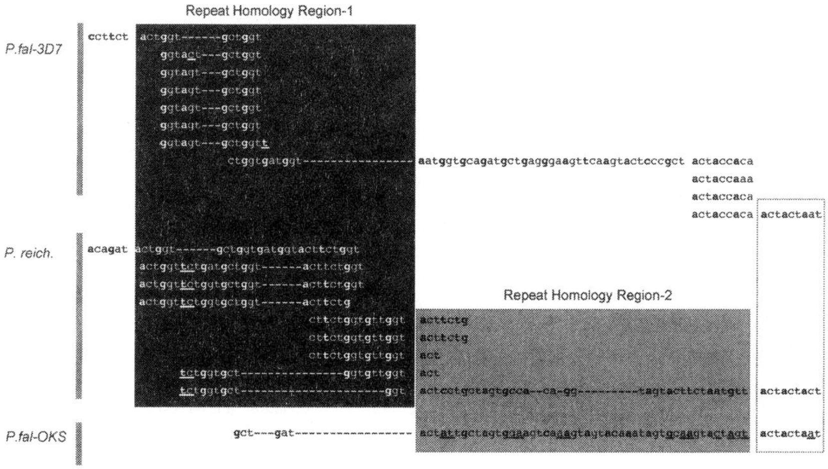

FIGURE 6. Partial nucleotide alignments of three Msp-2 gene sequences to manifest the repeats and the homologies between P. falciparum 3D7 and P. reichenowi (RHR1, black shading) and between P. falciparum OKS and P. reichenowi (RHR2, dark shading). Sequences read left to right and down where homologous repeats are present. The open box at the 3′ end of RHR2 shows a region of high similarity among all three alleles. Bold letters indicate the first nucleotide of each codon. Differences between aligned sequences are highlighted by underline. The alignment of repeats follows the convention of Fenton et al. (1991), so that repeats within and between sequences are aligned to show their homology.

the *P. reichenowi Msp-2* allelic repeats within this region. Although the conservation of these codons is clear among these two alleles, it appears that they have been lost altogether in the FC27-like alleles (represented by OKS in Figs. 5–7). However, the region adjacent to RHR1 in the *P. reichenowi Msp-2* sequence is similar to the first 21 aa of the 32-aa repeat found within the FC27 family, and this sequence is the basis for the inferred RHR2 (Fig. 5, dark gray shading). The last 9 nt of RHR2 also manifest the homology between all three sequences, including the short stretch following the $(actaccaa)_4$ repeat in 3D7. Note also the overlap between repeating nucleotides of *P. reichenowi Msp-2* in both RHR1 and RHR2.

A third RHR is located further downstream and shows the relationship between the 12-aa repeats of OKS and *P. reichenowi Msp-2* (Fig. 7). The repeat region in OKS is surrounded on either side by a 10-bp sequence (tacagaaagt), which occurs as only a single 5′ copy in the *P. reichenowi Msp-2* allele. Despite the lengthy repeat insertion in the OKS sequence, the homology of OKS and the *P. reichenowi Msp-2* in the region downstream of this repeat is apparent. And so it appears that the repeats were generated sometime after the split between *P. falciparum* and *P. reichenowi*.

Analysis of the single *P. reichenowi* sequence allows us to approximate the ancestral sequence of the two *P. falciparum Msp-2* allele families. Indeed, the comparison of the three RHRs discloses that whereas the precursor sequences for the various repeats probably were derived from the common *P. falciparum–P. reichenowi* ancestral species, the extant diversity among the *Msp-2* alleles has derived since the divergence of the two species. The distinctive dimorphism of the two *P. falciparum* alleles results from proliferation of repeats in two different regions of the molecule. Presumably because the overall MSP-2 molecule is constrained in size, the proliferation of repeats leads consequently to loss of nucleotides along the gene regions; i.e., the 3D7/Camp repeat precursors were lost in FC27 alleles, and the FC27 repeat precursors were lost in the 3D7 alleles.

As noted for *Csp*, the repetitive DNA sequences found within the

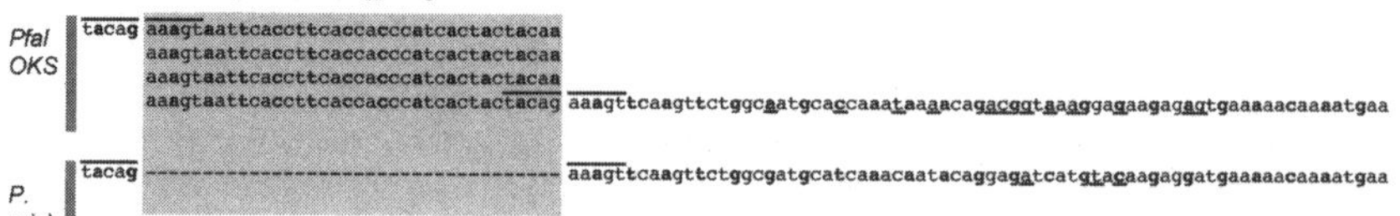

FIGURE 7. Nucleotide alignment of two Msp-2 gene sequences to manifest the repeats within RHR3 of P. falciparum OKS. This repeat region is not present in P. falciparum 3D7 or P. reichenowi. The repeat region of OKS continues contiguously from first to second to third to fourth row, left to right.

Msp-2 (and *Msp-1*) genes, as well as those in other *P. falciparum* antigenic determinants, are subject to much higher rates of mutation than NR sequences found within the same locus. Indeed, the paucity of silent substitutions within the NR regions indicates that intragenic recombination has generated repeat diversity in relatively short periods of time. Empirical estimates of mutation rates among repetitive DNA sequences, such as satellite DNA, are as high as 10^{-2} mutations/per generation and therefore several orders of magnitude greater than rates for point mutations (Schug *et al.*, 1998). The high mutation rates, coupled with strong selection for immune evasion, yield an extremely accelerated evolutionary rate for *P. falciparum* antigens.

ANTIGENIC POLYMORPHISM, INTRAGENIC RECOMBINATION, AND POPULATION STRUCTURE

Homologous comparisons among allelic variants of antigenic genes manifest that most of the variation is attributable to the rapid mutational processes associated with intragenic recombination. The increased rate of evolution among these genes reconciles the recent origin of extant *P. falciparum* populations with the abundance of antigenic diversity observed globally and locally. We have noted that nucleotide diversification can result from either intrahelical or interhelical events. An example of intrahelical recombination is that of mitotic, slipped-strand mismatch repair, which is considered to be the principal source of variation in repetitive units such as satellite DNA (Fig. 5). Interhelical recombination derives from the classical process of meiotic crossing over and recombination within or between loci on homologous chromosomes.

Both of these processes occur in *P. falciparum*. Kerr *et al.* (1994) have shown that meiotic, interhelical recombination occurs between mixed *Msp-2* genotype parasites passaged in laboratory animals. This process constitutes the basis for generating linkage maps of *P. falciparum* chromosomes (Su and Wellems, 1996). But we have shown that, despite the abundant intragenic recombination within *Csp* CR, there is an apparent absence of recombination between the 5′ and 3′ NR regions, suggesting that the duplication and deletion of RATs occur by mitotic processes such as the slipped-strand process modeled in Fig. 5 (Rich *et al.*, 1997). This process also has been implicated as the cause of repeat variation in *Msp-2* (Fenton *et al.*, 1991).

The debate over the relevance of sexual recombination between *P. falciparum* types may remain unsettled for some time. It is becoming increasingly clear that the population structure of *P. falciparum* may not be uniform throughout the species, but depends on local factors related to parasite, vector, and host biology (Paul *et al.*, 1995; Babiker and Walliker,

1997; Conway *et al.*, 1999; Sakihama *et al.*, 1999). An accurate determination of these factors is contingent on careful analysis of parasite genotypes and appropriate determination of homologous comparisons.

We are grateful to Benjamin Rosenthal and F. Ellis McKenzie for thoughtful insights and comments.

REFERENCES

Anderson, T. J. C., Su, X. Z., Bockarie, M. Lagog, M. & Day, K.P. (1999) Twelve microsatellite markers for characterization of *Plasmodium falciparum* from finger-prick blood samples. *Parasitology* 119, 113–125.

Ayala, F. J., Escalante, A. A., & Rich, S. M. (1999) Evolution of *Plasmodium* and the recent origin of the world populations of *Plasmodium falciparum. Parassitologia* 41, 55–68.

Ayala, F. J., Escalante, A. A., Lal, A. A., & Rich, S. M. (1998) Evolutionary relationships of human malaria parasites. In *Malaria: Parasite Biology, Pathogenesis, and Protection*, ed. Sherman, I. W. (ASM Press, Washington, DC), pp. 285–300.

Babiker, H. & Walliker, D. (1997) Current views on the population structure of *Plasmodium falciparum*: Implications for control. *Parasitol. Today* 13, 262–267.

Clyde, D. F., McCarthy, V. C., Miller, R. M., & Hornick, R. B. (1973) Specificity of protection of man immunized against sporozoite-induced falciparum malaria. *Am. J. Med. Sci.* 266, 398–403.

Coatney, R. G., Collins, W. E., Warren, M., & Contacos, P. G. (1971) *The Primate Malarias* (US Government Printing Office, Washington, DC).

Coluzzi, M. (1994) Malaria and the afrotropical ecosystems: impact of man-made environmental changes. *Parassitologia* 36, 223–227.

Coluzzi, M. (1997) Evoluzione Biologica i Grandi Problemi della Biologia. *Evoluzione Biologica i Grandi Problemi della Biologia* (Accademia dei Lincei, Rome), pp. 263–285.

Coluzzi, M. (1999) The clay feet of the malaria giant and its African roots: Hypotheses and inferences about origin, spread and control of *Plasmodium falciparum. Parassitologia* 41, 277–283.

Conway, D. J., Roper, C., Oduola, A. M. J., Arnot, D. E., Kremsner, P. G., Grobusch, M. P., Curtis, C. R., & Greenwood, B. M. (1999) High recombination rate in natural populations of *Plasmodium falciparum. Proc. Natl. Acad. Sci.* USA 96, 4506–4511.

de Zulueta, J. (1973) Malaria and Mediterranean history. *Parassitologia* 15, 1–15.

de Zulueta, J. (1994) Malaria and ecosystems: From prehistory to posteradication. *Parassitologia* 36, 7–15.

Dubbeld, M. A., Kocken, C. H., & Thomas, A. W. (1998) Merozoite surface protein 2 of *Plasmodium reichenowi* is a unique mosaic of *Plasmodium falciparum* allelic forms and species-specific elements. *Mol. Biochem. Parasitol.* 92, 187–192.

Escalante, A. A. & Ayala, F. J. (1994) Phylogeny of the malarial genus Plasmodium derived from rRNA gene sequences. *Proc. Natl. Acad. Sci.* USA 91, 11373–11377.

Escalante, A. A., Barrio, E., & Ayala, F. J. (1995) Evolutionary origin of human and primate malarias: Evidence from the circumsporozoite protein gene. *Mol. Biol. Evol.* 12, 616–626.

Escalante, A. A., Lal, A. A., & Ayala, F. J. (1998) Genetic polymorphism and natural selection in the malaria parasite *Plasmodium falciparum. Genetics* 149, 189–202.

Felger, I., Tavul, L., Kabintik, S., Marshall, V., Genton, B., Alpers, M., & Beck, H. P. (1994) Plasmodium falciparum—extensive polymorphism in merozoite surface antigen 2 alleles in an area with endemic malaria in Papua New Guinea. *Exp. Parasitol.* 79, 106–116.

Felger, I., Marshal, V. M., Reeder, J. C., Hunt, J. A., Mgone, C. S., & Beck, H. P. (1997) Sequence diversity and molecular evolution of the merozoite surface antigen 2 of *Plasmodium falciparum. J. Mol. Evol.* 45, 154–160.

Fenton, B., Clark, J. T., Khan, C. M. A., Robinson, J. V., Walliker, D., Ridley, R., Scaife, J. G., & McBride, J. S. (1991) Structural and antigenic polymorphism of the 35-kilodalton to 48-kilodalton merozoite surface antigen (MSA-2) of the malaria parasite *Plasmodium-falciparum. Mol. Cell. Biol.* 11, 963–974.

Frontali, C. (1994) Genome plasticity in Plasmodium. *Genetica* 94, 91–100.

Frontali, C. & Pizzi, E. (1991) Conservation and divergence of repeated structures in Plasmodium genomes the molecular drift. *Acta Leiden.* 60, 69–81.

Ho, K. F., Craddock, E. M., Piano, F., & Kambysellis, M. P. (1996) Phylogenetic analysis of DNA length mutations in a repetitive region of the Hawaiian Drosophila yolk protein gene YP2. *J. Mol. Evol.* 43, 116–124.

Hughes, A. L. (1992) Positive selection and interallelic recombination at the merozoite surface antigen-1 (MSA-1) locus of *Plasmodium falciparum. Mol. Biol. Evol.* 9, 381–393.

Kerr, P. J., Ranford-Cartwright, L. C., & Walliker, D. (1994) Proof of intragenic recombination in *Plasmodium falciparum. Mol. Biochem. Parasitol.* 66, 241–248.

Levinson, G. & Gutman, G. A. (1987) Slipped-strand mispairing a major mechanism for DNA sequence evolution. *Mol. Biol. Evol.* 4, 203–221.

Livingstone, F. B. (1958) Anthropological implications of sickle cell gene distribution in West Africa. *Am. Anthropol.* 60, 533–562.

López-Antuñano, F. & Schumunis, F. A. (1993) Plasmodia of humans. In *Parasitic Protozoa*, 2nd ed., vol. 5, ed. Kreier, J. P. (Academic Press Inc., New York), pp. 135–265.

Oron-Karni, V., Filon, D., Rund, D., & Oppenheim, A. (1997) A novel mechanism generating short deletion/insertions following slippage is suggested by a mutation in the human alpha(2)-globin gene. *Hum. Mol. Genet.* 6, 881–885.

Paul, R. E. L., Packer, M. J., Walmsley, M., Lagog, M., Ranford-Cartwright, L. C., Paru, R., & Day, K. P. (1995) Mating patterns in malaria parasite populations of Papua New Guinea. *Science* 269, 1709–1711.

Qari, S. H., Shi, Y. P., Pieniazek, N. J., Collins, W. E., & Lal, A. A. (1996) Phylogenetic relationship among the malaria parasites based on small subunit rRNA gene sequences: monophyletic nature of the human malaria parasite, *Plasmodium falciparum. Mol. Phylogenet. Evol.* 6, 157–165.

Rich, S. M., Hudson, R. R., & Ayala, F. J. (1997) *Plasmodium falciparum* antigenic diversity: Evidence of clonal population structure. *Proc. Natl. Acad. Sci.* USA 94, 13040–13045.

Rich, S. M., Licht, M. C., Hudson, R. R., & Ayala, F. J. (1998) Malaria's Eve: Evidence of a recent bottleneck in the global *Plasmodium falciparum* population. *Proc. Natl. Acad. Sci.* USA 95, 4425–4430.

Sakihama, N., Kimura, M., Hirayama, K., Kanda, T., Na-Bangchang, K., Jongwutiwes, S., Conway, D., & Tanabe, K. (1999) Allelic recombination and linkage disequilibrium within Msp-1 of *Plasmodium falciparum*, the malignant human malaria parasite. *Gene* 230, 47–54.

Schug, M. D., Hutter, C. M., Noor, M. A., & Aquadro, C. F. (1998) Mutation and evolution of microsatellites in *Drosophila melanogaster. Genetica* 102–103, 359–367.

Sherman, I. W. (1998) A brief history of malaria and discovery of the parasite's life cycle. In *Malaria: Parasite Biology, Pathogenesis, and Protection*, ed., Sherman, I. W. (ASM Press, Washington, DC), pp. 3–10.

Smythe, J. A., Coppel, R. L., Kay, K. P., Martin, R. K., Oduola, A. M. J., Kemp, D. J., & Anders, R. F. (1991) Structural diversity in the *Plasmodium-falciparum* merozoite surface antigen-2. *Proc. Natl. Acad. Sci.* USA 88, 1751–1755.

Su, X. & Wellems, T. E. (1996) Toward a high-resolution *Plasmodium falciparum* linkage map—polymorphic markers from hundreds of simple sequence repeats. *Genomics* 33, 430–444.

Tanabe, K., Mackay, M., Goman, M., & Scaife, J. G. (1987) Allelic dimorphism in a surface antigen gene of the malaria parasite *Plasmodium falciparum*. *J. Mol. Biol.* 195, 273–287.

Ukhayakumar, V., Shi, Y.-P., Kumar, S., Jue, D. L., Wohlhueter, R. M., & Lal, A. A. (1994) Antigenic diversity in the circumsporozoite protein of *Plasmodium falciparum* abrogates cytotoxic-T-cell recognition. *Infect. Immun.* 62, 1410–1413.

Waters, A. P., Higgins, D. G., & McCutchan, T. F. (1991) *Plasmodium falciparum* appears to have arisen as a result of lateral transfer between avian and human hosts. *Proc. Natl. Acad. Sci.* USA 88, 3140–3144.

Wiesenfeld, S. L. (1967) Sickle-cell trait in human biological and cultural evolution: development of agriculture causing increased malaria is bound to gene-pool changes causing malaria reduction. *Science* 157, 1134–1140.

World Health Organization (1995) *Tropical Disease Report, Twelfth Programme Report* (World Health Organization, Geneva).

Zevering, Y., Khamboonruang, C. & Good, M. F. (1994) Natural amino acid polymorphisms of the circumsporozoite protein of *Plasmodium falciparum* abrogate specific human CD4(+) T cell responsiveness. *Eur. J. Immunol.* 24, 1418–1425.

Part IV

POPULATION VARIATION

Nina Fedoroff ("Transposons and Genome Evolution in Plants," Chapter 10) notices that the publication 50 years ago of Stebbins' *Variation and Evolution in Plants* roughly coincides with the first reports by Barbara McClintock that there are genetic elements capable of transposing from one to another chromosomal location in maize. Today we know that transposable elements make up a large fraction of the DNA of agriculturally important plants, such as corn and wheat, and of animals such as mice and humans, and perhaps all species of mammals and many other vertebrates. Fedoroff reviews the history of the discovery of transposing elements and advances the hypothesis that the mechanisms controlling transposition are an instance of "the more general capacity of eukaryotic organisms to detect, mark, and retain duplicated DNA through regressive chromatin structures."

Grasses (family Poaceae) and their cultivated relatives encompass a gamut of genome size and structural complexity, that extends from rice at the lower end to wheat and sugarcane at the higher end, having nuclear DNAs more than 30 times larger than rice's. Maize is towards the middle, with about six times more nuclear DNA than rice, embodied in 10 pairs of chromosomes. The maize genome is replete with chromosomal duplications and repetitive DNA sequences, as Brandon S. Gaut and his collaborators tell us ("Maize as a Model for the Evolution of Plant Nuclear Genomes," Chapter 11). This complexity has motivated these authors to focus on maize as a model system for investigating the evolution of plant nuclear genomes. More than 11 million years ago, but after the sorghum and maize lineages had split, the maize genome became polyploid, which accounts for much of the difference in DNA content between these related species. The polyploid event was followed by diploidization and much rearrangement of the genome, so that maize is now a diploid. But there remains much "extra" DNA in maize, mostly consisting of multiple repetitions of retrotransposons that account for 50% of the nuclear genome. This multiplication has occurred within the last 5–6 million years and has also contributed to the genome differentiation between maize and sorghum. The evolutionary complexities of cultivated maize extend to individual genes that have been variously impacted by domestication and intensive breeding.

Michael T. Clegg and Mary L. Durbin ("Flower Color Variation: A Model for the Experimental Study of Evolution," Chapter 12) trace the development of flower color in the morning glory, from the molecular and genetic levels to the phenotype, as a model for analyzing adaptation. Most mutations determining phenotypic differences turn out to be due to transposon insertions. Insect pollinators discriminate against white flowers in populations where white flowers are rare. This would provide an advantage to white genes through self-fertilization in white maternal plants. The pattern of geographic distribution of white plants indicates that such advantage is counteracted by definite, but undiscovered disadvantages of the white phenotype. The authors conclude by proposing that floral color development is an area of special promise for understanding the complex gene interactions that impact the phenotype and its adaptation, precisely because "the translation between genes and phenotype is tractable . . . [and] the translation between environment and phenotype is more transparent for flower color than in most other cases."

Barbara A. Schaal and Kenneth M. Olsen point out, in "Gene Genealogies and Population Variation in Plants" (Chapter 13), that it was largely due to Stebbins that the investigation of individual variation within populations become part and parcel of the study of plant evolution. For many years beyond 1950, the focus of investigation was the phenotype: morphology, karyotype, and fitness components. Protein electrophoresis opened up the identification of allozyme variation and thus the study of allelic variation at individual genes. Restriction analysis and DNA sequencing have added the possibility of reconstructing the intraspecific genealogy of alleles. The mathematical theory of gene coalescence has provided the analytical tools for reconstruction and interpretation. Schaal and Olsen put all these tools to good use in several model cases: the recent rapid geographic expansion of *Arabidopsis thaliana*, with little differentiation between populations; the recolonization of European tree species from refugia created by the Pleistocene glaciation; the origin and domestication of cassava (manioc), the main carbohydrate source for 500 million people in the world tropics.

10

Transposons and Genome Evolution in Plants

NINA FEDOROFF

Although it is known today that transposons comprise a significant fraction of the genomes of many organisms, they eluded discovery through the first half century of genetic analysis and even once discovered, their ubiquity and abundance were not recognized for some time. This genetic invisibility of transposons focuses attention on the mechanisms that control not only transposition, but illegitimate recombination. The thesis is developed that the mechanisms that control transposition are a reflection of the more general capacity of eukaryotic organisms to detect, mark, and retain duplicated DNA through repressive chromatin structures.

The 50 years that have elapsed since the publication of Stebbins' *Variation and Evolution in Plants* have seen extraordinary changes in our understanding of how genomes are structured and how they change in evolution. The book's publication date roughly coincides with the first reports by Barbara McClintock that there are genetic

The Pennsylvania State University, University Park, PA 16802

This paper was presented at the National Academy of Sciences colloquium "Variation and Evolution in Plants and Microorganisms: Toward a New Synthesis 50 Years After Stebbins," held January 27–29, 2000, at the Arnold and Mabel Beckman Center in Irvine, CA.

Abbreviation: LTR, long terminal repeat.

elements capable of transposing to different chromosomal locations in maize plants (McClintock, 1945). The book contains a brief mention of Marcus Rhoades' observation that a standard recessive a_1 allele of a gene in the anthocyanin biosynthetic pathway can become unstable and revert at a high frequency to the dominant A_1 allele in a background containing a dominant *Dt* ("dotted") allele (Stebbins Jr., 1950). But transposable elements were not yet common fare, nor was it known that *Dt* is a transposon.

Today we know that transposons constitute a large fraction—even a majority—of the DNA in some species of plants and animals, among them mice, humans, and such agriculturally important plants as corn and wheat. Given what we now know about genome organization, it is paradoxical that the discovery of transposable elements lagged so far behind the discovery of the basic laws of genetic transmission. And it is equally curious that even when they were discovered, acceptance of their generality and recognition of their ubiquity came so slowly. It is perhaps an understatement to say that McClintock's early communications describing transposition were not widely hailed for their explanatory power. Indeed, McClintock commented in the introduction to her collected papers that the response to her first effort in 1950 to communicate her discovery of transposition in ". . . a journal with wide readership . . . ," specifically the *Proceedings of the National Academy of Sciences*, convinced her that ". . . the presented thesis, and evidence for it, could not be accepted by the majority of geneticists or by other biologists" (McClintock, 1987). By contrast, the explanatory power of Watson and Crick's 1953 *Nature* paper on the structure and mode of replication of nucleic acids was recognized immediately (Watson and Crick, 1953).

An informative parallel is provided by the contrast between the immediate recognition of the importance of Darwin's theory of evolution and the long delay between Mendel's articulation of the laws of heredity and their wide acceptance in evolutionary thinking (Carlson, 1966). It can be speculated that this was because Darwin's theory provided immediate explanations in the realm of the perceptible, whereas the hereditary mechanisms underlying variation were obscure. Variation, in Darwin's view, was continuous. Geneticists sharing his view formed the "biometrical school," devoted to the statistical analysis of inheritance. It was not at all clear how the simple rules derived by Mendel for the hereditary behavior of "differentiating characters" bore on the problem of evolution (Carlson, 1966). The relevance of discontinuous variation or the production of "sports," as morphological mutations were called, was even less obvious, because the biometric approach treated offspring as statistical combinations of parental traits. Thus the idea that the study of mutations was central to understanding evolution was close to unimaginable a century ago.

Equally unimaginable at mid-20th century was the idea that transposable elements are essential to understanding chromosome structure and evolution, much less organismal evolution. The efforts of Bateson and other geneticists had firmly established Mendelian "laws" as the central paradigm of genetics and the identification and mapping of genetic "loci" through the study of mutant alleles was proceeding apace. Because genetic mapping is predicated on the invariance of recombination frequencies, there was plentiful evidence that genes have fixed chromosomal locations. Written at this time, Stebbins' book in general and in particular his third chapter, titled "The Basis of Individual Variation," clearly acknowledges the existence of many chromosomal differences among organisms in a population, including duplications, inversions, translocations, and deletions. At the same time, the book reflects the prevailing view that these ". . . are not the materials that selection uses to fashion the diverse kinds of organisms which are the products of evolution" (Stebbins, Jr., 1950). Instead, Stebbins concludes that the majority of evolutionarily important changes in physiology and morphology are attributable to classical genetic "point" mutations.

Another half century has elapsed and the geneticist's "black box," sprung open, spills nucleotide sequences at an ever accelerating pace. Our computers sift through genomes in search of genes, knee-deep in transposons. How could we not have seen them before? The answer is as straightforward as it is mysterious and worthy of consideration: they are invisible to the geneticist. Well, almost invisible. And of course it depends on the geneticist.

THE DISCOVERY OF TRANSPOSITION

The study of unstable mutations that cause variegation dates back to De Vries, who formulated the concept of "ever-sporting varieties" and eventually came to the conclusion that these types of mutations do not obey Mendel's rules (de Vries, 1905). The first person to make substantial sense of their inheritance was the maize geneticist Emerson, who analyzed a variegating allele of the maize *P* locus during the first decades of this century (Emerson, 1914, 1917, 1929). His first paper on the subject opens with the statement that variegation ". . . is distinguished from other color patterns by its incorrigible irregularity" (Emerson, 1914). What follows is a brilliant analysis of "freak ears" containing large sectors in which the unstable *P* allele has either further mutated or reverted. Emerson was able to capture the behavior of unstable mutations in the Mendelian paradigm by postulating that variegation commenced with the temporary association of some type of inhibitor with a locus required for pigmentation. Emerson's suggestion was that normal pigmentation was restored upon loss of the inhibitor.

Several prominent geneticists, among them Correns and Goldschmidt, dismissed unstable mutations as a special category of "diseased genes" (Goldschmidt, 1938; Fedoroff, 1998). It was their view that little could be learned from the study of such mutations that was relevant to the study of conventional genes. But the drosophilist Demerec and the maize geneticist Rhoades shared Emerson's view that there was no difference in principle between stable and unstable mutations. Indeed, the Rhoades mutation cited in Stebbin's volume illustrates the important point that instability is conditional. Rhoades' experiments had revealed that a standard recessive allele of the maize *A1* locus, isolated decades earlier and in wide use as a stable null allele, could become unstable in a different genetic background. The key ingredient of the destabilizing background was the presence of the *Dt* locus, which caused reversion of the *a1* allele to wild type both somatically and germinally (Rhoades, 1936, 1938).

In the late 1930s, McClintock had begun to work with broken chromosomes and by the early 1940s she had devised a method for producing deletion mutations commencing with parental plants, each of which contributed a broken chromosome 9 lacking a terminal segment and the telomere. Searching for mutants in the progeny of such crosses, she observed a high frequency of variegating mutants of all kinds (McClintock, 1946). She noted that although reports of the appearance of new mutable genes were relatively rare in the maize literature, she already had isolated 14 new cases of such instability and observed more. She chose to follow the behavior of a locus, which she called *Dissociation* (*Ds*), for its propensity to cause the dissociation of the short arm of chromosome 9 at a position close to the centromere, although she soon appreciated that chromosome dissociation required the presence of another unlinked locus, which she designated the *Activator* (*Ac*) locus (McClintock, 1946, 1947).

By 1948, she had gained sufficient confidence that the *Ds* locus moves to report: "It is now known that the *Ds* locus may change its position in the chromosome" (McClintock, 1948; Fedoroff, 1998). The relationship between the chromosome-breaking *Ds* locus and variegation emerged as McClintock analyzed the progeny of a new variegating mutation of the *C* locus required for kernel pigmentation. She carried out an extraordinary series of painstakingly detailed cytological and genetic experiments on this new mutation, *c-m1*, whose instability was conditional and depended on the presence of the *Ac* locus (McClintock, 1948, 1949; Fedoroff, 1998). She showed that the origin of the unstable mutation coincided with the transposition of *Ds* from its original position near the centromere on chromosome 9 to a new site at the *C* locus and that when it reverted to a stable wild-type allele, *Ds* disappeared from the locus. Having established that

Ds could transpose into and out of the *C* locus germinally, she inferred that somatic variegation reflects the frequent transposition of *Ds* during development. Transposition explained both Emerson's and Rhoades' earlier observations. McClintock and Rhoades were good friends, of course, and it is evident from their correspondence that McClintock immediately saw the parallels between the behavior of the *c-m1* mutation and Rhoades' *a1* mutation (Lee Kass, personal communication).

The *Ac* and *Ds* elements are transposition-competent and transposition-defective members of a single transposon family. In the ensuing years, McClintock identified and studied a second transposon family, called *Suppressor-mutator* (*Spm*) (McClintock, 1951, 1954). Her studies on these element families were purely genetic, and she was able to make extraordinary progress in understanding the transposition mechanism because she studied the interactions between a single transposition-competent element and one or a small number of genes with insertions of cognate transposition-defective elements (Fedoroff, 1989). Two points about this early history of transposition merit emphasis. First, the active elements were denumerable and manageable as genetic entities, despite their propensity to move. Second, the number of different transposon families and family members uncovered genetically was (and still is) small. Hence the genetic impact of transposable elements was limited. McClintock recognized that the high frequency of new variegating mutations in her cultures was linked to the genetic perturbations associated with the presence of broken chromosomes (McClintock, 1946, 1978). Her inference, extraordinarily prescient, was that transposons are regular inhabitants of the genome, but genetically silent.

PLANT TRANSPOSONS IN THE AGE OF GENOMICS

With the cloning of the maize transposons, first the *Ac* element in my laboratory and later the cognate *En* and *Spm* elements in Heinz Saedler's and my laboratories, the picture began to change (Fedoroff *et al.*, 1983; Pereira *et al.*, 1985; Masson *et al.*, 1987). To begin with, it became obvious immediately that the maize genome contains more copies of a given transposon than there are genetically identifiable elements. Although most of these sequences are not complete transposons, there are nonetheless more complete transposons than can be perceived genetically (Fedoroff *et al.*, 1984). Importantly, it was clear almost immediately that a genetically active transposon could be distinguished from one that was genetically silent by its methylation pattern (Fedoroff *et al.*, 1984; Banks and Fedoroff, 1989). Both of these observations bear on the genetic visibility of transposons.

As maize genes and genome segments began to be cloned and sequenced, the discovery of new transposons accelerated. Although the transposons that McClintock identified and studied were DNA transposons, both *gypsy*-like and *copia*-like retrotransposons were soon identified in the maize genome and subsequently in many other plant genomes (Shepherd *et al.*, 1984; Flavell, 1992; Purugganan and Wessler, 1994; White *et al.*, 1994; Suoniemi *et al.*, 1997, 1998). It has also become evident that non-long terminal repeat (LTR) retrotransposons are abundant in maize, as well as other plant genomes (Schwarz-Sommer *et al.*, 1987; Noma *et al.*, 1999). Many additional maize transposon families have been identified through their sequence organization and their presence in or near genes (Spell *et al.*, 1988; Bureau and Wessler, 1992, 1994a, b; Wessler *et al.*, 1995; Bureau *et al.*, 1996). We now know that transposons and retrotransposons comprise half or more of the maize genome (Bennetzen *et al.*, 1998).

WHAT DO TRANSPOSONS DO?

Commencing with McClintock's elegant analyses of transposon-associated chromosomal rearrangements and extending into the literature of today, the range of transposon-associated genetic changes has continued to expand (McClintock, 1951). Insertion of plant transposons, like almost all known transposons, is accompanied by the duplication of a short flanking sequence of a few base pairs (Schwarz-Sommer *et al.*, 1985a). Plant transposons excise imprecisely, generally leaving part of the duplication at the former insertion site (Schwarz-Sommer *et al.*, 1985a). The consequences of insertion and excision of a transposon therefore depend on the location within the coding sequence and excision of an insertion from an exon commonly results in either an altered gene product or a frame-shift mutation. Transposon insertions can alter transcription and transcript processing, and there are cases in which transposons are processed out of transcripts by virtue of the presence of splice donor and acceptor sequences (Kim *et al.*, 1987; Wessler, 1989; Giroux *et al.*, 1994). Transposons also can promote the movement of large segments of DNA either by transposition or by illegitimate recombination (Courage-Tebbe *et al.*, 1983; Schwartz *et al.*, 1998).

THE PARADOX

One might think that given their abundance, transposable elements would rapidly randomize genome order. Yet the results of a decade of comparative plant genome studies has revealed that gene order is surprisingly conserved between species. Close relationships among genomes have been demonstrated in crop plants belonging to the *Solanaceae*, and the *Graminae*, between *Brassica* crops and *Arabidopsis*, among several legumes, and others (Tanksley *et al.*, 1992; Gale and Devos, 1998; Lagercrantz, 1998). The synteny among the genomes of economically important cereal grasses is so extensive that they are now represented by concentric circular maps (Gale and Devos, 1998). There are rearrangements, but a relatively small number of major inversions and transpositions is required to harmonize the present day maps. Such maps, of course, are crude representations of the genome, and rearrangements can emerge as the level of resolution increases (Tanksley *et al.*, 1988, 1992; Tikhonov *et al.*, 1999). The frequency of rearrangements also can differ markedly and there is evidence that rearrangements are more prevalent just after polyploidization (Song *et al.*, 1995; Gale and Devos, 1998). Even within a conservative lineage, however, some gene families are more heterogeneous in composition and map distribution than others (Leister *et al.*, 1998).

SYNTENY AND DIVERGENCE

What are the useful generalizations? First, synteny can extend down to a very fine level, but it is far from perfect. A detailed sequence comparison of the small region around the maize and sorghum *Adh1* loci reveals a surprising amount of change in a constant framework (Tikhonov *et al.*, 1999). The sorghum and maize genomes are 750 and 2,500 Mbp, respectively. The *Adh1* gene sequences are highly conserved, and complete sequencing revealed that there were seven and 10 additional genes in the homologous regions of maize and sorghum, respectively. The region of homology extends over about 65 kb of the sorghum genome, but occupies more than 200 kb in the maize genome. The gene order and orientation are conserved, although three of the genes found in the sorghum *Adh1* region are not in the maize *Adh1* region. The genes are located elsewhere in the maize genome, suggesting that they transposed away from the *Adh1* region (Tikhonov *et al.*, 1999). Although homology is confined largely to genes, there are also homologous intergenic regions. There are simple sequence repeats and small transposons, called MITES as a group, scattered throughout this region in both sorghum and maize. MITES are found primarily between genes, but several are in introns. The small MITE transposons are found neither in exons nor in retrotransposons. There are three non-LTR retrotransposons in the maize *Adh1* region and none in the sorghum *Adh1* region (Tikhonov *et al.*, 1999).

The major difference between the maize and sorghum *Adh* regions is the presence of very large continuous blocks of retrotransposons in maize that are not present in sorghum. Although most blocks are between genes, one appears to be inside a gene sequence. They are present in many, but not all intergenic regions. There is a relatively long stretch of almost 40 kb containing four genes in maize and seven genes in sorghum, which contains no retrotransposon blocks in maize and in which there is about 10 kb of extensive homology, some genic and some intergenic. Thus synteny extends down to a relatively fine level and includes both genic and intergenic sequences.

PLANT GENOMES EXPAND

A second generalization is that plant genomes grow. Genome sizes among flowering plants vary dramatically over almost 3 orders of magnitude, from the roughly 130 Mbp *Arabidopsis* genome to the 110,000 Mbp *Fritillaria assyriaca* genome (Bennett *et al.*, 1982). Genome size variation greatly exceeds estimates of differences in gene numbers (Bennetzen and Freeling, 1997). This, of course, is the celebrated C-value paradox (Thomas, 1971). Plant genomes expand by several mechanisms, including polyploidization, transposition, and duplication. Thus, for example, a fine-scale comparison of the *Arabidopsis thaliana* and *Brassica nigra* genomes reveals that the *Brassica* genome contains a triplication of the much smaller *Arabidopsis* genome, as well as chromosome fusions and rearrangements (Lagercrantz, 1998). There is evidence that the maize genome is a segmental allotetraploid (White and Doebley, 1998). It is estimated that up to 70% of flowering plants have polyploidy in their lineages (Leitch and Bennett, 1997). Thus replication of whole genomes or parts of genomes is a common and important theme in plant genome evolution.

TRANSPOSITION

Transposition is also a major cause of plant genome expansion. To begin with, transposition generates DNA. Retrotransposition results from transcription of genomic retrotransposons, followed by insertion of reverse transcripts into the genome at new sites (Howe and Berg, 1989). Plant transposons generate additional copies of themselves by virtue of excising from only one of two newly replicated sister chromatids and reinserting into as yet unreplicated sites (Fedoroff, 1989). Absent countering forces, genome expansion is an inevitable consequence of the properties of transposable elements. The accumulation of retrotransposon blocks between genes is a major factor in the size difference between the maize genome and those of its smaller relatives (SanMiguel *et al.*, 1996, 1998). Retrotransposon blocks occupy 74% of the recently sequenced 240-kb maize *Adh* region (Tikhonov *et al.*, 1999). These blocks contain 23 members of 11 different retrotransposon families, primarily as complete retrotransposons, but also occasionally as solo LTRs (Tikhonov *et al.*, 1999). Within these blocks, retrotransposons are commonly nested by insertion of retrotransposons into each other (SanMiguel *et al.*, 1996, 1998).

What is perhaps most surprising about the maize retrotransposon blocks that have been characterized is that they grow quite slowly. The transposition mechanism assures that retrotransposon ends are almost always identical when an element inserts, hence the divergence between the LTRs of a single element reflects the age of the insertion. Bennetzen and his colleagues found that the sequence difference between the LTRs of a given element is almost invariably less than the sequence difference between the LTRs of the element into which it is inserted. Using these differences to order and date the insertions, they inferred that all of the insertions have occurred within roughly the last 5 million years, well after the divergence of maize and sorghum (SanMiguel *et al.*, 1998). Importantly, no retrotransposons have been found in the corresponding *Adh1* flanking sequence in sorghum (SanMiguel *et al.*, 1998; Tikhonov *et al.*, 1999). This raises the possibility that retrotransposon activity may differ between closely related lineages.

AMPLIFICATION AND REARRANGEMENT

New copies of transposons and retrotransposons provide new sites of homology for unequal crossing over. Evidence that transposable elements are central to the evolutionary restructuring of genomes has accumulated in every organism for which sufficient sequence data exist. Exceptionally detailed examples of the role of transposition, retrotransposition, amplification, and transposon-mediated rearrangements in the evolution of a contemporary chromosome are provided by recent studies on the human Y chromosome (Saxena *et al.*, 1996; Schwartz *et al.*, 1998; Lahn and Page, 1999a, b). Although the level of resolution is not yet sufficient in many cases to determine the molecular history of each duplication, it is evident that many, if not a majority of plant genes belong to gene families ranging in size from a few members to hundreds (Michelmore and Meyers, 1998; Riechmann and Meyerowitz, 1998; Martienssen and Irish, 1999; Rabinowicz *et al.*, 1999). *R* genes, for example, comprise a superfamily of similar *myc*-homologous, helix–loop–helix transcriptional activators of genes in anthocyanin biosynthesis (Ludwig *et al.*, 1989; Perrot and Cone, 1989; Consonni *et al.*, 1993). Detailed analysis of the *R-r* complex, a well-studied member of the *R* superfamily, reveals a history of transposon-catalyzed rearrangement and duplication (Walker *et al.*, 1995).

There also may be other genetic mechanisms that drive genome expansion. A recent analysis of the behavior of maize chromosomal knobs reveals that the pattern of segregation under the influence of a "meiotic drive" locus of as yet unknown function results in the preferential transmission of chromosomes with larger knobs over chromosomes with smaller knobs (Buckler *et al.*, 1999). Maize knobs are blocks of similar short tandemly repeated sequences, ranging from as few as 100 copies to as many as 25,000 per site (Ananiev *et al.*, 1998a). Their structure and dispersed occurrence further suggest that they are transposable (Ananiev *et al.*, 1998a, b; Buckler *et al.*, 1999). The combination of transposability and preferential transmission of chromosomes with expanded knobs thus provides an additional mechanism for genome expansion.

GENOME CONTRACTION

Are there genetic mechanisms that contract genomes? Careful analysis of the relative deletion frequency and length in drosophilid non-LTR retrotransposons supports the inference that there are more deletions per point mutation in Drosophila than in mammals and that the average deletion size is almost eight times larger (Petrov and Hartl, 1998). Thus mechanisms that contract genomes by preferential deletion may exist, as well. Bennetzen and Kellogg have argued that despite ample evidence for the operation of mechanisms that expand genomes in plants, there is little evidence that plant genomes contract (Bennetzen and Kellogg, 1997). The maize intergenic regions that have been analyzed, for example, comprise predominantly intact retrotransposons, rather than solo LTRs, which can arise by unequal crossovers between the repeats at retrotransposon ends and are common in other genomes (Bennetzen and Kellogg, 1997). However, it also is known that both the Ac and Spm transposons of maize frequently give rise to internally deleted elements, and Ac ends are very much more abundant in the maize genome than are full-length elements, suggesting deletional decay of transposon sequences (Fedoroff *et al.*, 1983, 1984; Schwarz-Sommer *et al.*, 1985b; Masson *et al.*, 1987). So it would not be surprising to find mechanisms that preferentially eliminate sequences. And indeed, preferential loss of nonredundant sequences early after polyploidization has been detected in wheat (Feldman *et al.*, 1997).

CONTROLLING TRANSCRIPTION, RECOMBINATION, AND TRANSPOSITION

Despite our growing awareness of the abundance of plant transposable elements and the role they have played in shaping contemporary chromosome organization, the fact is they eluded discovery for the first half century of intensive genetic analysis. Thus what is perhaps the most striking observation about transposable elements is not their instability, but precisely the opposite: their stability. Not only are insertion mutations in genes infrequent, but retrotransposition events are so widely separated that the time interval between insertions in a particular region of the genome can be counted in hundreds of thousands to millions of years (SanMiguel *et al.*, 1998). Chromosomes containing many hundreds of thousands of transposable elements are as stable as chromosomes containing few. By what means are such sequences prevented from transposing, recombining, deleting, and rearranging?

The transposon problem can be viewed as one aspect of a larger problem in genome evolution: why does duplicated DNA persist? Duplications are a by-product of the properties of the DNA replication and recombination machinery. Short stretches of homology suffice to give rise to duplications by slippage during replication, homology-dependent unequal crossing-over, and double-strand breakage/repair (Gorbunova and Levy, 1997; Liang *et al.*, 1998). But duplications are problematical. Once a duplication exists, the mechanisms that generated it also permit unequal crossing over between identical repeats (Anderson and Roth, 1977; Perelson and Bell, 1977; Koch, 1979). Prokaryotes readily duplicate genetic material, but do not retain duplications (Perelson and Bell, 1977; Romero and Palacios, 1997). Thus the ability of genomes to expand by duplication is predicated on their ability to sequester homologous sequences from the cell's recombination machinery and retain them, which may necessitate the invention of mechanisms to recognize and differentially mark duplications. Some lower eukaryotes, including *Neurospora crassa* (Selker and Garrett, 1988; Selker, 1997) and *Ascobolus immersus* (Rossignol and Faugeron, 1994, 1995), have the capacity to recognize and mark duplicated sequences by methylating them. Sequence methylation silences transcription, enhances the mutability of the duplicated sequence, and inhibits recombination (Selker, 1997; Maloisel and Rossignol, 1998).

Some years ago, Adrian Bird pointed out that there are two evolutionary discontinuities in the average number of genes per genome (Bird, 1995). The first is an increase between prokaryotes and eukaryotes and the second is between invertebrates and vertebrates. He suggests that with a given cellular organization there may be an upper limit on the tolerable gene numbers imposed by the imprecision of the biochemical mechanisms controlling gene expression. He suggested that the transcriptional "noise reduction" mechanisms that arose at the prokaryote/eukaryote boundary were the nuclear envelope, chromatin, and separation of the transcriptional and translational machinery, as well as RNA processing, capping, and polyadenylation to discriminate authentic from spurious transcripts. He proposed that genome-wide DNA methylation is the novel "noise reduction" mechanism that has permitted the additional quantal leap in gene numbers characteristic of vertebrates.

HOMOLOGY-DEPENDENT GENE SILENCING

The results of both classical and contemporary studies on the silencing of redundant gene copies in plants suggests that both methylation and other epigenetic mechanisms reflect a much more fundamental ability to recognize and regulate gene dosage (Kooter *et al.*, 1999). McClintock understood that transposable elements exist in a genetically intact, but cryptic form in the genome and she carried out genetic analyses of *Spm* transposons undergoing epigenetic changes in their ability to transpose (McClintock, 1962). We later found that the genetically inactive *Spm* transposons are methylated in critical regulatory sequences (Banks and Fedoroff, 1989). It also has been reported that the large intergenic retrotransposon blocks in maize are extensively methylated (Bennetzen *et al.*, 1994).

The discovery that the introduction of a transgene can lead to the transcriptional silencing and methylation of both the introduced gene and its endogenous homolog brought gene silencing mechanisms under intense study (Park *et al.*, 1996; Kooter *et al.*, 1999). Genes can be silenced both transcriptionally and posttranscriptionally consequent on the introduction of additional copies. Posttranscriptional silencing appears to be caused by RNA destabilization, whereas transcriptional gene silencing involves DNA methylation (Vaucheret *et al.*, 1998; Kooter *et al.*, 1999). There is also some evidence that posttranscriptional silencing triggers DNA methylation (Wassenegger *et al.*, 1994). The results of recent studies on the classical epigenetic phenomenon of *R* locus paramutation in maize have revealed that local endoreduplication of a chromosomal segment both triggers silencing and can render the endoreduplicated locus capable of silencing an active allele of the gene on a homolog (Kermicle *et al.*, 1995). Similar observations have been made with transgenes, as well as endogenous gene duplications at different chromosomal locations in tobacco and *Arabidopsis* (Matzke *et al.*, 1994; Luff *et al.*, 1999).

A connection between gene silencing and chromatin structure has come from the analysis of mutants altered in methylation and in transcriptional gene silencing (Jeddeloh *et al.*, 1998; Kooter *et al.*, 1999). Both approaches have identified alleles of the *ddm1* locus, which encodes a protein with homology to known chromatin remodeling proteins. This suggests that the repressive mechanisms of DNA methylation and chromatin structure are linked in plants, as they are in animal cells (Ng *et al.*, 1999; Wade *et al.*, 1999). Evidence also is accumulating that double-stranded RNA mediates gene silencing, both in plants and in a variety of other organisms (Waterhouse *et al.*, 1998; Fire, 1999). Analyses of mutants altered in posttranscriptional gene silencing in *Neurospora* have identified an RNA-dependent RNA polymerase, as well as a RecQ helicase-like protein, homologs of which are known to be involved in DNA repair and recombination (Cogoni and Macino, 1999a, b).

THE ORIGIN OF TRANSPOSONS AND METHYLATION

Although it is popular to assert that transposons are genomic "parasites" and that DNA methylation evolved to control them, I suggest that the evidence supports neither notion (Yoder *et al.*, 1997). The idea that transposons as parasitic, selfish DNA comes from a couple of essays written two decades ago, one by Doolittle and Sapienza (1980) and one by Orgel and Crick (1980). These essays sought rightly to free us from the then prevalent notion that genome structure is optimized by phenotypic selection. But the persistence of the moniker "selfish DNA" has become an impediment to further understanding of the origin, historical contribution, and contemporary role of transposons in chromosome structure.

Transposons may be an inevitable by-product of the evolution of sequence-specific endonucleases. Complete transposons have been shown to arise from a single cleavage site and an endonuclease gene (Morita *et al.*, 1999). Although the successful constitution of a transposon from the recognition sequences used in Ig gene rearrangement and the RAG1 and RAG2 proteins was interpreted as evidence that the V(D)J recombination system evolved from an ancient mobile DNA element, the fact is that the critical components of a transposon and a site-specific rearrangement system are the same (Hiom *et al.*, 1998). Thus questions about the origin of certain kinds of transposons may devolve to questions about the association of sequence-specific DNA binding domains with endonuclease domains.

Although the majority of methylated sequences in a genome can be transposable elements, the view that DNA methylation evolved to control transposons seems implausible in the light of evidence that duplications of any kind trigger methylation in organisms that methylate DNA (Yoder *et al.*, 1997; Garrick *et al.*, 1998; Selker, 1999). And organisms that do not methylate DNA also have mechanisms for detecting duplications and sequestering repeats (Pirrotta, 1997; Sherman and Pillus, 1997; Henikoff, 1998). Genome expansion by duplication is predicated on preventing illegitimate recombination between duplicated sequences. Although different eukaryotic lineages appear to have invented different mechanisms, what is common to repeat-induced silencing in all eukaryotes is the stable packaging of DNA into "repressive" chromatin. It may be that the evolution of mechanisms that recognize, mark, and sequester duplications into repressive chromatin structures, among which some involve DNA methylation, were the prerequisites for expansion of genomes by endoreduplication at all scales. The additional benefit of such "repressive" mechanisms in minimizing spurious transcription could be secondary sequelae. Because sequence duplication is inherent in transposition, the ability to recognize and repress duplications would serve to minimize both the activity and the adverse impacts of transposons, rendering them genetically invisible and favoring their gradual accumulation.

An important and as yet underappreciated property of compacted, inactive genomic regions is their ability to impose their organization on adjacent, as well as nonadjacent, active regions, often in a homology-dependent manner. This is evidenced in position effect variegation in *Drosophila*, an organism that does not methylate its DNA, as well as in plant paramutation, which involves DNA methylation (Kermicle *et al.*, 1995; Henikoff, 1998). What has been learned recently from analyzing gene silencing and paramutation suggests that it does not take many tandem duplications to trigger the formation of a compacted, silenced region. A silenced region then may become a "sink" for insertions within it, as well as a silencer for homologous sequences located adjacent to it or elsewhere in the genome (Henikoff, 1998; Jakowitsch *et al.*, 1999).

CONCLUSIONS

The key to understanding the prevalence of transposons in contemporary genomes, as well as their genetic invisibility, therefore may lie not in transposons themselves, but in the much more fundamental capacity of eukaryotic organisms to recognize and sequester duplications. Whether transposons, retrotransposons, and other repetitive elements accumulate extensively in a given evolutionary lineage may depend on several factors, among them the efficiency of repressive mechanisms and the rate at which the sequences undergo mutational and deletional decay. For example, methylation of C residues enhances the mutability of CG base pairs, hence methylation accelerates the divergence rate of newly arising duplications. This happens in an extreme form in *Neurospora*, in which many methylated CGs are mutated in the span of a single generation, and at more measured rates in plants and mammals, in which the mutability can be detected by virtue of a marked deficiency of the base pairs and triplets that are normally methylated (Selker, 1990; SanMiguel *et al.*, 1998; Wang *et al.*, 1998).

The burgeoning analyses of genomes also makes it evident that repressive mechanisms are imperfect. However slowly, genomes are inexorably restructured by transposition and rearrangements arising from ectopic interactions between dispersed transposons. Thus there is little remaining doubt that transposons are central to genome evolution. What is less clear is the relationship between genome restructuring and morphological change. We know that the magnitude of the morphological differences between species does not necessarily reflect the magnitude of the genetic or chromosomal differences between them. It recently has become evident, for example, that the marked morphological and developmental differences between teosinte and maize are attributable to a very small number of genes and that for some genes, the differences are regulatory, rather than structural (Doebley *et al.*, 1997; White and Doebley, 1998). It is also well known that genes are expressed differently depending on their chromosomal position. But what remains to be discovered is the extent to which chromosomal restructuring contributes to organismal evolution.

REFERENCES

Ananiev, E. V., Phillips, R. L. & Rines, H. W. (1998a) Complex structure of knob DNA on maize chromosome 9. Retrotransposon invasion into heterochromatin. *Genetics* 149, 2025–2037.

Ananiev, E. V., Phillips, R. L. & Rines, H. W. (1998b) A knob-associated tandem repeat in maize capable of forming fold-back DNA segments: are chromosome knobs megatransposons? *Proc. Natl. Acad. Sci. USA* 95, 10785–10790.

Anderson, R. P. & Roth, J. R. (1977) Tandem genetic duplications in phage and bacteria. *Annu. Rev. Microbiol.* 31, 473–505.

Banks, J. A. & Fedoroff, N. (1989) Patterns of developmental and heritable change in methylation of the *Suppressor-mutator* transposable element. *Dev. Genet.* 10, 425–437.

Bennett, M. D., Smith, J. B. & Heslop-Harrison, J. S. (1982) Nuclear DNA amounts in angiosperms. *Proc. Natl. Acad. Sci. USA* 216, 179–199.

Bennetzen, J. L. & Freeling, M. (1997) The unified grass genome: synergy in synteny. *Genome Res.* 7, 301–306.

Bennetzen, J. L. & Kellogg, E. A. (1997) Do plants have a one-way ticket to genomic obesity? *Plant Cell* 9, 1507–1514.

Bennetzen, J. L., SanMiguel, P, Chen, M., Tikhonov, A., Francki, M. & Avramova, Z. (1998) Grass genomes. *Proc. Natl. Acad. Sci. USA* 95, 1975–1978.

Bennetzen, J. L., Schrick, K., Springer, P. S., Brown, W. E. & SanMiguel, P. (1994) Active maize genes are unmodified and flanked by diverse classes of modified highly repetitive DNA. *Genome* 37, 565–576.

Bird, A. P. (1995) Gene number, noise reduction and biological complexity. *Trends Genet.* 11, 94–100.

Buckler, E. S. T., Phelps-Durr, T. L., Buckler, C. S., Dawe, R. K., Doebley, J. F. & Holtsford, T. P. (1999) Meiotic drive of chromosomal knobs reshaped the maize genome. *Genetics* 153, 415–426.

Bureau, T. E., Ronald, P. C. & Wessler, S. R. (1996) A computer-based systematic survey reveals the predominance of small inverted-repeat elements in wild-type rice genes. *Proc. Natl. Acad. Sci. USA* 93, 8524–8529.

Bureau, T. E. & Wessler, S. R. (1992) Tourist: a large family of small inverted repeat elements frequently associated with maize genes. *Plant Cell* 4, 1283–1294.

Bureau, T. E. & Wessler, S. R. (1994a) Mobile inverted-repeat elements of the Tourist family are associated with the genes of many cereal grasses. *Proc. Natl. Acad. Sci. USA* 91, 1411–1415.

Bureau, T. E. & Wessler, S. R. (1994b) Stowaway: a new family of inverted repeat elements associated with the genes of both monocotyledonous and dicotyledonous plants. *Plant Cell* 6, 907–916.

Carlson, E. A. (1966) *The Gene: A Critical History* (W. B. Saunders Co., Philadelphia).

Cogoni, C. & Macino, G. (1999a) Gene silencing in *Neurospora crassa* requires a protein homologous to RNA-dependent RNA polymerase. *Nature* 399, 166–169.

Cogoni, C. & Macino, G. (1999b) Posttranscriptional gene silencing in *Neurospora* by a RecQ DNA helicase. *Science* 286, 2342–2344.

Consonni, G., Geuna, F., Gavazzi, G. & Tonelli, C. (1993) Molecular homology among members of the R gene family in maize. *Plant J.* 3, 335–346.

Courage-Tebbe, U., Doring, H. P., Fedoroff, N. & Starlinger, P. (1983) The controlling element *Ds* at the Shrunken locus in *Zea mays*: structure of the unstable *sh-m5933* allele and several revertants. *Cell* 34, 383–393.

De Vries, H. (1905) *Species and Varieties: Their Origin by Mutation* (Open Court Publishing Co., Chicago), 2nd Ed.

Doebley, J., Stec, A. & Hubbard, L. (1997) The evolution of apical dominance in maize. *Nature* 386, 485–488.

Doolittle, W. F. & Sapienza, C. (1980) Selfish genes, the phenotype paradigm and genome evolution. *Nature* 284, 601–603.

Emerson, R. A. (1914) The inheritance of a recurring somatic variation in variegated ears of maize. *Am. Naturalist* 48, 87–115.

Emerson, R. A. (1917) Genetical studies of variegated pericarp in maize. *Genetics* 2, 1–35.

Emerson, R. A. (1929) The frequency of somatic mutation in variegated pericarp of maize. *Genetics* 14, 488–511.

Fedoroff, N. (1989) Maize transposable elements. In *Mobile DNA*, eds. Howe, M. & Berg, D. (American Society for Microbiology, Washington), pp. 375–411.

Fedoroff, N., Furtek, D. & Nelson, O. (1984) Cloning of the *Bronze* locus in maize by a simple and generalizable procedure using the transposable controlling element *Ac*. *Proc. Natl. Acad. Sci. USA* 81, 3825–3829.

Fedoroff, N., Wessler, S. & Shure, M. (1983) Isolation of the transposable maize controlling elements *Ac* and *Ds*. *Cell* 35, 243–251.

Fedoroff, N. V. (1998) The discovery of transposable elements. In *Discoveries in Plant Biology* (World Scientific Publishing Co., Singapore), pp. 89–104.

Feldman, M., Liu, B., Segal, G., Abbo, S., Levy, A. A. & Vega, J. M. (1997) Rapid elimination of low-copy DNA sequences in polyploid wheat: a possible mechanism for differentiation of homoeologous chromosomes. *Genetics* 147, 1381–1387.

Fire, A. (1999) RNA-triggered gene silencing. *TIG* 15, 358–363.

Flavell, A. J. (1992) Ty1-copia group retrotransposons and the evolution of retroelements in the eukaryotes. *Genetica* 86, 203–214.

Gale, M. D. & Devos, K. M. (1998) Plant comparative genetics after 10 years. *Science* 282, 656–659.

Garrick, D., Fiering, S., Martin, D. I. & Whitelaw, E. (1998) Repeat-induced gene silencing in mammals. *Nat. Genet.* 18, 56–59.

Giroux, M. J., Clancy, M., Baier, J., Ingham, L., McCarty, D. & Hannah, L. C. (1994) De novo synthesis of an intron by the maize transposable element *Dissociation*. *Proc. Natl. Acad. Sci. USA* 91, 12150–12154.

Goldschmidt, R. (1938) *Physiological Genetics* (McGraw-Hill, New York).

Gorbunova, V. & Levy, A. A. (1997) Non-homologous DNA end joining in plant cells is associated with deletions and filler DNA insertions. *Nucleic Acids Res.* 25, 4650–4657.

Henikoff, S. (1998) Conspiracy of silence among repeated transgenes. *Bioessays* 20, 532–535.

Hiom, K., Melek, M. & Gellert, M. (1998) DNA transposition by the RAG1 and RAG2 proteins: a possible source of oncogenic translocations. *Cell* 94, 463–470.

Howe, M. & Berg, D. (1989) *Mobile DNA* (American Soc. Microbiol., Washington).

Jakowitsch, J., Papp, I., Moscone, E. A., van der Winden, J., Matzke, M. & Matzke, A. J. (1999) Molecular and cytogenetic characterization of a transgene locus that induces silencing and methylation of homologous promoters in trans. *Plant J.* 17, 131–140.

Jeddeloh, J. A., Bender, J. & Richards, E. J. (1998) The DNA methylation locus *DDM1* is required for maintenance of gene silencing in *Arabidopsis. Genes Dev.* 12, 1714–1725.

Kermicle, J. L., Eggleston, W. B. & Alleman, M. (1995) Organization of paramutagenicity in *R-stippled* maize. *Genetics* 141, 361–372.

Kim, H. Y., Schiefelbein, J. W., Raboy, V., Furtek, D. B. & Nelson, O., Jr. (1987) RNA splicing permits expression of a maize gene with a defective *Suppressor-mutator* transposable element insertion in an exon. *Proc. Natl. Acad. Sci. USA* 84, 5863–5867.

Koch, A. L. (1979) Selection and recombination in populations containing tandem multiple genes. *J. Mol. Evol.* 14, 273–285.

Kooter, J. M., Matzke, M. A. & Meyer, P. (1999) Listening to the silent genes: transgene silencing, gene regulation and pathogen control. *Trends Plant Sci.* 4, 340–347.

Lagercrantz, U. (1998) Comparative mapping between *Arabidopsis thaliana* and *Brassica nigra* indicates that *Brassica* genomes have evolved through extensive genome replication accompanied by chromosome fusions and frequent rearrangements. *Genetics* 150, 1217–1228.

Lahn, B. T. & Page, D. C. (1999a) Four evolutionary strata on the human X chromosome. *Science* 286, 964–967.

Lahn, B. T. & Page, D. C. (1999b) Retroposition of autosomal mRNA yielded testis-specific gene family on human Y chromosome. *Nat. Genet.* 21, 429–433.

Leister, D., Kurth, J., Laurie, D. A., Yano, M., Sasaki, T., Devos, K., Graner, A. & Schulze-Lefert, P. (1998) Rapid reorganization of resistance gene homologues in cereal genomes. *Proc. Natl. Acad. Sci. USA* 95, 370–375.

Leitch, I. J. & Bennett, M. D. (1997) Polyploidy in angiosperms. *Trends Plant Sci.* 2, 470–476.

Liang, F., Han, M., Romanienko, P. J. & Jasin, M. (1998) Homology-directed repair is a major double-strand break repair pathway in mammalian cells. *Proc. Natl. Acad. Sci. USA* 95, 5172–5177.

Ludwig, S. R., Habera, L. F., Dellaporta, S. L. & Wessler, S. R. (1989) *Lc*, a member of the maize R gene family responsible for tissue-specific anthocyanin production, encoded a protein similar to transcriptional activators and contains the *myc*-homology region. *Proc. Natl. Acad. Sci. USA* 86, 7092–7096.

Luff, B., Pawlowski, L. & Bender, J. (1999) An inverted repeat triggers cytosine methylation of identical sequences in *Arabidopsis. Mol. Cell* 3, 505–511.

Maloisel, L. & Rossignol, J. L. (1998) Suppression of crossing-over by DNA methylation in *Ascobolus. Genes Dev.* 12, 1381–1389.

Martienssen, R. & Irish, V. (1999) Copying out our ABCs: the role of gene redundancy in interpreting genetic hierarchies. *Trends Genet.* 15, 435–437.

Masson, P., Surosky, R., Kingsbury, J. A. & Fedoroff, N. V. (1987) Genetic and molecular analysis of the *Spm*-dependent *a-m2* alleles of the maize *a* locus. *Genetics* 117, 117–137.

Matzke, A. J., Neuhuber, F., Park, Y. D., Ambros, P. F. & Matzke, M. A. (1994) Homology-dependent gene silencing in transgenic plants: epistatic silencing loci contain multiple copies of methylated transgenes. *Mol. Gen. Genet.* 244, 219–229.

McClintock, B. (1945) Neurospora: I. Preliminary observations of the chromosomes of *Neurospora crassa. Amer. J. Bot.* 32, 671–678.

McClintock, B. (1946) Maize genetics. *Carnegie Inst. Wash. Yr. Bk.* 45, 176–186.

McClintock, B. (1947) Cytogenetic studies of maize and *Neurospora. Carnegie Inst. Wash. Yr. Bk.* 46, 146–152.

McClintock, B. (1948) Mutable loci in maize. *Carnegie Inst. Wash. Yr. Bk.* 47, 155–169.

McClintock. B. (1949) Mutable loci in maize. *Carnegie Inst. Wash. Yr. Bk.* 48, 142–154.

McClintock, B. (1951) Mutable loci in maize. *Carnegie Inst. Wash. Yr. Bk.* 50, 174–181.

McClintock, B. (1954) Mutations in maize and chromosomal aberrations in *Neurospora. Carnegie Inst. Wash. Yr. Bk.* 53, 254–260.

McClintock, B. (1962) Topographical relations between elements of control systems in maize. *Carnegie Inst. Wash. Yr. Bk.* 61, 448–461.

McClintock. B. (1978) Mechanisms that rapidly reorganize the genome. *Stadler Genet. Symp.* 10, 25–48.

McClintock. B. (1987) *The Discovery and Characterization of Transposable Elements* (Garland Publishing, Inc., New York).

Michelmore, R. W. & Meyers, B. C. (1998) Clusters of resistance genes in plants evolve by divergent selection and a birth-and-death process. *Genome Res.* 8, 1113–1130.

Morita, M., Umemoto, A., Watanabe, H., Nakazono, N. & Sugino, Y. (1999) Generation of new transposons in vivo: an evolutionary role for the "staggered" head-to-head dimer and one-ended transposition. *Mol. Gen. Genet.* 261, 953–957.

Ng, H.-H., Zhang, Y., Hedrich, B., Johnson, C. A., Turner, B. M., Erdjument-Bromage, H., Tempst, P., Reinberg, D. & Bird, A. (1999) MBD2 is a transcriptional repressor belonging to the MeCP1 histone deacetylase complex. *Nature Genet.* 23, 58–61.

Noma, K., Ohtsubo, E. & Ohtsubo, H. (1999) Non-LTR retrotransposons (LINEs) as ubiquitous components of plant genomes. *Mol. Gen. Genet.* 261, 71–79.

Orgel, L. E. & Crick, F. H. C. (1980) Selfish DNA: the ultimate parasite. *Nature* 284, 604–607.

Park, Y. D., Papp, I., Moscone, E. A., Iglesias, V. A., Vaucheret, H., Matzke, A. J. & Matzke, M. A. (1996) Gene silencing mediated by promoter homology occurs at the level of transcription and results in meiotically heritable alterations in methylation and gene activity. *Plant J.* 9, 183–194.

Pereira, A., Schwarz-Sommer, Z., Gierl, A., Bertram, I., Peterson, P. A. & Saedler, H. (1985) Genetic and molecular analysis of the *Enhancer (En)* transposable element system of *Zea mays*. *EMBO J.* 4, 17–23.

Perelson, A. S. & Bell, G. I. (1977) Mathematical models for the evolution of multigene families by unequal crossing over. *Nature* 265, 304–310.

Perrot, G. H. & Cone, K. C. (1989) Nucleotide sequence of the maize *R-S* gene. *Nucleic Acids Res.* 17, 8003.

Petrov, D. A. & Hartl, D. L. (1998) High rate of DNA loss in the *Drosophila melanogaster* and *Drosophila virilis* species groups. *Mol. Biol. Evol.* 15, 293–302.

Pirrotta, V. (1997) Chromatin-silencing mechanisms in *Drosophila* maintain patterns of gene expression. *Trends Genet.* 13, 314–318.

Purugganan, M. D. & Wessler, S. R. (1994) Molecular evolution of *magellan*, a maize Ty3/gypsy-like retrotransposon. *Proc. Natl. Acad. Sci. USA* 91, 11674–11678.

Rabinowicz, P. D., Braun, E. L., Wolfe, A. D., Bowen, B. & Grotewold, E. (1999) Maize *R2R3 Myb* genes: Sequence analysis reveals amplification in the higher plants. *Genetics* 153, 427–444.

Rhoades, M. M. (1936) The effect of varying gene dosage on aleurone colour in maize. *J. Genet.* 33, 347–354.

Rhoades, M. M. (1938) Effect of the *Dt* gene on the mutability of the a_1 allele in maize. *Genetics* 23, 377–397.

Riechmann, J. L. & Meyerowitz, E. M. (1998) The AP2/EREBP family of plant transcription factors. *Biol. Chem.* 379, 633–646.

Romero, D. & Palacios, R. (1997) Gene amplification and genome plasticity in prokaryotes. *Annu. Rev. Genet.* 31, 91–111.

Rossignol, J. L. & Faugeron, G. (1994) Gene inactivation triggered by recognition between DNA repeats. *Experientia* 50, 307–317.

Rossignol, J. L. & Faugeron, G. (1995) MIP: an epigenetic gene silencing process in *Ascobolus immersus*. *Curr. Top. Microbiol. Immunol.* 197, 179–191.

SanMiguel, P., Gaut, B. S., Tikhonov, A., Nakajima, Y. & Bennetzen, J. L. (1998) The paleontology of intergene retrotransposons of maize. *Nat. Genet.* 20, 43–45.

SanMiguel, P. Tikhonov, A., Jin, Y. K., Motchoulskaia, N., Zakharov, D., Melake-Berhan, A., Springer, P. S., Edwards, K. J., Lee, M., Avramova, Z, & Bennetzen, J. L. (1996) Nested retrotransposons in the intergenic regions of the maize genome. *Science* 274, 765–768.

Saxena, R., Brown, L. G., Hawkins, T., Alagappan, R. K., Skaletsky, H., Reeve, M. P., Reijo, R., Rozen, S., Dinulos, M. B., Disteche, C. M. & Page, D. C. (1996) The DAZ gene cluster on the human Y chromosome arose from an autosomal gene that was transposed, repeatedly amplified and pruned. *Nat. Genet.* 14, 292–299.

Schwartz, A., Chan, D. C., Brown, L. G., Alagappan, R., Pettay, D., Disteche, C., McGillivray, B., de la Chapelle, A. & Page, D. C. (1998) Reconstructing hominid Y evolution: X-homologous block, created by X-Y transposition, was disrupted by Yp inversion through LINE-LINE recombination. *Hum. Mol. Genet.* 7, 1–11.

Schwarz-Sommer, Z., Gierl, A., Cuypers, H., Peterson, P. A. & Saedler, H. (1985a) Plant transposable elements generate the DNA sequence diversity needed in evolution. *EMBO J.* 4, 591–597.

Schwarz-Sommer, Z., Gierl, A., Berndtgen, R. & Saedler, H. (1985b) Sequence comparison of 'states' of *a1-m1* suggest a model of *Spm (En)* action. *EMBO J.* 4, 2439–2443.

Schwarz-Sommer, Z., Leclercq, L., Gobel, E. & Saedler, H. (1987) *CIN4*, an insert altering the structure of the *A1* gene in *Zea mays*, exhibits properties of nonviral retrotransposons. *EMBO J.* 6, 3873–3880.

Selker, E. U. (1990) Premeiotic instability of repeated sequences in *Neurospora crassa. Annu. Rev. Genet.* 24, 579–613.

Selker, E. U. (1997) Epigenetic phenomena in filamentous fungi: useful paradigms or repeat-induced confusion? *Trends Genet.* 13, 296–301.

Selker, E. U. (1999) Gene silencing: repeats that count. *Cell* 97, 157–160.

Selker, E. U. & Garrett, P. W. (1988) DNA sequence duplications trigger gene inactivation in *Neurospora crassa. Proc. Natl. Acad. Sci. USA* 85, 6870–6874.

Shepherd, N. S., Schwarz-Sommer, Z., Blumberg vel Spalve, J., Gupta, M., Wienand, U. & Saedler, H. (1984) Similarity of the *Cin1* repetitive family of Zea mays to eukaryotic transposable elements. *Nature* 307, 185–187.

Sherman, J. M. & Pillus, L. (1997) An uncertain silence. *Trends Genet.* 13, 308–313.

Song, K., Lu, P., Tang, K. & Osborn, T. C. (1995) Rapid genome change in synthetic polyploids of *Brassica* and its implications for polyploid evolution. *Proc. Natl. Acad. Sci. USA* 92, 7719–7723.

Spell, M. L., Baran, G. & Wessler, S. R. (1988) An RFLP adjacent to the maize waxy gene has the structure of a transposable element. *Mol. Gen. Genet.* 211, 364–366.

Stebbins, G. L., Jr. (1950) *Variation and Evolution in Plants* (Columbia Univ. Press, New York).

Suoniemi, A., Schmidt, D. & Schulman, A. H. (1997) *BARE-1* insertion site preferences and evolutionary conservation of RNA and cDNA processing sites. *Genetica* 100, 219–230.

Suoniemi, A. Tanskanen, J. & Schulman, A. H. (1998) Gypsy-like retrotransposons are widespread in the plant kingdom. *Plant. J.* 13, 699–705.

Tanksley, S. D., Bernatzky, R., Lapitan, N. L. & Prince, J. P. (1988) Conservation of gene repertoire but not gene order in pepper and tomato. *Proc. Natl. Acad. Sci. USA* 85, 6419–6423.

Tanksley, S. D., Ganal, M. W., Prince, J. P., de Vicente, M. C., Ponierbale, M. W., Broun, P., Fulton, T. M. Giovannoni, J. J., Grandillo, S., Martin, G. B., Messeguer, R., Miller, J. C., Miller, L., Paterson, A.H., Pineda, O., Roder, M. S., Wing, R. A., Wu, W. & Young, N. D. (1992) High density molecular linkage maps of the tomato and potato genomes. *Genetics* 132, 1141–1160.

Thomas, C. A. (1971) The genetic organisation of chromosomes. *Annu. Rev. Genet.* 5, 237–256.

Tikhonov, A. P., SanMiguel, P. J., Nakajima, Y., Gorenstein, N. M., Bennetzen, J. L. & Avramova, Z. (1999) Colinearity and its exceptions in orthologous *adh* regions of maize and sorghum. *Proc. Natl. Acad. Sci. USA* 96, 7409–7414.

Vaucheret, H., Beclin, C., Elmayan, T., Feuerbach, F., Godon, C., Morel, J. B., Mourrain, P., Palauqui, J. C. & Vernhettes, S. (1998) Transgene-induced gene silencing in plants. *Plant J.* 16, 651–659.

Wade, P. A., Gegonne, A., Jones, P. L. Ballestar, E., Aubry, F. & Wolffe, A. P. (1999) Mi-2 complex couples DNA methylation to chromatin remodeling and histone deacetylation. *Nat. Genet.* 23, 62–66.

Walker, E. L., Robbins, T. P., Bureau, T. E., Kermicle, J. & Dellaporta, S. L. (1995) Transposon-mediated chromosomal rearrangements and gene duplications in the formation of the maize *R-r* complex. *EMBO J.* 14, 2350–2363.

Wang, D. G., Fan, J. B., Siao, C. J., Berno, A., Young, P., Sapolsky, R., Ghandour, G., Perkins, N., Winchester, E., Spencer, J., Kruglyak, L., Stein, L., Hsie, L., Topaloglou, T., Hubbell, E., Robinson, E., Mittmann, M., Morris, M. S., Shen, N. P., Kilburn, D., Rioux, J., Nusbaum, C., Rozen, S., Hudson, T. J., Lipshutz, R., Chee, M. & Lander, E. S. (1998) Large-scale identification, mapping, and genotyping of single-nucleotide polymorphisms in the human genome. *Science* 280, 1077–1082.

Wassenegger, M., Heimes, S., Riedel, L. & Sanger, H. L. (1994) RNA-directed *de novo* methylation of genomic sequences in plants. *Cell* 76, 567–576.

Waterhouse, P. M., Graham, M. W. & Wang, M. B. (1998) Virus resistance and gene silencing in plants can be induced by simultaneous expression of sense and antisense RNA. *Proc. Natl. Acad. Sci. USA* 95, 13959–13964.

Watson, J. D. & Crick, F. H. C. (1953) Molecular structure of nucleic acids. *Nature* 171, 737–738.

Wessler, S. R. (1989) The splicing of maize transposable elements from pre-mRNA—a minireview. *Gene* 82, 127–133.

Wessler, S. R., Bureau, T. E. & White, S. E. (1995) LTR-retrotransposons and MITEs: important players in the evolution of plant genomes. *Curr. Opin. Genet. Dev.* 5, 814–821.

White, S. & Doebley, J. (1998) Of genes and genomes and the origin of maize. *Trends Genet.* 14, 327–332.

White, S. E., Habera, L. F. & Wessler, S. R. (1994) Retrotransposons in the flanking regions of normal plant genes: a role for *copia*-like elements in the evolution of gene structure and expression. *Proc. Natl. Acad. Sci. USA* 91, 11792–11796.

Yoder, J. A., Walsh, C. P. & Bestor, T. H. (1997) Cytosine methylation and the ecology of intragenomic parasites. *Trends Genet.* 13, 335–340.

11

Maize as a Model for the Evolution of Plant Nuclear Genomes

BRANDON S. GAUT, MAUD LE THIERRY D'ENNEQUIN, ANDREW S. PEEK, AND MARK C. SAWKINS

The maize genome is replete with chromosomal duplications and repetitive DNA. The duplications resulted from an ancient polyploid event that occurred over 11 million years ago. Based on DNA sequence data, the polyploid event occurred after the divergence between sorghum and maize, and hence the polyploid event explains some of the difference in DNA content between these two species. Genomic rearrangement and diploidization followed the polyploid event. Most of the repetitive DNA in the maize genome is retrotransposable elements, and they comprise 50% of the genome. Retrotransposon multiplication has been relatively recent—within the last 5–6 million years—suggesting that the proliferation of retrotransposons has also contributed to differences in DNA content between sorghum and maize. There are still unanswered questions about repetitive DNA, including the distribution of repetitive DNA throughout the genome, the relative impacts of retrotransposons and chromosomal duplication in plant genome evolution, and the hypothesized correlation of duplication events with transposition. Population genetic processes also affect the evolution of genomes. We discuss how centromeric genes should, in theory,

Department of Ecology and Evolutionary Biology, University of California, Irvine, CA 92697-2525

This paper was presented at the National Academy of Sciences colloquium "Variation and Evolution in Plants and Microorganisms: Toward a New Synthesis 50 Years After Stebbins," held January 27–29, 2000, at the Arnold and Mabel Beckman Center in Irvine, CA.

Abbreviations: mya, million years ago; LTR, long terminal repeat.

contain less genetic diversity than noncentromeric genes. In addition, studies of diversity in the wild relatives of maize indicate that different genes have different histories and also show that domestication and intensive breeding have had heterogeneous effects on genetic diversity across genes.

Genomic technologies have produced a wealth of data on the organization and structure of genomes. These data range from extensive marker-based genetic maps to "chromosome paintings" based on fluorescent *in situ* hybridization to complete genomic DNA sequences. Although genomic approaches have changed the amount and type of data, the challenges of interpreting genomic data in an evolutionary context have changed little from the challenges faced by Stebbins (1950) and the coauthors of the evolutionary synthesis. The challenges are to infer the mechanisms of evolution and to construct a comprehensive picture of evolutionary change.

In this paper, we will focus on the processes that contribute to the evolution of plant nuclear genomes by using maize (*Zea mays*) as a model system. In some respects, it is premature to discuss the evolution of plant genomes, because the pending completion of the *Arabidopsis* (*Arabidopsis thaliana*) genome, with rice (*Oryza sativa*) following, is sure to unlock many mysteries about plant genome evolution. However, it must be remembered that *Arabidopsis* and rice are being sequenced precisely because their genomes are atypically small and streamlined. Even after these genomes are sequenced, it will still be a tremendous challenge to understand the evolution of plant nuclear genomes, like the maize genome, for which entire DNA sequences will not be readily available.

Maize is a member of the grass family (Poaceae). The grasses represent a range of genome size and structural complexity, with rice on one extreme. A diploid with 12 chromosomes ($2n = 24$), rice has one of the smallest plant genomes, with only 0.9 pg of DNA per 2C nucleus (Fig. 1). Other grass species exhibit far larger genomes. Wheat, for example, is a hexaploid with 21 chromosomes ($2n = 42$) and a haploid DNA content of 33.1 pg (Bennett and Leitch, 1995). Genera like *Saccharum* (sugarcane) and *Festuca* are even more complicated, displaying wide variation in ploidy level and over 100 chromosomes in some species. As a diploid with 10 chromosomes ($2n = 20$) and a 2C genome content roughly 6-fold larger than rice, maize lies somewhere in the middle of grass genome size and structural complexity (Fig. 1).

This paper focuses on the impact of chromosomal duplication, transposition, and nucleotide substitution on the evolution of the maize genome. We will discuss chromosomal duplication and transposition separately and will pay particular attention to their effects on DNA content. Nucleotide

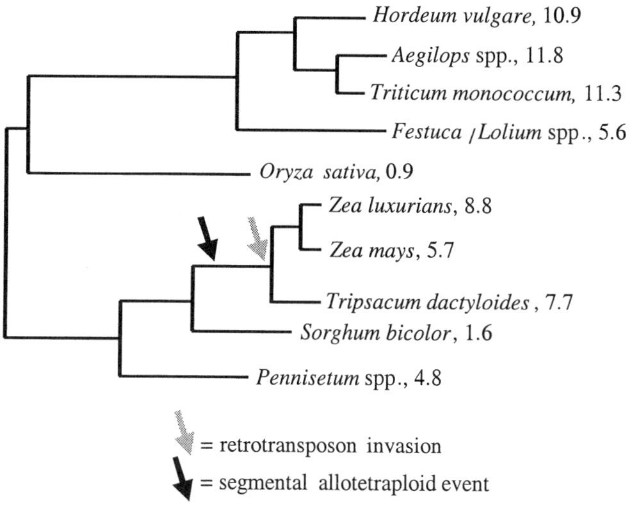

FIGURE 1. A phylogeny of diploid grass species. Numerical values next to species names represent the 2C genome content of the species, measured in picograms. The phylogeny and genome content information is taken from figure 1 of Bennetzen and Kellogg (1997). The arrows represent the hypothesized timing of evolutionary events.

substitution will be discussed in the context of genetic diversity. Patterns of genetic diversity provide insight into the population genetic processes that act on different regions of the genome and thus uncover the evolutionary forces that act on genomes. We focus on maize throughout the paper but also generalize to other species when appropriate.

POLYPLOIDY AND CHROMOSOMAL DUPLICATION

An Ancient Polyploid Origin

The first hints of the complex organization of the maize genome came from cytological studies. Although maize is diploid, early studies by McClintock (McClintock 1930, 1933) demonstrated the association of non-homologous chromosomes during meiosis. Later studies documented the formation of bivalents and multivalents in maize haploids (Snope, 1967; Ting, 1966). Altogether, cytological observations suggested that the maize genome contains extensive regions of homology, probably reflecting chromosomal duplications.

Evidence for chromosomal duplication also came from linkage information. In 1951, Rhoades (1951, 1955) noted that some regions of linkage maps did not contain mutants, and he proposed that the lack of mutants reflected genetic redundancy caused by chromosomal duplication. Rhoades' proposal has since been supported by molecular data. For example, isozyme studies have documented the presence of duplicated, linked loci in maize (Goodman *et al.*, 1980; McMillin and Scandalios, 1980; Wendel *et al.*, 1986, 1989), and restriction fragment length polymorphism mapping studies have shown that many markers map to two or more chromosomal locations (Davis, 1999; Helentjaris *et al.*, 1988). These mapping studies have established that some chromosomes—e.g., chromosomes 1 and 5 and chromosomes 2 and 7—share duplicated segments. Perhaps the most surprising information about the extent of gene duplication in maize is that 72% of single-copy rice genes are duplicated in maize (Ahn and Tankley, 1993).

Extensive chromosomal duplication in maize has been interpreted as evidence for a polyploid origin of the genome (Anderson, 1945; Rhoades, 1951), but until recently there had been no estimation of the timing and mode of this polyploid event. In 1997, Gaut and Doebley (1997) inferred the timing and mode of the polyploid event by studying DNA sequences from maize duplicated genes. To infer the mode of origin, Gaut and Doebley first modeled patterns of genetic divergence under three different types of polyploid formation: autopolyploidy, genomic allopolyploidy, and segmental allopolyploidy. (Briefly, allopolyploids are created by hybridization between species, with a genomic allopolyploid based on species that have fully differentiated chromosomes and a segmental allopolyploid based on species that have only partially differentiated chromosomes. Autopolyploidy refers to a polyploid event based on an intraspecific event. Stebbins contributed a great deal toward the definition and use of these terms, and precise definitions can be found in Stebbins, 1950.) The models' predictions were then compared with patterns of DNA sequence divergence in 14 pairs of maize duplicated genes. The sequence data were consistent with a segmental allotetraploid model of origin but inconsistent with the other two models of polyploid formation. Hence, the authors concluded that the maize genome was the product of a segmental allotetraploid event. They estimated the timing of the event by applying a molecular clock to the sequence data.

The hypothesized origin of the maize genome is detailed in Fig. 2 (Gaut and Doebley, 1997). Briefly, this hypothesis states that (*i*) maize is the product of a segmental allotetraploid event, (*ii*) the two diploid progenitors (or "parents") of maize diverged ≈20.5 mya, (*iii*) the tetraploid event occurred between 16.5 and 11.4 mya, sometime after the divergence of *Sorghum* from one of the progenitor lineages, and (*iv*) the genome "rediploidized" before 11.4 mya. Although valuable, there are at least three reasons to be cautious about the hypothesis. The first reason is that

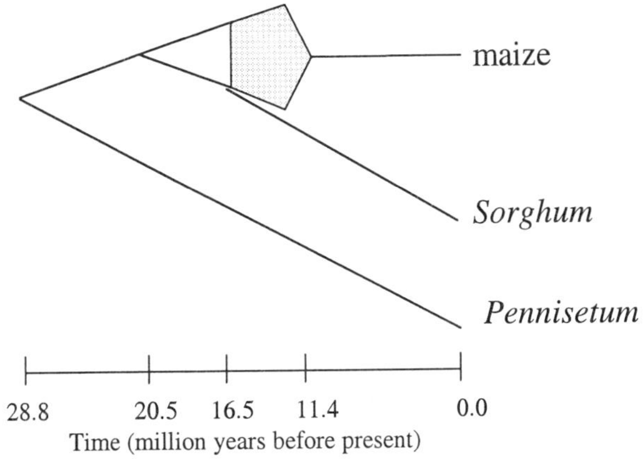

FIGURE 2. A hypothesis for the origin of the maize genome (Gaut and Doebley, 1997). Under this hypothesis, Pennisetum and maize diverged ≈29 million years ago (mya), followed ≈9 million years later by the divergence of the two diploid progenitors of maize. Sorghum diverged from one of these progenitor lineages (≈16.5 mya) before the two diploid progenitors united to form allopolyploid maize. The polyploid event occurred sometime between 16.5 mya and 11.4 mya, with subsequent diploidization completed by 11.4 mya. Gray shading represents the period in which allotetraploidy and diploidization occurred.

the hypothesis is based on a relatively small number of DNA sequences—i.e., only 14 pairs of duplicated sequences. The second reason is that some of the sequences were not mapped to a chromosomal location. Ideally, these analyses should be based on a far greater number of sequences, all of which are known to reside in regions of known chromosomal duplication. Finally, it was not possible to test molecular clock assumptions rigorously for all of the sequence data, and thus some of the clock-based time estimates are subject to an unknown amount of error. Despite the need for caution, the study of Gaut and Doebley (1997) provides the first glimpse into the mode and timing of an ancient plant polyploid event, and it also proposes a hypothesis that is testable with additional data.

The Polyploid Event and the Divergence of Maize and Sorghum

Fig. 1 places the segmental allotetraploid event in a phylogenetic context, and this context raises three important points about the comparison of maize to sorghum. First, if the allotetraploid event occurred after maize and sorghum diverged, then the maize genome should be duplicated more extensively than the sorghum genome. A corollary prediction is that maize and sorghum should not share common chromosomal duplications. Ultimately, these predictions can be tested with comparative genetic maps. At this point, however, it is unclear from comparative genetic maps as to whether the two genomes share extensive duplications in common, largely because published sorghum maps lack sufficient coverage (Berhan *et al.*, 1993; Chittenden *et al.*, 1994; Pereira *et al.*, 1994; Whitkus *et al.*, 1992). However, mapping information indicates that a higher proportion of markers is duplicated in maize than in sorghum. For example, Pereira *et al.* (1994) found that 44% of restriction fragment length polymorphism markers detected more bands in maize than in sorghum; conversely, only 7% of markers detected more bands in sorghum than in maize. This information is consistent with the phylogenetic placement of the allotetraploid event (Fig. 1).

The second point centers on chromosome number. Maize and sorghum (*Sorghum bicolor*) have the same number of chromosomes ($2n = 20$). If maize underwent an allotetraploid event after the divergence of maize from sorghum, why do these plants have an identical number of chromosomes? At present, there is no suitable answer to this question, but there has been discussion about the evolution of chromosome number. Traditionally, it has been assumed that the basal haploid chromosome number of the tribe Andropogoneae, which encompasses maize, sorghum, and *Tripsacum*, was $n = 5$ (Celarier, 1956; Molina and Naranjo, 1987). More recently, it has been suggested that the basal haploid chromosome of the tribe was $n = 10$ (Spangler *et al.*, 1999). If the basal number was 10, one can hypothesize both that the chromosome number of *S. bicolor* has remained unchanged and that maize was the product of an allopolyploid event between two species with a reduced number of chromosomes ($n = 5$). This scenario is plausible, because the tribe contains diploid taxa with $n = 5$ (e.g., *Elionurus* and *Sorghum* species; Spangler *et al.*, 1999) and because comparative maps provide support that maize consists of two $n = 5$ subgenomes (Devos and Gale, 1997; Moore *et al.*, 1995).

Wilson *et al.* (1999) have asserted that maize came from an ancestor with neither 5 nor 10 chromosomes. Based on genetic map data, they argued that the chromosome number of maize before the allotetraploid event was $n = 8$. The chromosome number was doubled subsequently to $n = 16$ ($2n = 32$) during the maize allotetraploid event and then reduced further by diploidization and fusion to the current number ($n = 10$; $2n = 20$). Unfortunately, however, the argument of Wilson *et al.* contains errors regarding the timing and phylogenetic context of the allotetraploid event. For example, they suggest that the allotetraploid event occurred after the divergence of maize and *Tripsacum*, whereas most evidence suggests that the allotetraploid event occurred *before* the divergence of maize and *Tripsacum*. When these errors are taken into account, their arguments for the evolution of chromosome number seem unlikely. In short, there are no definitive answers either as to the evolution of chromosome number in this group or as to why *S. bicolor* and maize have the same number of chromosomes.

The third and final point about maize and sorghum centers on the difference in genome content between the two species. The segmental allotetraploid event predicts 2-fold variation in DNA content between sorghum and maize, but it does not account for the actual 3.5-fold variation in DNA content (Fig. 1). Based on this information, differences in DNA content probably reflect the allopolyploid event *and* additional evolutionary changes, such as the accumulation of repetitive DNA.

Genome Rearrangement After an Allopolyploid Event

It must be remembered that extant maize is a diploid, and thus the segmental allotetraploid hypothesis presumes that the maize genome rearranged and diploidized. Is this presumption reasonable? Is genome rearrangement common after allopolyploid events?

Thus far, studies of synthetic plant polyploids suggest that genomes rearrange rapidly after allopolyploid events (reviewed in Wendel, 2000). In one study, Song *et al.* (1995) created four synthetic allopolyploids. After recovery of F_2 polyploids, each line was selfed until the F_5 generation. Plants from the F_2 and each subsequent generation were subjected to Southern hybridization with a panel of 89 probes. Southern blotting revealed remarkable differences in fragment profiles from generation to generation. In one synthetic polyploid, 66% of the probes detected fragment loss, fragment gain, or a change in fragment size, demonstrating that extensive rearrangement can occur rapidly after allopolyploid formation. Feldman and coworkers (Feldman *et al.*, 1997; Liu *et al.*, 1998; Liu, 1998) performed similar studies in *Triticum* and *Aegilops*. Their results suggest that allopolyploids lose noncoding sequences in a directed, nonrandom fashion and that coding sequences are modified extensively (Feldman *et al.*, 1997; Liu *et al.*, 1998; Liu, 1998).

Empirical studies detect rapid rearrangement of allopolyploid genomes, but rapid rearrangement is not equivalent to a complete diploidization. However, there is growing evidence that many plant, animal, and fungal genomes are the products of ancient polyploid events that were followed by rearrangement and a reduction in ploidy level. Yeast is one example. The DNA sequence of the yeast genome contains numerous blocks of duplicated genes. The phase (or direction) of the blocks are nonrandomly associated with centromeres, suggesting that the blocks were produced by the process of chromosomal duplication (Wolfe and Shields, 1997). Altogether, the data suggest that the yeast genome is the product of an ancient tetraploid event followed by rearrangement and diploidization (Seoighe and Wolfe, 1998). Vertebrates are another example of diploidized ancient polyploids; it is believed that vertebrates are degenerate polyploids owing to two polyploid events before the radiation of fish and mammals (Postlethwait *et al.*, 1998). Similar examples come from plants; for example, both *Glycine* (soybean) (Shoemaker *et al.*, 1996) and *Brassica* species (Bohuon *et al.*, 1996; Cavell *et al.*, 1998) seem to be degenerate polyploids. Based on this information, one can conclude that diploidization after polyploidy is evolutionarily common.

For maize, it should be possible to garner insights into the processes of rearrangement and diploidization from extant patterns of chromosomal duplication. Mapping studies have documented regions of chromosomal duplication in maize (Table 1). (It is important to note that Table 1 includes *only* those chromosomes that were explicitly defined as duplicated by the authors; Table 1 does *not* include all of the chromosome pairs on which markers are known to crosshybridize.) As Table 1 demonstrates, there is some disagreement among studies about chromosomal duplications, for two reasons. First, different studies use different data, leading to different conclusions. Second, and perhaps more importantly, researchers rarely denote their criteria for defining chromosomal duplications, and thus criteria likely differ among studies. Ultimately, chromosomal duplications should be defined by objective statistical criteria.

Nonetheless, there is a consensus about some chromosomal pairs. For example, it is now well established that portions of chromosome 1 are duplicated on chromosomes 5 and 9 (Table 1). The evolutionary implication for these pairings is that the process of diploidization rearranged one copy of chromosome 1. (Alternatively, chromosome 1 could be an amalgamation of regions from different parental chromosomes.) Chromosome

TABLE 1. Duplicated chromosomes in maize and the studies that identified them

Duplicated chromosomes	References
1–5	Helentjaris et al., 1988; Wilson et al., 1999; Gale and Devos, 1998
1–9	Helentjaris et al., 1988; Wilson et al., 1999; Gale and Devos, 1998
2–4	Helentjaris et al., 1988
2–7	Helentjaris et al., 1988; Wilson et al., 1999; Gale and Devos, 1998
2–10	Helentjaris et al., 1988; Ahn and Tankley, 1993; Wilson et al., 1999; Gale and Devos, 1998
3–8	Helentjaris et al., 1988; Ahn and Tankley, 1993; Wilson et al., 1999; Gale and Devos, 1998
3–10	Gale and Devos, 1998
4–5	Wilson et al., 1999; Gale and Devos, 1998
6–8	Helentjaris et al., 1988; Wilson et al., 1999; Gale and Devos, 1998
6–9	Wilson et al., 1999; Gale and Devos, 1998

2 had a similar fate in that portions of chromosome 2 are also found on chromosomes 7, 10, and perhaps 4 (Table 1). More extensive evaluation of these duplications will provide an indication as to whether there has been any bias in rearrangements. For example, there is a strong bias for paracentric inversions, as opposed to translocations and pericentric inversions, between potato and tomato. It was reasoned that the bias toward paracentric inversions reflects the relatively low effect of paracentric inversions on fitness (Bonierbale *et al.*, 1988). Additional studies of chromosomal duplications in maize could provide additional insights into the kind of rearrangements that are most evolutionarily stable.

The Importance of Chromosomal Duplication in Genome Evolution

Is maize typical with regard to its polyploid history and prevalent chromosomal duplication? There is no doubt that polyploidy is common in plants, with up to 70% of angiosperms owing their history to polyploidy (Masterson, 1994; Stebbins, 1950). Furthermore, genetic maps demonstrate that a great number of species contain chromosomal duplications. Even species with streamlined genomes contain chromosomal duplications; for example, rice has a large duplication between chromosomes 11 and 12 (Harushima *et al.*, 1998) and *Arabidopsis* also has at least one large chromosomal duplication (Mayer *et al.*, 1999). Other plant genomes with chromosomal duplications include sorghum (Chittenden *et al.*, 1994), cotton (Reinisch *et al.*, 1994), soybean (Shoemaker *et al.*, 1996), and *Brassica* species (Bohuon *et al.*, 1996; Cavell *et al.*, 1998). Some of these genomes are degenerate polyploids like maize, but others may owe their chromosomal duplications to independent segmental events.

It is important to note that chromosomal duplications are usually inferred from genetic maps, but most (if not all) genetic maps are based on low copy-number markers. Low copy-number markers are systematically biased against detecting duplicated chromosomal segments, and hence the extent of chromosomal duplication is likely grossly underestimated for most plant taxa. In addition, the resolution of most genetic maps is low, such that relatively small areas of chromosomal duplication cannot be detected. The result is that we do not have a realistic understanding of either the extent to which chromosomes are duplicated or the extent to which genomes contain functional redundancies. We can, however, look to *Arabidopsis* sequence data as preliminary examples of the extent of chromosomal duplication. Based on the sequences of chromosomes 2 and 4 (Lin *et al.*, 1999; Mayer *et al.*, 1999), it is estimated that 10–20% of the low-copy regions of the *Arabidopsis* genome lie within duplicated chromosomal regions (Mayer *et al.*, 1999). Given that the *Arabidopsis* genome is streamlined, this percentage is undoubtedly much higher in complex genomes. It is possible that most genes in most plant genomes reside in duplicated chromosomal regions.

MULTIPLICATION OF REPEAT SEQUENCES

Extent and Identification of Repetitive DNA

Repetitive DNA constitutes a high proportion of plant genomes. This fact has been confirmed experimentally by reassociation (or C_0t) kinetics. For example, Flavell *et al.* (1974) found that repetitive DNA (defined, in this case, as DNA with more than 100 copies per genome) constitutes ≈80% of genomes with a haploid DNA content >5 pg. In contrast, small genomes of <5 pg contain 62% repetitive DNA on average. Maize falls into this range; reassociation experiments indicate that the genome contains from 60% to 80% repetitive DNA (Flavell *et al.*, 1974; Hake and Walbot, 1980). The repetitive DNA of maize can be categorized further as 20% highly repetitive (over 800,000 copies per genome) and 40% middle repetitive (over 1,000 copies per genome; Hake and Walbot, 1980).

It is obvious that repetitive DNA is a large component of the maize genome, and thus the proliferation of repeat sequences has had important evolutionary implications. However, reassociation studies alone cannot answer two important questions about repetitive DNA in maize: what is the repetitive DNA, and when did it arise?

To date, the most complete answers to these two questions come from studies of the maize *Adh1* region by Bennetzen and coworkers (SanMiguel *et al.*, 1996, 1998; Springer *et al.*, 1994; Tikhonov *et al.*, 1999). They isolated a 280-kilobase yeast artificial chromosome clone of the *Adh1* region and characterized the composition of the repetitive intergenic DNA. Retrotransposons comprise roughly 62% of the 240 kilobases analyzed, with an additional 6% of the clone consisting of miniature inverted-repeat transposable elements, remnants of DNA transposons, and other low-copy repeats. In total, the region contained 23 retrotransposons representing 10 distinct families. Of the 23 retroelements, 10 inserted within another element, resulting in a nested or "layered" structure of intergenic DNA within maize (Fig. 3). The architecture of this region suggests that retrotransposons preferentially target other retroelements for insertion.

Perhaps the most interesting feature of the *Adh1* region is that it seems to be a representative region of the maize genome. Three observations support this contention. First, Southern blot and other analyses suggest that the retrotransposon families in the *Adh1* region comprise at least 50% of the maize genome; altogether, just three of the retroelement families found in the *Adh1* region constitute a full 25% of the genome (SanMiguel *et al.*, 1996). Second, 85% of repetitive DNAs from other regions were also

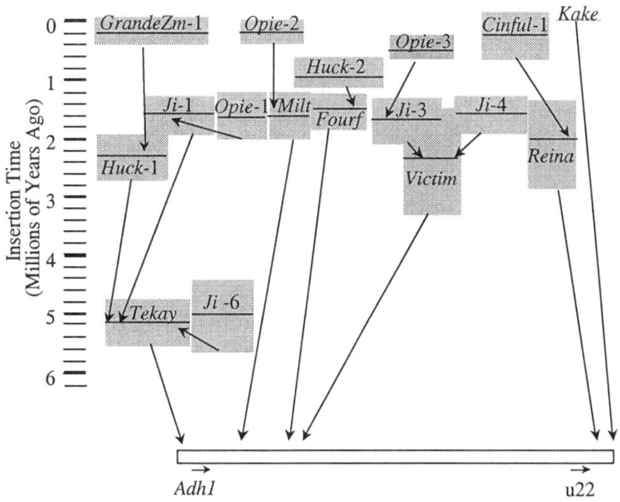

FIGURE 3. The estimated insertion times of retrotransposons in the Adh1 region (SanMiguel *et al.*, 1998). Each gray box represents a retrotransposon. The horizontal line through the box is the estimate of insertion time, and the height of the box represents the standard deviation of the estimate. Arrows between boxes indicate the order of insertion. For example, Huck-2 inserted into Fourf ≈1 mya.

present in the *Adh1* region (although it should be noted that the sample of repetitive DNAs from other regions was small and thus this estimate may not be robust). Finally, a more recent study suggests that retrotransposons hybridize fairly uniformly to maize bacterial artificial chromosome clones, suggesting that the distribution of retrotransposons is reasonably homogeneous throughout the genome (B. Meyers, personal communication).

The Timing of Retrotransposon Multiplication

Maize repetitive DNA seems to be primarily retrotransposons, but the second question remains: when did these retroelements multiply? To answer this question, SanMiguel *et al.* (1998) sequenced the long terminal repeat (LTR) of retrotransposons in the *Adh1* region. The rationale was as follows: when a single retrotransposon inserts into genomic DNA, both copies of the LTR are identical. Over time, the LTRs accumulate nucleotide substitutions and diverge in sequence. If the accumulation of nucleotide substitutions occurs at a regular pace, the number of nucleotide differences between the two LTRs provide insight into the date of LTR divergence and hence the date of retrotransposon insertion.

SanMiguel *et al.* (1998) applied this approach to estimate the insertion time for 17 LTRs from the *Adh1* region (Fig. 3). The results show that the oldest retrotransposon insertion is ≈5.2 mya and that most (15 of 17) retrotransposons inserted within the last 3.0 million years. The question arises as to whether these time estimates are reasonable. One feature that supports the results is that the time estimates correspond to the layering of retrotransposons (Fig. 3). In other words, in most cases (10 of 11) the insertion date for a retrotransposon is less than the insertion date for the retrotransposon into which it inserted. (The one exception is an instance in which the insertion dates are statistically indistinguishable.) Another observation that supports these results is that the sorghum *Adh1* region lacks retrotransposons (Tikhonov *et al.*, 1999). Based on this information and ignoring the possibility of extensive retrotransposon loss in sorghum (Bennetzen and Kellogg, 1997), retrotransposons in the maize *Adh1* region must have amassed in the ≈16 million years since the divergence of sorghum and maize.

The implications of the study are important. If the *Adh1* region is representative and the retrotransposons in this region constitute 50% of the genome, the maize genome has doubled in size in the last 5–6 million years. Like the polyploid event, retrotransposon proliferation represents a doubling of genome content over a relatively short evolutionary time scale.

Fig. 1 indicates that retrotransposon multiplication likely began in the evolutionary lineage leading to maize and *Tripsacum*, which diverged roughly ≈4.5–4.8 mya (Hilton and Gaut, 1998). Thus, most maize retrotransposon activity postdates the divergence of genera, but the oldest retrotransposons in the maize *Adh1* region likely predate the split between *Zea* and *Tripsacum*. This discussion underscores the importance of studying *Tripsacum* to understand evolutionary events in maize better; if Fig. 1 is accurate, *Tripsacum* should share both chromosomal duplications and some retrotransposon activity in common with maize. It is known that *Zea* and *Tripsacum* share at least one low-copy retrotransposon that is absent from other closely related genera (Vicient and Martinez-Izquierdo, 1997), but there is generally little information about chromosomal duplications or retrotransposons in *Tripsacum*.

Based on the available information, two large events differentiate the maize lineage from the sorghum lineage. The first event, segmental allotetraploidy, resulted in a 2-fold increase in maize DNA content. The second event, retrotransposon proliferation, produced another 2-fold increase in maize DNA content. Together, these events adequately explain the 3.5-fold difference in DNA content between maize and sorghum. However, it should be noted that there is also substantial variation in genomic DNA content among *Zea* and *Tripsacum* species (Fig. 1) (Bennett and

Leitch, 1995; Bennett and Smith, 1991); this variation may reflect different amounts of retrotransposon proliferation or independent chromosomal duplications.

Remaining Questions

Studies of the *Adh1* region by Bennetzen and coworkers (Springer *et al.*, 1994; SanMiguel *et al.*, 1996, 1998; Tikhonov, 1999) have provided invaluable insight into the structure and dynamics of maize intergenic DNA, but at least three important questions remain.

Question 1. Are retrotransposons distributed homogeneously among genomic regions? The *Adh1* studies, as well as other studies (B. Meyers, personal communication), suggest that retrotransposon distribution may be roughly homogenous among regions of the maize genome. However, other lines of evidence suggest that such homogeneity is unlikely. For example, evolutionary theory predicts that transposable elements should gather in regions of low recombination, such as centromeres (Charlesworth *et al.*, 1986, 1994). This prediction holds in *Arabidopsis*, where sequence data from chromosomes 2 and 4 indicate an increase in the frequency of transposable elements near centromeres (Copenhaver, 1999).

There are other reasons to suggest that retrotransposon distribution may not be homogeneous throughout the maize genome. One obvious reason is that there are heterogeneities in chromosomal structure, such as euchromatin, heterochromatin, nucleolus organizing regions, telomeres, centromeres, and knobs. Nonetheless, recent research indicates that retrotransposons constitute a substantial fraction of both heterochromatic centromeres and heterochromatic knobs (Ananiev *et al.*, 1998a, b); for one chromosome 9 knob, retroelements comprise roughly one-third of knob-specific clones (Ananiev *et al.*, 1998c). Many of the retrotransposons in knob and centromeric DNA belong to the element families found in the *Adh1* region. Despite these commonalties, there are also substantive differences among knobs, centromeres, and the *Adh1* region. For example, centromeres contain a centromere-specific retrotransposon (CentA; Ananiev *et al.*, 1998b). Similarly, chromosomal knobs associate with 180-bp and 350-bp repeat elements that are otherwise sparse in the genome (Ananiev *et al.*, 1998a). Altogether, the emerging picture is one in which some retroelement families are fairly ubiquitous, and other repetitive DNAs are heterogeneous in their distribution (e.g., Zhang *et al.*, 2000).

The work of Bernardi and coworkers (Barakat *et al.*, 1997; Carels *et al.*, 1995) is an intriguing addition to this picture. They fractionated DNA by G:C content and hybridized each G:C fraction to 38 coding-region probes. The coding genes hybridize almost exclusively to a DNA fraction of very narrow G:C content (1% of the total range), and this narrow fraction corresponds to 17% of the DNA content of the genome. To explain this hybridization pattern, Bernardi and coworkers (Barakat *et al.*, 1997; Carels *et al.*, 1995) reasoned that maize coding genes must be located in "gene-rich" regions and that these gene-rich regions must be flanked by DNA with highly homogeneous G:C contents. They proposed that this flanking DNA could consist of retrotransposons like those flanking the *Adh1* gene (San Miguel *et al.*, 1996).

The results from G:C fractionation experiments and studies of the *Adh1* region are inconsistent. On the one hand, the study of the *Adh1* region, coupled with studies of centromeres and knobs, suggest that retrotransposon distribution is widespread, representing 50% of the genome. On the other hand, Bernardi and coworkers' work implicitly suggests that retrotransposon distributions are heterogeneous, with a higher concentration of retroelements in the 17% of the genome that represents coding DNA. Ultimately, there may be a resolution to differences implied by different studies, but such a resolution will require more sequencing of large chromosomal clones representing diverse genomic regions.

Question 2. What contributes more to the evolution of DNA content: multiplication of repetitive DNA or chromosomal duplication? The evolutionary history of maize suggests that retrotransposon multiplication and chromosomal duplication (by way of polyploidy) each have generated a 2-fold increase in DNA content within the last 16 million years. Hence, the net effect of these two evolutionary processes is similar in maize. In contrast, it seems that the multiplication of repeat sequences is the primary contributor to differences in DNA content between many taxa (Flavell *et al.*, 1974). For example, barley and rice have similar complements of low-copy genes (Saghai-Maroof *et al.*, 1996) but a 12-fold difference in DNA content (Fig. 1). The difference in DNA content is thus probably attributable to differences in the amount of repetitive DNA (Saghai-Maroof *et al.*, 1996).

It is premature to make the general statement that repeat proliferation contributes more to the evolution of DNA content than chromosomal duplications for two reasons. First, as mentioned previously, mapping studies are biased against the discovery of duplications, and for this reason, there is as yet no accurate indication of the extent of chromosomal duplication in complex genomes. Second, duplication and repeat proliferation are not independent. Duplication plays a role in repeat proliferation, because duplication doubles repetitive DNA as well as low-copy DNA.

Question 3. Are chromosomal duplication events correlated with an increase in the rate of transposition? This question originates from the work of Matzke, Matzke, and colleague (Matzke and Matzke, 1998; Matzke *et al.*, 1999). They argue that polyploid genomes contain duplications of all genes and thus are relatively well buffered against mutations caused by transposon insertion. As a consequence, transposable elements multiply and are maintained in polyploid genomes. For maize, the fact that two major events (polyploidy and retrotransposon multiplication) are located on the same phylogenetic lineage gives credence to the idea that these phenomena are biologically correlated (Fig. 1), but it is not yet known whether this correlation is widely observed.

GENETIC VARIATION IN GENES ALONG CHROMOSOMES

Genetic Diversity as a Function of Recombination, Natural Selection, and Chromosomal Position

Genomes are dynamic entities that can be modified extensively by polyploidy and transposon multiplication. However, ongoing evolutionary processes like mutation, recombination, natural selection, and migration also shape the genome. The effect of these extant processes on the genome can be inferred from careful study of genetic diversity.

Diversity throughout the genome is affected strongly by the interplay of recombination and natural selection. In *Drosophila*, for example, genetic diversity varies along the chromosome as a function of recombination rate (Begun and Aquadro, 1992; Hamblin and Aquadro, 1999). Loci near centromeres tend to have low recombination rates and also tend to have low levels of genetic diversity, but both recombination rate and genetic diversity increase toward the tip of chromosomes. This relationship is not because recombination is mutagenic; rather, it reflects an interdependence between natural selection and recombination (Begun and Aquadro, 1992; Charlesworth *et al.*, 1995). In regions of low recombination, for example, linkage between nucleotide sites ensures that selection for or against a single nucleotide substitution will affect a large region of the genome. In regions of high recombination, nucleotide sites are nearly independent; thus, selection on a single site affects a much smaller region of the genome. The result of the interdependence between selection and recombination is that (*i*) levels of genetic diversity can be a function of chromosomal position and (*ii*) large chromosomal regions can be depauperate of genetic diversity.

The correlation between chromosomal position and genetic diversity has been confirmed in plants (Dvorak *et al.*, 1998; Stephan and Langley, 1998), but it is not yet clear whether recombination in maize follows a simple pattern along chromosomes. For example, it has been documented that maize single-copy regions act as recombination hot spots, but recombination rates also vary among single-copy regions (Civardi *et al.*, 1994; Okagaki and Weil, 1997; Timmermans *et al.*, 1996). Altogether, these studies suggest that the relationship between chromosomal position and recombination rate may not be as straightforward in maize as in *Drosophila*. More thorough elucidation of recombination rates in maize requires comparisons between genetic and physical maps; such physical maps are being produced but are not yet completed.

Nonetheless, we have a goal to quantify patterns of genetic diversity more accurately in the maize genome. To make this quantification, we have begun a long-term study of 100 maize genes along chromosomes 1 and 3. To measure genetic diversity in each gene, we will sample DNA sequences from ≈70 individuals representing maize, its progenitor, and two other wild *Zea* taxa. The project has many long-term goals, including (*i*) to investigate the relationship between chromosomal position and genetic diversity, (*ii*) to examine the impact of domestication on genetic diversity in maize, (*iii*) to compare the evolutionary history among species across genes, and (*iv*) to create a public single-nucleotide-polymorphism database.

The first stage of this ongoing project is to measure genetic diversity in 25 chromosome 1 genes from 16 maize individuals representing Mexican and South American land races and 9 individuals representing U.S. inbred lines. The results of this first stage will be reported in detail elsewhere, but we can make a preliminary contrast of diversity in centromeric vs. noncentromeric genes. Average diversity per base pair in four genes within 5 centimorgans of the centromere is $\theta = 0.0144$, as determined by using Watterson's estimator (Watterson, 1975). This level of diversity is slightly lower than average diversity in 11 noncentromeric genes (average Watterson's $\theta = 0.0170$), but the centromeric genes do not have extremely low levels of diversity. For example, all four centromeric genes contain more diversity than 3 of the 11 noncentromeric genes. Thus, we report that there is as yet no clear evidence for a strong reduction in genetic diversity near the centromere of chromosome 1.

Discordant Evolutionary Histories Among Genes

One interesting feature of genetic diversity studies of maize and its wild relatives is that evolutionary histories differ among loci. As an example, consider Fig. 4, which summarizes sequence data from four genes. The genes *Adh1* and *Glb1* provide very similar pictures of the relationship of the wild species *Z. luxurians* to other members of the genus *Zea* (Eyre-Walker *et al.*, 1998; Hilton and Gaut, 1998); in short, for both of these genes, *Z. luxurians* sequences comprise a separate, well defined clade. In contrast, *Z. luxurians* individuals contain sequences that are very similar

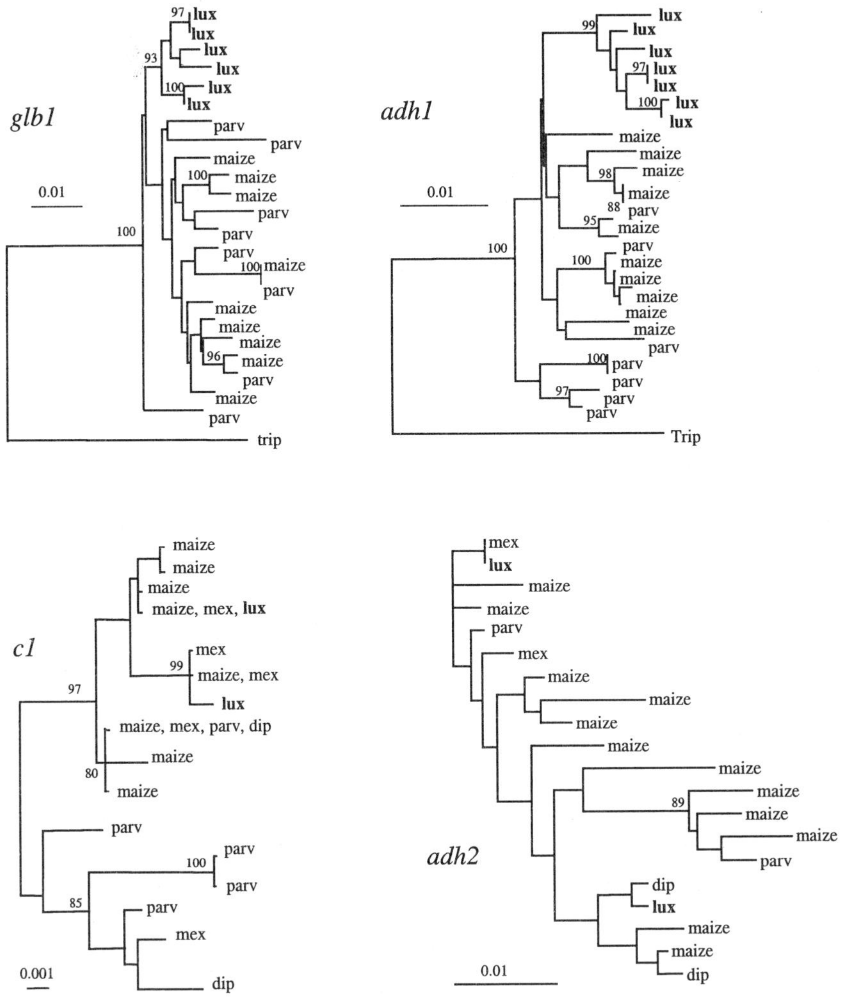

FIGURE 4. Genealogies of four genes, based on the neighbor-joining method (Saitou and Nei, 1987) with Kimura 2-parameter distances (Kimura, 1980). Taxa are abbreviated as follows: maize, domesticated maize; parv, ancestor of domesticated maize (Z. mays subsp. parviglumis); mex, Z. mays subsp. mexicana; lux, Zea luxurians; dip, Zea diploperennis; trip, Tripsacum dactyloides. Sequences from Z. luxurians are shown in bold. The data are from Eyre-Walker *et al.*, 1998; Goloubinoff *et al.*, 1993; Hanson *et al.*, 1996; Hilton and Gaut, 1998. Scale bars indicate level of divergence among sequences; bootstrap values >80% are shown.

(or even identical) to sequences from other *Zea* taxa for *Adh2* (Goloubinoff *et al.*, 1993) and *c1* (Hanson *et al.*, 1996). Thus, the picture of evolutionary history from *Adh1* and *Glb1* is not consistent with information from *c1* and *Adh2*. (Fig. 4 focuses on genealogical or phylogenetic information for ease of presentation, but sequence statistics also suggest that these genes have different evolutionary histories.) One interesting feature of Fig. 4 is that *Adh1* and *Glb1* are located within a 12-centimorgan region of chromosome 1; *Adh2* and *c1* are found on chromosomes 4 and 9, respectively.

We have sampled extensively from the wild relatives of maize for only a handful of genes, but discordant patterns, such as those demonstrated in Fig. 4, continue to be identified. The challenge of these data will be to infer the evolutionary processes that contribute to discordant evolutionary histories among genes. Several possibilities exist, including differences in nucleotide substitution rates, introgression (migration) rates, and natural selection among genes. One interesting possibility is that genealogical patterns among genes may correlate with chromosomal location.

In this context, it is worth noting that studies of *Drosophila* species have also demonstrated discordant patterns of genetic diversity among loci. For example, Wang *et al.* (1997) studied three loci in three *Drosophila* species. Two of the loci (*Hsp82* and *period*) yielded very similar pictures of genetic divergence among taxa. At these two loci, sequences were well differentiated among taxa. However, the pattern of genetic diversity in the third *Drosophila* locus (*Adh*) was incongruent with data from the first two loci. In this last locus, DNA sequences from different taxa were not highly diverged. Wang *et al.* (1997) used population genetic tools to contrast genealogical information among *Drosophila* loci, and they concluded that introgression among species has occurred at a much higher rate at one locus (*Adh*) than at the other two loci (*Hsp82* and *period*). In short, *Drosophila* studies strongly suggest that the processes affecting genetic diversity can vary among loci and also demonstrate the importance of comparing genealogical information across species and across loci.

In crops, artificial selection can cause discordant patterns of genetic diversity among loci. Thus far, levels of nucleotide sequence diversity have been measured in maize and its wild progenitor (*Z. mays* subsp. *parviglumis*) for six genes (summarized in White and Doebley, 1999). All six genes indicate that maize has reduced genetic diversity relative to its wild progenitor, probably reflecting a genetic bottleneck during domestication (Eyre-Walker *et al.*, 1998; Hilton and Gaut, 1998). However, the level of reduction in genetic diversity varies substantially among genes. For four of the six genes, maize retains at least half of the genetic diversity of its wild progenitor. For the remaining two genes (*c1* and *tb1*), maize contains less than 20% of the level of diversity of its wild progenitor (Hanson *et al.*, 1996; Wang *et al.*, 1999). Low diversity in *c1* and *tb1* likely reflects artificial selection by the early domesticators of maize. The *tb1* gene was probably selected to affect morphological changes in branching pattern (Wang *et al.*, 1999), and *c1* may have been selected for production of purple pigment in maize kernels (Hanson *et al.*, 1996).

Just as domestication has had a heterogeneous effect across loci, so has the process of maize breeding. For nine genes that we have sampled extensively thus far, U.S. inbred lines average roughly 65% the level of genetic diversity of the broader sample of maize. This level of reduction from maize land races to U.S. maize is commensurate with the original reduction in genetic diversity from wild progenitor to domesticated maize (Hilton and Gaut, 1998). Altogether, owing to reductions in diversity caused by initial domestication and subsequent intensive breeding, our initial estimates indicate that U.S. inbreds contain only ≈40% of the level of genetic diversity of the wild ancestor of maize.

Thus far, studies of genetic diversity have shown that maize genes have different levels of genetic diversity, and diversity in some genes has been affected strongly by artificial selection. In addition, studies of wild *Zea* taxa indicate that genes differ in their evolutionary histories among taxa. Our ongoing study of 100 genes will help determine whether patterns of evolutionary history among genes are, in fact, correlated with chromosomal location and will also contribute to the overall understanding of the evolutionary forces acting on plant genomes.

The authors acknowledge National Science Foundation Grants DBI-9872631 and DEB-9815855 and U.S. Department of Agriculture Grant 98-35301-6153.

REFERENCES

Ahn, S. & Tankley, S. D. (1993) Comparative linkage maps of the rice and maize genomes. *Proc. Natl. Acad. Sci. USA.* 90, 7980–7984.

Ananiev, E. V., Phillips, R. L. & Rines, H. W. (1998a) A knob-associated repeat in maize capable of forming fold-back DNA segments: Are chromosome knobs megatransposons? *Proc. Natl. Acad. Sci. USA* 95, 10785–10790.

Ananiev, E. V., Phillips, R. L. & Rines, H. W. (1998b) Chromosome-specific molecular organization of maize (*Zea mays* L.) centromeric regions. *Proc. Natl. Acad. Sci. USA* 95, 13073–13078.

Ananiev, E. V., Phillips, R. L. & Rines, H. W. (1998c) Complex structure of knob DNA on maize chromosome 9: Retrotransposon invasion into heterochromatin. *Genetics* 149, 2025–2037.

Anderson, E. (1945) What is *Zea mays*? A report of progress. Chron. Bot. 9: 88–92.

Barakat, A., N. Carels and G. Bernardi. (1997) The distribution of genes in the genomes of Gramineae. *Proc. Natl. Acad. Sci. USA.* 94, 6857–6861.

Begun, D. J. & Aquadro, C. F. (1992) Levels of naturally occurring DNA polymorphism correlate with recombination rates in *Drosophila melanogaster*. *Nature* 356, 519–520.

Bennett, M. D. & Leitch, I. J. (1995) Nuclear DNA amounts in angiosperms. *Ann. Bot.* 76, 113–176.

Bennett, M. D. & Smith, J. B. (1991) Nuclear DNA amounts in angiosperms. *Phil. Trans. Roy. Soc. Lond. B* 334, 309–345.

Bennetzen, J. L. & Kellogg, E. A. (1997) Do plants have a one-way ticket to genomic obesity? *Plant Cell* 9, 1509–1514.

Berhan, A. M., Hulbert, S. H., Butler, L. G. & Bennetzen, J. L. (1993) Structure and evolution of the genome of *Sorghum bicolor* and *Zea mays*. *Theor. Appl. Genet.* 86, 598–604.

Bohuon, E. J. R., Keith, D. J., Parkin, I. A. P., Sharpe, A. G. & Lydiate, D. J. (1996) Alignment of the conserved C genomes of *Brassica oleracea* and *Brassica napus*. *Theor. Appl. Genet.* 93, 833–839.

Bonierbale, M. W., Plaisted, R. L. & Tanksley, S. D. (1988) RFLP maps based on a common set of clones reveal modes of chromosomal evolution in potato and tomato. *Genetics* 120, 1095–1103.

Carels, N., Barakat, A. & Bernardi, G. (1995) The gene distribution of the maize genome. *Proc. Natl. Acad. Sci. USA* 92, 11057–11060.

Cavell, A. C., Lydiate, D. J., Parkin, I. A. P., Dean, C. & Trick, M. (1998) Collinearity between a 30-centimorgan segment of *Arabidopsis thaliana* chromosome 4 and duplicated regions within the *Brassica napus* genome. *Genome* 41, 62–69.

Celarier, R. P. (1956) Additional evidence of five as the basic chromosome number of the Andropoganeae. *Rhodora* 58, 135–143.

Charlesworth, B., Langley, C. H. & Stephan, W. (1986) The evolution of restricted recombination and the accumulation of repeated DNA sequences. *Genetics* 112, 947–962.

Charlesworth, B., Sniegowski, P. & Stephan, W. (1994) The evolutionary dynamics of repetitive DNA in eukaryotes. *Nature* 371, 215–220.

Charlesworth, D., Charlesworth, B. & Morgan, M. T. (1995) The pattern of neutral molecular variation under the background selection model. *Genetics* 141, 1619–1632.

Chittenden, L. M., Schertz, K. F., Lin, Y. R., Wing, R. A. & Paterson, A. H. (1994) A detailed RFLP map of Sorghum bicolor X S. propinquum, suitable for high density mapping, suggests ancestral duplication of sorghum chromosomes or chromosomal segments. *Theor. Appl. Genet.* 87, 925–933.

Civardi, L., Xia, Y., Edwards, K. J., Schnable, P. S. & Nikolau, B. J. (1994) The relationship between genetic and physical distances in the clones *a1-sh2* interval of the *Zea mays* L. genome. *Proc. Natl. Acad. Sci. USA* 91, 8268–8272.

Copenhaver, G. N. K, Kuromori, T, Benito, M. I., Kaul, S, Lin, X. Y., Bevan, M., Murphy, G., Harris, B., Parnell, L. D., McCombie, W. R., Martienssen, R. A., Marra, M., & Preuss, D. (1999) Genetic definition and sequence analysis of *Arabidopsis* centromeres. *Science* 286, 2468–2474.

Davis, G. M., Baysdorfer, C., Musket, T., Grant, D., Staebell, M., Xu, G., Polacco, M., Koster, L., Melia-Hancock, S., Houchins, K., Chao, S., & Coe, E. H.. (1999) A maize map standard with sequenced core markers, grass genome reference points and 932 expressed sequence tagged sites (ESTs) in a 1736-locus map. *Genetics* 152, 1137–1172.

Devos, K. M. & Gale, M. D. (1997) Comparative genetics in the grasses. *Pl. Mol. Biol.* 35, 3–15.

Dvorak, J., Luo, M.-C. & Yang, J.-L. (1998) Restriction fragment length polymorphism and divergence in the genomic regions of high and low recombination in self-fertilizing and cross-fertilizing *Aegilops* species. *Genetics* 148, 423–434.

Eyre-Walker, A., Gaut, R. L., Hilton, H., Feldman, D. L. & Gaut, B. S. (1998) Investigation of the bottleneck leading to the domestication of maize. *Proc. Natl. Acad. Sci. USA* 95, 4441–4446.

Feldman, M., Liu, B., Segal, G., Abbo, S., Levy, A. A. & Vega, J. M. (1997) Rapid elimination of low-copy DNA sequences in polyploid wheat: A possible mechanism for differentiation of homoeologous chromosomes. *Genetics* 147, 1381–1387.

Flavell, R. B., Bennett, M. D., Smith, J. B. & Smith, D. B. (1974) Genome size and the proportion of repeated nucleotide sequence DNA in plants. *Biochem. Genet.* 12, 257–269.

Gale, M. D. & Devos, K. M. (1998) Comparative genetics in the grasses. *Proc. Natl. Acad. Sci. USA* 95, 1971–1974.

Gaut, B. S. & Doebley, J. F. (1997) DNA sequence evidence for the segmental allotetraploid origin of maize. *Proc. Natl. Acad. Sci. USA* 94, 6809–6814.

Goloubinoff, P., Paabo, S. & Wilson, A. C. (1993) Evolution of maize inferred from sequence diversity of an *Adh2* gene segment from archaelogical specimens. *Proc. Natl. Acad. Sci. USA* 90, 1997–2001.

Goodman, M. M., Stuber, C. W., Newton, K. & Weissinger, H. H. (1980) Linkage relationships of 19 enzyme loci in maize. *Genetics* 96, 697–710.

Hake, S. & Walbot, V. (1980) The genome of *Zea mays*, its organization and homology to related grasses. *Chromosoma* 79, 251–270.

Hamblin, M. T. & Aquadro, C. F. (1999) DNA sequence variation and the recombinational landscape in Drosophila pseudoobscura: A study of the second chromosome. *Genetics* 153, 859–869.

Hanson, M. A., Gaut, B. S., Stec, A. O., Fuerstenberg, S. I., Goodman, M. M., Coe, E. H. & Doebley, J. (1996) Evolution of anthocyanin biosynthesis in maize kernels: the role of regulatory and enzymatic loci. *Genetics* 143, 1395–1407.

Harushima, Y., Yano, M., Shomura, A., Sato, M., Shimano, T., Kuboki, Y., Yamamoto, T., Lin, S.-Y., Antonio, B. A., Parco, A. Kajiya H., Huang, N., Yamamoto, K., Nagamura, Y., Kurata, N., Khush, G. S., & Sasaki, T. (1998) A high-density rice genetic linkage map with 2275 markers using a single F_2 population. *Genetics* 148, 479–494.

Helentjaris, T., Weber, D. & Wright, S. (1988) Identification of the genomic locations of duplicate nucleotide sequences in maize by analysis of restriction fragment length polymorphism. *Genetics* 118, 353–363.

Hilton, H. & Gaut, B. S. (1998) Speciation and domestication in maize and its wild relatives: evidence from the *Globulin-1* gene. *Genetics* 150, 863–872.

Kimura, M. (1980) A simple method for estimating evolutionary rates of base substitutions through comparative studies of nucleotide sequences. *J. Mol. Evol.* 16, 111–120.

Lin, X. Y., S. Kaul, S. S., Rounsley, S., Shea, T. P., Benito M. I., Town, C. D., Fujii, C. Y., Mason, T., Bowman, C. L., Barnstead, M., Feldblyum, T. V., Buell, C. R., Ketchum, K. A., Lee, J., Ronning, C. M., Koo, H. L., Moffat, K. S., Cronin, L. A., Shen, M., Pai, G., Van Aken, S., Umayam, L., Tallon, L. J., Gill, J. E., Adams, M. D., Carrera, A. J., Creasy, T. H., Goodman, H. M., Somerville, C. R., Copenhaver, G. P., Preuss, D., Nierman, W. C., White, O., Eisen, J. A., Salzberg, S. L., Fraser, C. M., & Venter, J. C. (1999) Sequence and analysis of chromosome 2 of the plant *Arabidopsis thaliana*. *Nature* 402, 761–768.

Liu, B., Vega, J. M., Segal, G., Abbo, S., Rodova, H. & Feldman, M. (1998) Rapid genomic changes in newly synthesized amphiploids of *Triticum* and *Aegilops*. I. Changes in low-copy noncoding DNA sequences. *Genome* 41, 272–277.

Liu, B., Vega, J. M. & Feldman, M. (1998) Rapid genomic changes in newly synthesized amphiploids of *Triticum* and *Aegilops*. II. Changes in low-copy coding DNA sequences. *Genome* 41, 535–542.

Masterson, J. (1994) Stomatal size in fossil plants: evidence for polyploidy in majority of angiosperms. *Science* 264, 421–423.

Matzke, M. A. & Matzke, A. J. M. (1998) Polyploid and Transposons. *TREE* 13, 241.

Matzke, M. A., Mittelsten-Scheid, O. & Matzke, A. J. M. (1999) Rapid structural and epigenetic changes in polyploid and aneuploid genomes. *BioEssays* 21, 761–767.

Mayer, K., Schuller, C., Wambutt, R., Murphy, G., Volckaert, G., Pohl, T., Dusterhoft, A., Stiekema, W., Entian, K. D., Terryn, N., Harris, B., Ansorge, W., Brandt, P., Grivell, L., Rieger, M., Weichselgartner, M., de Simone, V., Obermaier, B., Mache, R., Muller, M.,

Kreis, M., Delseny, M., Puigdomenech, P., Watson, M., Schmidtheini, T., Reichert, B., Portatelle, D., Perez-Alonso, M., Boutry, M., Bancroft, I., Vos, P., Hoheisel, J., Zimmermann, W., Wedler, H., Ridley, P., Langham, S. A., McCullagh, B., Bilham, L., Robben, J., Van der Schueren, J., Grymonprez, B., Chuang, Y. J., Vandenbussche, F., Braeken, M., Weltjens, I., Voet, M., Bastiaens, I., Aert, R., Defoor, E, Weitzenegger, T., Bothe, G., Ramsperger, U., Hilbert, H., Braun, M., Holzer, E., Brandt, A., Peters, S., van Staveren, M., Dirkse, W., Mooijman, P., Lankhorst, R. K., Rose, M., Hauf, J., Kotter, P., Berneiser, S., Hempel, S., Feldpausch, M., Lamberth, S., Van den Daele, H., De Keyser, A., Buysshaert, C., Gielen, J., Villarroel, R., De Clercq, R., Van Montagu, M., Rogers, J., Cronin, A., Quail, M., Bray-Allen, S., Clark, L., Doggett, J., Hall, S., Kay, M., Lennard, N., McLay, K., Mayes, R., Pettett, A., Rajandream, M. A., Lyne, M., Benes, V., Rechmann, S., Borkova, D., Blocker, H., Scharfe, M., Grimm, M., Lohnert, T. H., Dose, S., de Haan, M., Maarse, A., Schafer, M, Muller-Auer, S., Gabel, C., Fuchs, M., Fartmann, B., Granderath, K., Dauner, D., Herzl, A., Neumann, S., Argiriou, A., Vitale, D., Liguori, R., Piravandi, E., Massenet, O., Quigley, F., Clabauld, G., Mundlein, A., Felber, R., Schnabl, S., Hiller, R., Schmidt, W., Lecharny, A., Aubourg, S., Chefdor, F., Cooke, R., Berger, C., Montfort, A., Casacuberta, E., Gibbons, T., Weber, N., Vandenbol, M., Bargues, M, Terol, J., Torres, A., Perez-Perez, A., Purnelle, B., Bent, E., Johnson, S., Tacon, D., Jesse, T., Heijnen, L., Schwarz, S., Scholler, P., Heber, S., Francs, P., Bielke, C., Frishman, D., Haase, D., Lemcke, K., Mewes, H. W., Stocker, S., Zaccaria, P., Bevan, M., Wilson, R. K., de la Bastide, M, Habermann, K., Parnell, L., Dedhia, N., Gnoj, L., Schutz, K., Huang, E., Spiegel, L., Sehkon, M., Murray, J., Sheet, P., Cordes, M., Abu-Threideh, J., Stoneking, T., Kalicki, J., Graves, T., Harmon, G., Edwards, J., Latreille, P., Courtney, L., Cloud, J., Abbott, A., Scott, K., Johnson, D., Minx, P., Bentley, D., Fulton, B., Miller, N., Greco, T., Kemp, K., Kramer, J., Fulton, L., Mardis, E., Dante, M., Pepin, K., Hillier, L., Nelson, J., Spieth, J., Ryan, E., Andrews, S., Geisel, C., Layman, D., Du, H., Ali, J., Berghoff, A., Jones, K., Drone, K., Cotton, M., Joshu, C., Antonoiu, B., Zidanic, M., Strong, C., Sun, H., Lamar, B., Yordan, C., Ma, P., Zhong, J., Preston, R., Vil, D, Shekher, M., Matero, A., Shah, R., & Swaby, I. (1999) Sequence and analysis of chromosome 4 of the plant Arabidopsis thaliana. *Nature* 402, 769–777.

McClintock, B. (1930) A cytological demonstration of the location of an interchange between two non-homologous chromosomes of *Zea mays*. *Proc. Natl. Acad. Sci, USA* 16, 791–796.

McClintock, B. (1933) The association of non-homologous parts of chromosomes in the mid-prophase of meiosis in *Zea mays*. *Zeitchrift fur Zellforschung und mikroskopische Anatomie* 19, 191–237.

McMillin, D. E. & Scandalios, J. G. (1980) Duplicated cytosolic malate dehydrogenase genes in *Zea mays*. *Proc. Natl. Acad. Sci. USA* 77, 4866–4870.

Molina, M. D. & Naranjo, C. A. (1987) Cytogenetic studies in the genus *Zea:* 1. Evidence for five as the basic chromosome number. *Theor. Appl. Genet*. 73, 542–550.

Moore, G., Devos, K. M., Wang, Z. & Gale, M. D. (1995) Cereal genome evolution—grasses, line up and form a circle. *Curr. Biol.* 5, 737–739.

Okagaki, R. J. & Weil, C. F. (1997) Analysis of recombination sites within the maize *waxy* locus. *Genetics* 147, 815–821.

Pereira, M. G., Lee, M., Bramel-Cox, P., Woodman, W., Doebley, J. & Whitkus, R. (1994) Construction of an RFLP map in sorghum and comparative mapping in maize. *Genome* 37, 236–243.

Postlethwait, J. H., Yan, Y.-L., Gates, M. A., Horne, S., Arnores, A., Brownlie, A., Donovan, A., Egan, E. S., Force, A., Gong, Z. Y., Goutel, C., Fritz, A., Kelsh, R., Knapik, E., Liao, E., Paw, B., Ransom, D., Singer, A., Thomson, M., Abduljabbar, T. S., Yelick, P., Beier, D., Joly, J. S., Larhammar, D., Rosa, F., Westerfield, M., Zon, L. I., Johnson, S. L., & Talbot, W. S. (1998) Vertebrate genome evolution and the zebrafish gene map. *Nature Genetics* 18, 345–349.

Reinisch, A. J., Dong, J., Brubaker, C. L., Stelly, D. M., Wendel, J. F. & Paterson, A. H. (1994) A detailed RFLP map of cotton, *Gossypium hirsutum* X *Gossypium barbadense*-Chromosome organization and evolution in a disomic polyploid genome. *Genetics* 138, 829–847.

Rhoades, M. M. (1951) Duplicated genes in maize. *Am. Nat.* 85, 105–110.

Rhoades, M. M. (1955) The cytogenetics of maize. In *Corn and Corn Improvement*, ed. Sprague, G. F. (Academic Press, NY), pp. 123–219.

Saghai-Maroof, M. A., Yang, G. P., Biyashev, R. M., Maughan, P. J. & Zhang, Q. (1996) Analysis of the barley and rice genomes by comparative RFLP linkage mapping. *Theor. Appl. Genet.* 92, 541–551.

Saitou, N. & Nei, M. (1987) The neighbor-joining method: a new method for reconstructing phylogenetic trees. *Mol. Biol. Evol.* 4, 406–425.

SanMiguel, P., Tickhonov, A., Jin, Y.-K., Melake-Berhan, A., Springer, P. S., Edwards, K. J., Avramova, Z. & Bennetzen, J. L. (1996) Nested retrotransposons in the intergenic regions of the maize genome. *Science* 274, 765–768.

SanMiguel, P. J., Gaut, B. S., Tikhonov, A., Nakajima, Y. & Bennetzen, J. L. (1998) The paleontology of intergene retrotransposons of maize: dating the strata. *Nature Genetics* 20, 43–45.

Seoighe, C. & Wolfe, K. H. (1998) Extent of genomic rearrangement after genome duplication in yeast. *Proc. Natl. Acad. Sci. USA* 95, 4447–4452.

Shoemaker, R. C., Polzin, K., Labate, J., Specht, J., Brummer, E. C., Olson, T., Young, N., Concibido, V., Wilcox, J., Tamulonis, J. P., Kochert, G., & Boerma, H. R. (1996) Genome duplication in soybean (*Glycine* subgenus *soja*). *Genetics* 144, 329–338.

Snope, A. J. (1967) The relationship of abnormal chromosome 10 to b-chromosomes in maize. *Chromosoma* 21, 243–349.

Song, K., Lu, P., Tang, K. & Osborn, T. C. (1995) Rapid genome change in synthetic polyploids of *Brassica* and its implication for polyploid evolution. *Proc. Natl. Acad. Sci. USA* 92, 7719–7723.

Spangler, R., Zaitchik, B., Russo, E. & Kellogg, E. A. (1999) Andropogoneae evolution and generic limits in *Sorghum* (Poaceae) using *ndh*F sequences. *Syst. Bot.* 24, 267–281.

Springer, P. S., Edwards, K. J. & Bennetzen, J. L. (1994) DNA class organization on maize adh1 yeast artificial chromosomes. *Proc. Natl. Acad. Sci. USA* 91, 863–867.

Stebbins, G. L. (1950) *Variation and Evolution in Plants* (Columbia University Press, New York, NY).

Stephan, W. & Langley, C. H. (1998) DNA polymorphism in Lycopersicon and crossing-over per physical length. *Genetics* 150, 1585–1593.

Tikhonov, A. P., SanMiguel, P. J., Nakajima, Y., Gorenstein, N. M., Bennetzen, J. L. & Avramova, Z. (1999) Colinearity and its exceptions in orthologous *adh* regions of maize and sorghum. *Proc. Natl. Acad. Sci. USA* 96, 7409–7414.

Timmermans, M. C. P., Das, O. P. & Messing, J. (1996) Characterization of a meiotic crossover in maize identified by a restriction fragment length polymorphism-based method. *Genetics* 143, 1771–1783.

Ting, Y. C. (1966) Duplications and meiotic behavior of the chromosomes in haploid maize (*Zea mays* L.). *Cytologia* 31, 324–329.

Vicient, C. M. & Martinez-Izquierdo, J. A. (1997) Discovery of a Zde1 transposable element in *Zea* species as a consequence of retrotransposon insertion. *Gene* 184, 257–261.

Wang, R. L., Stec, A., Hey, J., Lukens, L. & Doebley, J. (1999) The limits of selection during maize domestication. *Nature* 398, 236–239.

Wang, R. L., Wakeley, J. & Hey, J. (1997) Gene flow and natural selection in the origin of *Drosophila pseudoobscura* and close relatives. *Genetics* 147, 1091–1106.

Watterson, G. A. (1975) On the number of segregating sites in genetical models without recombination. *Theor. Popul. Biol.* 7, 188–193.

Wendel, J. F. (2000) Genome evolution in polyploids. *Pl. Mol. Biol.* 42, 225–249.

Wendel, J. F., Stuber, C. W., Edwards, M. D. & Goodman, M. M. (1986) Duplicated chromosomal segments in *Zea mays* L.: further evidence from Hexokinase isozymes. *Theor. Appl. Genet.* 72, 178–185.

Wendel, J. F., Stuber, C. W., Goodman, M. M. & Beckett, J. B. (1989) Duplicated plastid and triplicated cytosolic isozymes of triosphosphate isomerase in maize (*Zea mays* L.). *J. Hered.* 80, 218–228.

White, S. E. & Doebley, J. F. (1999) The molecular evolution of terminal ear l, a regulatory gene in the genus Zea. *Genetics* 153, 1455–1462.

Whitkus, R., Doebley, J. & Lee, M. (1992) Comparative genome mapping of sorghum and maize. *Genetics* 132, 1119–1130.

Wilson, W. A., Harrington, S. E., Woodman, W. L., Lee, M., Sorrells, M. E. & McCouch, S. R. (1999) Inferences on the genome structure of progenitor maize through comparative analysis of rice, maize and the domesticated panicoids. *Genetics* 153, 453–473.

Wolfe, K. H. & Shields, D. C. (1997) Molecular evidence for an ancient duplication of the entire yeast genome. *Nature* 387, 708–713.

Zhang, Q., Arbuckle, J. & Wessler, S. R. (2000) Recent, extensive, and preferential insertion of members of the miniature inverted-repeat transposable element family Heartbreaker into genic regions of maize. *Proc. Natl. Acad. Sci. USA* 97, 1160–1165.

12
Flower Color Variation: A Model for the Experimental Study of Evolution

MICHAEL T. CLEGG AND MARY L. DURBIN

We review the study of flower color polymorphisms in the morning glory as a model for the analysis of adaptation. The pathway involved in the determination of flower color phenotype is traced from the molecular and genetic levels to the phenotypic level. Many of the genes that determine the enzymatic components of flavonoid biosynthesis are redundant, but, despite this complexity, it is possible to associate discrete floral phenotypes with individual genes. An important finding is that almost all of the mutations that determine phenotypic differences are the result of transposon insertions. Thus, the flower color diversity seized on by early human domesticators of this plant is a consequence of the rich variety of mobile elements that reside in the morning glory genome. We then consider a long history of research aimed at uncovering the ecological fate of these various flower phenotypes in the southeastern U.S. A large body of work has shown that insect pollinators discriminate against white phenotypes when white flowers are rare in populations. Because the plant is self-compatible, pollinator bias causes an increase in self-fertilization in white

---

Department of Botany and Plant Sciences, University of California, Riverside, CA 92521-0124

This paper was presented at the National Academy of Sciences colloquium "Variation and Evolution in Plants and Microorganisms: Toward a New Synthesis 50 Years After Stebbins," held January 27–29, 2000, at the Arnold and Mabel Beckman Center in Irvine, CA.

Abbreviations: CHS, chalcone synthase; CHI, chalcone isomerase; F3H, flavanone 3b-hydroxylase; DFR, dihydroflavonol reductase; ANS, anthocyanidin synthase.

maternal plants, which should lead to an increase in the frequency of white genes, according to modifier gene theory. Studies of geographical distributions indicate other, as yet undiscovered, disadvantages associated with the white phenotype. The ultimate goal of connecting ecology to molecular genetics through the medium of phenotype is yet to be attained, but this approach may represent a model for analyzing the translation between these two levels of biological organization.

The study of adaptation is a problem that intersects all disciplines of biology. The study of adaptation is also among the most difficult and challenging areas of experimental research because a complete causal analysis of adaptation involves a translation between different levels of biological organization, the ecological, the phenotypic, and the molecular level (Clegg, 2000). In this article we review more than 20 years of research on flower color polymorphisms in the common morning glory [*Ipomoea purpurea* (L.) Roth] that was aimed at exploring this interface.

When the morning glory research program was initiated in the late 1970s, flower color polymorphisms appeared to be a natural starting point because (*i*) they represented simple discrete phenotypes that were susceptible to genetic analysis; (*ii*) a substantial body of work existed on the biochemistry of plant secondary metabolism; (*iii*) flower color was known to be important in insect pollinator behavior; and (*iv*) selection on reproductive performance should be among the most effective forms of selection, and, as a consequence, it should be the component of selection most likely to yield to experimental analysis. Virtually nothing was known in the late 1970s about the molecular biology of the genes that determine the anthocyanin pigments responsible for flower color, so nothing was known about the specific mutational changes associated with different flower color polymorphisms. Since that time a great deal of progress has been made in describing the molecular biology of the genes of flavonoid biosynthesis that determine flower color, but we are still some distance from a complete causal analysis that connects ecology to phenotype to genes.

We begin by discussing the natural history of the morning glory and then turn to a brief account of the genetics of flower color variation in the common morning glory. Next, we describe the flavonoid biosynthetic pathway that determines flower color, and we review pertinent work on the molecular genetics of the genes that encode enzymes within this pathway. Finally, we consider progress in the analysis of selection on flower color variation in natural and experimental populations of the common morning glory.

NATURAL HISTORY OF *I. PURPUREA*

The genus *Ipomoea* includes approximately 600 species distributed on a worldwide scale (Austin and Huaman, 1996) that are characterized by a diversity of floral morphologies and pigmentation patterns. In addition, a wide variety of growth habits, ranging from annual species to perennial vines to longer-lived arborescent forms, are represented in the genus. The common morning glory is an annual bee-pollinated self-compatible vine with showy flowers that is a native of the highlands of central Mexico. As the name morning glory suggests, the flowers open early in the morning and are available for fertilization for a few hours, after which the flower wilts and abscises from the vine. The plant is also a common weed in the southeastern U.S., where it is found in association with field corn and soybean plantings, as well as in roadside and disturbed habitats. The common morning glory is characterized by a series of flower color polymorphisms that include white, pink, and blue (or dark blue) phenotypes. A primary pollinator in the southeastern U.S. is the bumblebee (*Bombus pennsylvanicus* and *Bombus impatiens*), but occasionally plants are visited by honey bees and some lepidopterans (Epperson and Clegg, 1987a). The flower color phenotypes are thought to have been selected by pre-Columbian peoples, perhaps in association with maize culture (Glover *et al.*, 1996). At some point, the plant was introduced into the southeastern U.S., although the routes and times of introduction remain uncertain. Early floras of the southeastern U.S. indicate the presence of *I. purpurea* populations by the late 1600s, providing a minimum estimate of the residence time in this geographical region.

GENETICS OF FLOWER COLOR VARIANTS IN *I. PURPUREA*

At least 21 floral phenotypes are determined by five genetic loci (Barker, 1917; Ennos and Clegg, 1983; Epperson and Clegg, 1988). Most of these phenotypes have analogous forms in the Japanese morning glory (*Ipomoea nil*), and the genetics of the floral variants in both species appear to be similar, but not identical (Imai, 1927). A widespread polymorphism determines pink versus blue flowers (P/p locus), but the genotype at two other loci modifies the intensity of expression of the P/p locus. One modifier is an intensifier locus (I/i) that doubles the anthocyanin pigmentation in the recessive ii genotype (Schoen *et al.*, 1984). The second modifier locus is a regulatory locus that determines the patterning and degree of floral pigmentation (W/w locus). A fourth locus, the A/a locus, is epistatic to the P/p, I/i, and W/w loci in that the recessive albino phenotype yields a white floral limb independent of genotypic state at the other loci. The A/a locus is also characterized by unstable alleles (denoted a^* or a^f)

that exhibit pigmented sectors on an otherwise albino floral limb (Epperson and Clegg, 1987b, 1992). The pigmented sectors display the color associated with the P/p genotype. Finally, the pigmentation in the floral tube appears to be controlled separately from the outer floral limb, but the genetics of floral tube variation have not been analyzed. Fig. 1 displays the flower color phenotypes determined by these genetic loci.

Flavonoid Biosynthetic Pathway

To put the phenotypic variation into a biochemical context, it is useful to sketch the main outlines of the flavonoid biosynthetic pathway (Fig. 2), which culminates in the production of anthocyanins, the main pigments responsible for flower color. The presence or absence of these pigments affects the coloration of the floral display, which attracts pollinators. The

FIGURE 1. Flower color variation in I. purpurea. Loci are described in the text. The locus that determines the phenotype shown is highlighted in bold. Dashes indicate that the phenotype is dominant and only the dominant allele is therefore indicated. In the aa genotype, for example, the A/a locus is epistatic to the P/p and I/i loci; therefore, the albino phenotype determined by the recessive aa is the same regardless of the state of the other loci.

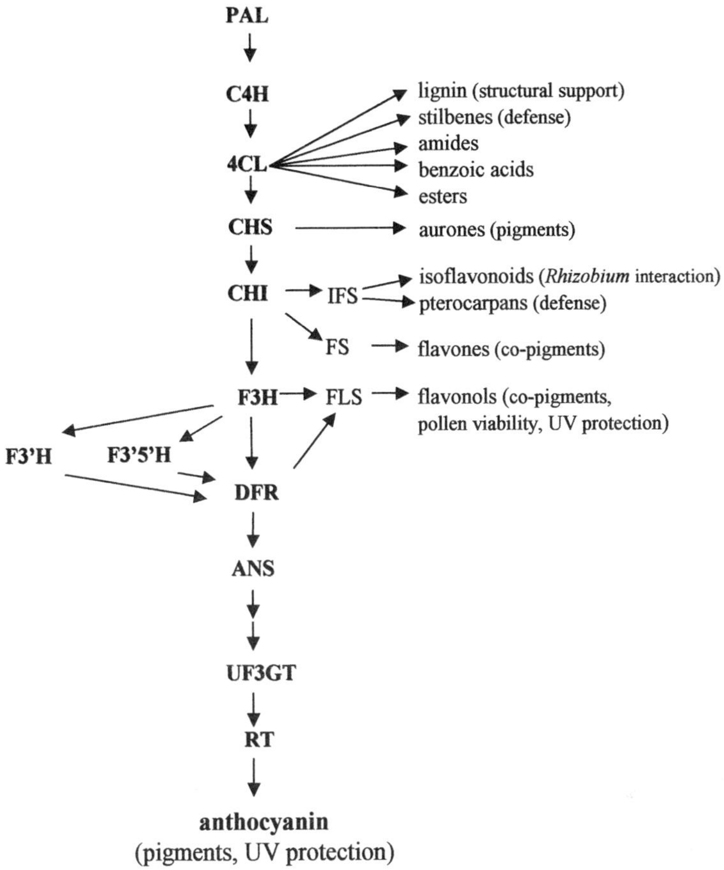

FIGURE 2. Flavonoid biosynthetic pathway. Enzymes involved in anthocyanin biosynthesis and side branches leading to related flavonoid pathways are shown. PAL, phenylalanine ammonialyase; C4H, cinnamate 4-hydroxylase; 4Cl, coumarate: CoA ligase; CHS, chalcone synthase; CHI, chalcone isomerase; F3H, flavanone 3β-hydroxylase; F3'H, dihydroflavonol 3' hydroxylase; F3'5'H, dihydroflavonol 3'5' hydroxylase; DFR, dihydroflavonol reductase; ANS, anthocyanidin synthase; UF3GT, UDP-glucose flavonol 3-0-glucosyl transferase; RT, rhamnosyl transferase.

anthocyanin pigments are therefore important to reproductive success and hence to gene transmission. In addition to pigment production, several side branches of the pathway also produce compounds that are important in plant disease defense, pollen viability, microbial interactions, and UV protection (Koes *et al.*, 1994). The flavonoid pathway has a pleiotropic role in plants, and one must consider that a single mutation in the pathway may have multiple phenotypic effects. The pleiotropic role of the flavonoid pathway is a complication in the effort to link phenotype to molecular changes, but, in addition, many of the genes of the flavonoid pathway are now known to consist of small multigene families (Table 1), so it is essential to associate a mutation in a particular gene with a phenotype of interest. That is, one must identify the gene family member responsible for the observed phenotype.

MOLECULAR CHARACTERIZATION OF THE GENES OF FLAVONOID BIOSYNTHESIS IN *IPOMOEA*

The first committed step in the flavonoid biosynthetic pathway is encoded by the enzyme chalcone synthase (CHS), which catalyzes the formation of naringenin chalcone from three molecules of malonyl-CoA and one molecule of *p*-coumaroyl-CoA (Kreuzaler and Hahlbrock, 1975). Several lines of evidence suggested that the A/a locus might encode a CHS gene. First, the albino phenotype is epistatic to all other flower color variants, suggesting an early point of action in the pathway, and, second, the albino phenotype is consistent with a blockage at CHS. In an attempt to show that the A/a locus encodes a chalcone synthase enzyme, we initiated efforts to clone chalcone synthase genes from a genomic library of *I. purpurea*. Multiple screenings of the library by using a heterologous probe from parsley (Reinhold *et al.*, 1983) were unsuccessful. Subsequent screening of the library with a tomato CHS clone (O'Neill *et al.*, 1990) resulted in the cloning of four genes (CHS-A, -B, -C, and a pseudogene) initially identified as CHS on the basis of nucleotide sequence similarity to other published CHS gene sequences (Durbin *et al.*, 1995).

To provide a comparative context for *Ipomoea* CHS gene family evolution, one can look at the *Petunia* CHS gene family. *Petunia* is important because both *Ipomoea* and *Petunia* are in the same flowering plant order (Solanales), and the *Petunia* CHS gene family had been extensively characterized, with as many as 12 genes provisionally identified (Koes *et al.*, 1987). The four *I. purpurea* genes appear to share a common line of descent with an unusual *Petunia* gene (CHS-B) that is relatively distant in nucleotide sequence from other *Petunia* CHS genes.

Subsequently, Fukada-Tanaka *et al.* (1997) cloned and characterized two more CHS genes from *Ipomoea* (CHS-D and -E) by using differential display and AFLP-based mRNA fingerprinting (Habu *et al.*, 1997). These genes proved to be more closely related in nucleotide sequence to the majority of CHS genes characterized in *Petunia*. Biochemical analysis of *Ipomoea* CHS genes A, B, D, and E revealed that only CHS-D and -E are capable of catalyzing the condensation reaction that results in naringenin chalcone (Shiokawa *et al.*, 2000). The CHS-A and -B genes appear to en-

TABLE 1. Estimated copy number of flavonoid genes in the *I. purpurea* genome

Genes	Number of known family members
CHS	6
CHI	1
F3H	2
F3′H	1
F3′5′H	?
DFR	3
ANS	1
UF3GT	1
RT	?

UF3GT, UDP-glucose flavonol 3-0-glucosyl transferase; RT, rhamnosyl transferase.

code enzymes that produce bisnoryangonin but not naringenin chalcone (H. Noguchi, personal communication). All of the *Ipomoea* CHS genes (except the pseudogene) are expressed, but each displays differential regulatory and developmental control (Durbin *et al.*, 2000; Johzuka-Hisatomi *et al.*, 1999). CHS-D is the most abundantly expressed transcript and is now known to be the one CHS gene solely responsible for anthocyanin production in the floral limb (Durbin *et al.*, 2000; Habu *et al.*, 1998). As discussed in greater detail below, the identification of CHS-D as the A/a locus followed the identification of a transposon (Habu *et al.*, 1998) in CHS-D, which accounted for the sectoring phenotype and loss of pigment. The CHS-E gene is also expressed in the limb but at a much lower level than CHS-D (Durbin *et al.*, 2000; Johzuka-Hisatomi *et al.*, 1999). CHS-E is now believed to be responsible for pigment production in the tube of the flower (Johzuka-Hisatomi *et al.*, 1999).

The next step in the pathway is encoded by chalcone isomerase (CHI), which catalyzes the isomerization of the naringenin chalcone product of CHS to the flavanone naringenin. In the absence of functional CHI, spontaneous isomerization may still occur *in vivo* because this is known to occur *in vitro* under physiological conditions (Miles and Main, 1985). CHI appears to be a single copy gene in *I. purpurea,* as multiple screenings of the genomic library have yielded only a single gene (unpublished data).

Flavanone 3β-hydroxylase (F3H) follows CHI in the pathway and hydroxylates flavonones at the 3 position to form dihydroflavonols, which are required for the synthesis of anthocyanidins and flavonols. F3H consists of at least two copies in the *I. purpurea* genome, both of which are expressed (McCaig, 1998). It is not known at this time whether one of the F3H genes is solely responsible for anthocyanin production in the limb (as is the case for CHS-D). Also, because they were both identified as F3H on the basis of nucleotide similarity, it is not known whether both are even capable of performing the hydroxylation step. One functional copy of F3H and one pseudogene have so far been identified in *I. nil* (Hoshino *et al.*, 1997a; S. Iida, personal communication).

Another hydroxylase, dihydroflavanol 3′ hydroxylase (F3′H), hydroxylates the 3′ position of the dihydroflavonol produced by F3H. This results in the eventual production of the red/magenta cyanidin. F3′H has been characterized in both *I. purpurea* and *I. nil* (Morita *et al.*, 1999) (S. Iida, personal communication).* Yet another hydroxylase, dihydroflavonol 3′5′-hydroxylase (F3′5′H), hydroxylates the 3′ and 5′ position of the dihydroflavonol produced by F3H. This product ultimately leads to the production of the blue/purple delphinidin. F3′5′H has not yet been characterized in *Ipomoea*.

The next step in the flavonoid pathway is dihydroflavonol reductase (DFR) which reduces dihydroflavonols to leucoanthocyanidins. In *I. purpurea*, DFR is a small gene family consisting of at least three tandemly arranged copies (DFR-A, -B, and -C) (Inagaki *et al.*, 1999). DFR-B has been identified as the gene responsible for anthocyanin production in the floral limb based on work from *I. nil* in which a transposon disrupts the DFR-B gene, resulting in a sectoring phenotype and loss of pigment (Inagaki *et al.*, 1994; see below). The function of the two other DFR genes in *I. purpurea* is not known, nor is it known whether they are capable of performing the reductase reaction.

Anthocyanidin synthase (ANS) encodes a dioxygenase and appears to be single copy in *I. purpurea*. UDP-glucose flavonol 3-0-glucosyl transferase glycosylates anthocyanidins and flavonols on the 3 position. This gene appears to be single copy in *I. purpurea*. Rhamnosyl transferase adds rhamnosyl to glucose to form rutinoside. This gene is as yet uncharacterized in *I. purpurea*.

Most Mutant Phenotypes Appear To Be the Result of Transposon Insertions

A wide variety of mobile elements (Table 2) have been identified in the *Ipomoea* genome, largely because of work from the laboratory of Shigeru Iida at the National Institute for Basic Biology in Okasaki, Japan. Some of these mobile element insertions cause phenotypic changes, including those responsible for several flower color variants (Table 3). Much of this work has concentrated on the Japanese morning glory (*I. nil*), where the rich history of morning glory genetics in Japan has provided an extensive research foundation. Considerable work has also been done both in

*Morita, Y., Hoshino, A., Tanaka, Y. Kusumi, T., Saito, N. & Iida, S. (1999) Plant Cell Physiol. 40, Suppl., 124 (abstr.).

TABLE 2. Transposons and mobile elements in *I. purpurea* and *I. nil* associated with genes of the flavonoid pathway

Element	Location	Species	Type	Reference
Tpn1	DFR-B	*I. nil*	En/Spm	Inagaki *et al.* (1994)
Tpn2	CHI	*I. nil*	En/Spm	Hoshino and Iida (1997b)
Tpn3	CHS-D	*I. nil*	En/Spm	Hoshino and Iida (1999)
Tip100	CHS-D	*I. purpurea*	Ac/Ds	Habu *et al.* (1998)
Tip201	F3′H	*I. purpurea*	Ac/Ds	S. Iida, personal communication; Morita *et al.* (1999)
RTipl	ANS flanking	*I. purpurea*	Ty3 gypsy	Hisatomi *et al.* (1997b)
MiniSip1 MiniSip2	ANS flanking	*I. purpurea*	Mini-satellite	Hisatomi *et al.* (1997b)
MELS	DFR flanking	*I. nil*	Mites	Hisatomi *et al.* (1997a)
MELS	CHS-D flanking	*I. purpurea*	Mites	Johzuka-Hisatomi *et al.* (1999)
Sinelp	CHS-D flanking	*I. purpurea*	Sine	unpublished data

TABLE 3. Mutations linked to phenotype in *I. purpurea* and *I. nil*

Phenotype	Mutant allele	Location	Reference
Mutations identified in *I. purpurea*			
Numerous pink sectors on white corolla	$a^{f}1$	*Tip100* in intron of CHS-D	Habu *et al.* (1998)
Few purple sectors on white corolla	a*	*Tip100* in 5′ flanking and intron of CHS-D	unpublished data
Stable white	a12	Two copies of *Tip100* in intron of CHS-D	Johzuka-Hisatomi *et al.* (1999)
Stable white	a	Genomic rearrangement of exon 1 of CHS-D	unpublished data
Pink corolla	p	*Tip201* in exon III of F3′H	S. Iida, personal communication; Morita *et al.* (1999)
Mutations identified in *I. nil*			
Magenta corolla	p	Point mutation at F3′H	S. Iida, personal communication; Morita *et al.* (1999)
Round spots on unpigmented corolla	*speckled*	*Tpn2* insertion in CHI	Hoshino and Iida (1997b)
Colored flecks and sectors	$a\text{-}3^{f}$	*Tpn1* insertion in DFR-B	Inagaki *et al.* (1994)
Double corolla	dp	*Tpn-botan* into C class MADS-box gene	Nitasaka (1997)
White corolla	r-1	*Tpn3* in CHS-D	Hoshino and Iida (1999)

the U.S. and in Japan on *I. purpurea*, which is a member of the same subgenus as *I. nil*. Both species appear to be quite closely related at the DNA sequence level. Molecular clock calculations indicate that the two species diverged roughly three million years ago based on synonymous divergence at DFR and CHS genes (unpublished data). These two closely related species provide a useful comparative framework for the analysis of genetic change over a relatively short period of evolutionary time.

A 3.9-kb Ac/Ds-like element (Tip100) has been identified in the intron of CHS-D near the 5′ junction in flaked (a^f) mutants of *I. purpurea* (Habu *et al.*, 1998). The a^f flower is white with pigmented sectors on the corolla. Another CHS-D mutant phenotype (a12), which has white flowers and no pigmented sectors, has been shown to carry two copies of Tip100 in the intron in the opposite orientation (Habu *et al.*, 1998). In another stable white phenotype (a), a rearrangement of DNA sequence between exon I and an adjacent Tip100 element has occurred (unpublished data). In yet another mutant (a*) of *I. purpurea*, which has a white corolla with very few pigmented sectors, an additional copy of Tip100 was found in the 5′ flanking region of CHS-D (unpublished data). The discovery that disruption of the CHS-D gene results in an unpigmented phenotype confirms that CHS-D is the only CHS gene family member that is responsible for pigment production in the floral limb (Durbin *et al.*, 2000; Johzuka-Hisatomi *et al.*, 1999). Another Ac/Ds-like element, Tip201, has also been discovered in *I. purpurea* (Morita *et al.*, 1999) (S. Iida, personal communication).[‡] Tip201 is found inserted in exon III of F3′H, resulting in a pink phenotype rather than the wild-type blue color corresponding to the P/p locus (Morita *et al.*, 1999) (S. Iida, personal communication).[‡] A point mutation in F3′H in *I. nil* also results in a red phenotype instead of blue (S. Iida, personal communication). Interestingly, this is the only phenotypic change involving flower color that has been characterized at the molecular level in either *I. nil* or *I. purpurea* that is not attributable to the presence of a transposable element.

An En/Spm-like mobile element termed Tpn1 was characterized by Inagaki *et al.* (1994) in the *I. nil* genome. Tpn1 is a 6.4-kb non-autonomous element inserted within the second intron of DFR-B in a mutant termed flecked ($a\text{-}3^{flecked}$). This phenotype has predominantly white flowers with colored sectors. An unusual feature of Tpn1 is that it contains a segment of DNA consisting of four exons that encode part of an HMG-box (High Mobility Group DNA-binding proteins) (Takahashi *et al.*, 1999). This finding is consistent with the observation that transposable elements can cause

---

[‡]Morita, Y., Hoshino, A., Tanaka, Y. Kusumi, T., Saito, N. & Iida, S. (1999) Plant Cell Physiol. 40, Suppl., 124 (abstr.).

rearrangements of the genome (Lönnig and Saedler, 1997). Not only did Inagaki *et al.* (1994) establish that the observed sectoring phenotype was caused by the movement of this new transposable element, but the finding also established that, of the three DFR genes characterized, only one gene family member (DFR-B) is responsible for pigment production in the floral limb.

Another En/Spm-like element, Tpn2, has also been identified in *I. nil* (Hoshino and Iida, 1997b, c).§ Tpn2 is a 6.5-kb element with similarity to Tpn1. Tpn2 is inserted in the second intron of CHI in the speckled mutant. This phenotype is pale yellow with round speckles of pigment on the corolla (Abe *et al.*, 1997). In addition to the En/Spm-like transposons, mobile element-like motifs were identified in the flanking regions of the DFR genes. These elements are similar to miniature inverted-repeat transposable elements (Hisatomi *et al.*, 1997a; Johzuka-Hisatomi *et al.*, 1999; Wessler *et al.*, 1995). Mobile element-like motifs are also found in the DFR region of *I. purpurea* (Johzuka-Hisatomi *et al.*, 1999) and in the CHS-D region of both I. *purpurea* and *I. nil*. Three copies of one of the elements (mobile element-like motif 6) from the CHS-D region were also found in the DFR region (Johzuka-Hisatomi *et al.*, 1999).

Furthermore, three copies of a directly repeated sequence of about 193 bp are found at the 3′ end of the intron in CHS-D in some lines of *I. purpurea* (Johzuka-Hisatomi *et al.*, 1999). Other lines of *I. purpurea* contain four copies of this repeat (unpublished data). In contrast, *I. nil* contains only one copy of this repeat, but it is interrupted by a 529-bp insertion (Johzuka-Hisatomi *et al.*, 1999).

Another En/Spm-related element termed Tpn3 has been found in the CHS-D gene of *I. nil* in the r-1 mutant (Hoshino and Iida, 1999).¶ This mutant bears white flowers with colored tubes. The insertion of this 5.57-kb element into CHS-D results in the accumulation of abnormal sizes of CHS-D mRNAs in the floral tissue (Hoshino and Iida, 1999).¶

A long terminal repeat retrotransposon, RTip1, has been reported associated with the ANS gene region in *I. purpurea* (Hisatomi *et al.*, 1997b). The element is 12.4 kb, contains two long terminal repeat sequences of about 590 kb, and appears to be a defective Ty3 gypsy-like element. The RTip1 element resides within yet another element (MiniSip1), which is described as a minisatellite. Another minisatellite, MiniSip2, has also been described and is located within the Rtip1 element. Thus, there are three elements piggybacked on one another in the ANS 3′ flanking region. Although no phenotypic changes have been linked to these elements,

§Hoshino, A. & Iida, S. (1997b) Genes Genet. Syst. 72, 422 (abstr.).

¶Hoshino, A. & Iida, S. (1999) Plant Cell Physiol. 40, Suppl., 26 (abstr.).

DNA rearrangements in the vicinity of ANS are attributed to their presence (Hisatomi *et al.*, 1997b).

A sine-like element (SineIp) has also been identified in the 5′ flanking region of CHS-D (unpublished data) in some lines of *I. purpurea*. This element is 236 bp and contains the Pol III promoters at the 5′ end of the element. It has 15 bp direct terminal repeats. No phenotypic changes have yet been associated with this element.

Another floral mutation in *I. nil*, not related to the flavonoid pathway, but which is caused by a transposon insertion, is the duplicated mutant. In this mutant phenotype, sexual organs are replaced by perianth organs (petals and sepals) resulting in a double floral whorl. An En/Spm-like insertion, termed Tpn-botan, was found in a C class MADS-box-like gene (Nitasaka, 1997)* that is evidently responsible for this phenotype. A similar phenotype has also been described in *I. purpurea* (Epperson and Clegg, 1992), although the molecular basis for the mutation in *I. purpurea* is unknown.

The vast majority of the phenotypic variation in *Ipomoea* characterized to date at the molecular level appears to be caused by the insertion or deletion of transposable elements (Table 3). It is apparent that a wide variety of mobile elements exist in the *Ipomoea* genome, and these are evidently quite active based on the relatively modest period of evolutionary time that separates *I. nil* and *I. purpurea*. We now turn to studies of the population genetics of some flower color phenotypes in *I. Purpurea*.

GEOGRAPHIC DISTRIBUTION OF FLOWER COLOR POLYMORPHISMS IN *I. PURPUREA*

The geographic distribution of genetic diversity in *I. purpurea* appears paradoxical. Levels of flower color polymorphism are high in the southeastern U.S. whereas Mexican populations are frequently monomorphic for the blue color form (Epperson and Clegg, 1986; Glover *et al.*, 1996). The situation is reversed for biochemical and molecular variation. Mexican populations have levels of isozyme polymorphism that are similar to other annual plants (Glover *et al.*, 1996; Hamrick and Godt, 1989), but U.S. populations are depauperate in isozyme variation. Surveys of ribosomal DNA restriction fragment variation and samples of gene sequence data for the (CHS-A) locus also reveal reduced levels of variation in U.S. populations relative to Mexican populations (Glover *et al.*, 1996; Huttley *et al.*, 1997). We speculate that this pattern is a consequence of the introduction of horticultural forms selected for flower color diversity into the U.S. However, we do not know the source of these introductions.

---

*Nitasaka, E. (1997) *Genes Genet. Syst.* 72, 421 (abstr.).

There are at least two plausible introduction scenarios. One scenario posits a northward migration of the common morning glory along with maize culture over 1,000 years ago. A second scenario posits an introduction of the common morning glory into the southeastern U.S. by European settlers of this region as a horticultural plant. The genetic evidence points to a strong founder effect associated with the introduction of the common morning glory into the southeastern U.S. Maize does not appear to have experienced so extreme a founder effect, and it is not obvious why the two species would experience different population restrictions during a common migration process, so we regard the first scenario as less likely. Under either scenario, we may regard the southeastern U.S. populations as a crude series of experiments in the microevolution of flower color determining genes.

Epperson and Clegg (1986) conducted geographic surveys of flower color variation within the southeastern U.S. at three different spatial scales. The smallest spatial scale (the intrapopulation scale) was analyzed via spatial autocorrelation statistics (Sokal and Oden, 1978); second, the sub regional scale (defined as local populations that range from 0.8 to 32 km apart) was analyzed via gene frequency distances between populations (Nei, 1972); and third, the regional scale ranging from 80 to 560 km between populations, and including much of the southeastern U.S., was also analyzed by using genetic distance statistics. A strong result of these analyses is that there is no correlation between genetic distance and geographic distance at subregional or regional scales for either the W/w or the P/p loci. Populations separated by a few kilometers are as differentiated from one another with respect to the P/p and W/w flower color determining loci as those separated by hundreds of kilometers. Such a pattern is consistent with the hypothesis that the flower color variants were randomly introduced into multiple locations in the southeastern U.S. Analyses of spatial distributions within local populations led to two major conclusions: first, spatial autocorrelation statistics were heterogeneous between the W/w and P/p loci within the same local populations; and, second, analyses of spatial correlograms for the P/p locus revealed genetic neighborhood sizes consistent with an isolation-by-distance model of population structure (Epperson and Clegg, 1986). We begin by elaborating on the second conclusion; we then turn to the importance of the first conclusion in establishing that the W/w locus is subject to selection within local populations.

For clarity, it is useful to review a few elementary definitions in spatial statistics. A spatial autocorrelation measures the correlation in state of a system at two points that are separated by x distance units. For example, in the common morning glory case, the state may be the flower color determined by the P/p locus x distance units apart. The autocorrelation is often graphed as a function of distance (x), and the resulting graph is called the correlogram [$y = f(x)$]. For discrete characters, the convention is to transform the autocorrelation into a standard normal deviate with mean 0 and variance 1, and this transformed function is graphed as the correlogram. Let us again take the concrete example of the common morning glory P/p locus. Three graphs result from the autocorrelation of blue with blue (y_{bb}), pink with pink (y_{pp}), and blue with pink (y_{bp}), each as a function of distance. When graphed as a function of distance, the data from many local *I. purpurea* populations in the southeastern U.S. revealed a common pattern for the P/p locus y_{pp} or $y_{bb} > 2$ and $y_{bp} < -2$ initially (for the smallest values of x). This result indicates a positive autocorrelation over short spatial distances among like phenotypes (blue with blue or pink with pink) and a negative autocorrelation at short spatial scales among unlike phenotypes (blue with pink). The distance (x), where the autocorrelation first crosses the x axis ($y = 0$), provides an operational definition of patch size. These analyses revealed substantial patchiness for the P/p locus and yielded a minimum estimate of about 120 pink (recessive homozygous) plants per patch. The estimates of patch size are highly consistent with simulations of spatial distributions based on a pattern of pollinator flight distances that is strongly biased toward nearest neighbor moves (Turner *et al.*, 1982; Sokal and Wartenberg, 1983).

In contrast to the P/p locus pattern, the W/w locus genotypes show little or no spatial patchiness (that is, $y = 0$ at the smallest values of x, in populations in which $y > 2$ or < -2 initially for the P/p locus). Because both loci are transmitted to successive generations within populations through the same mating process, we expect homogeneous spatial patterns. Heterogeneous spatial patterns argue that isolation-by-distance is not the sole factor governing spatial patterns within local populations. Other forces must be invoked to explain the W/w locus pattern, and the most likely appears to be selection against white homozygous plants. Epperson (1990) has carried out extensive simulation studies that tend to validate this conclusion, and, further, the simulations suggest an intensity of selection against white homozygotes (ww) of approximately 10%. Despite this conclusion, the frequency of white alleles varies from a low of 0% to a high of 43% across local populations with a mean value of about 10% (Epperson and Clegg, 1986). Because the various local populations can be thought of as quasi-independent experiments in which selection has operated for a number of generations, the persistence of white alleles argues strongly for countervailing selective forces favoring the retention of white phenotypes.

The spatial pattern of albino alleles, including unstable alleles (a*), is of interest because this locus also determines a white recessive phenotype that may experience selective pressures common to those experienced by the ww phenotype. Limited sampling across the southeastern U.S. indicates very low frequencies of albino alleles (<1%) in most local populations (Epperson and Clegg, 1992). In addition, unstable alleles also appear to be rare in most local populations. An exceptional local population near Athens, GA is the only population sampled with moderate frequencies of both albino (a) and unstable (a*) alleles (Epperson and Clegg, 1992).

SELECTION ON FLOWER COLOR PHENOTYPES

Because flowering plants often depend on insect pollinators for reproductive success, a natural question to investigate is whether pollinators discriminate among the various flower color phenotypes in morning glory populations. A number of experiments conducted in different years and by different investigators in both natural and experimental populations agree in revealing a bias by bumblebee pollinators against visiting white flowers when white is less than 25% of the population (Brown and Clegg, 1984; Epperson and Clegg, 1987a; Rausher *et al.*, 1993). In contrast, there is no evidence of pollinator discrimination among P/p (blue/pink) or I/i (dark/intense) locus phenotypes (Brown and Clegg, 1984; Schoen and Clegg, 1985). The under visitation of white phenotypes is correlated with an increased frequency of self-fertilization by white maternal parents based on estimates using isozyme marker loci (Brown and Clegg, 1984; Epperson and Clegg, 1987a). However, Schoen and Clegg (1985) discovered that pollinator visitation rates and outcrossing estimates did not differ between white and pigmented phenotypes when the two forms were in equal frequency. Later, Epperson and Clegg (1987a) and Rausher *et al.* (1993) showed that the pollinator discrimination against whites, and the reduced outcrossing rate of white maternal plants, was frequency-dependent and disappeared when the frequency of the white phenotype approached 50%. These observations and experiments indicate that white loci act as mating system modifier loci and consequently bias their own transmission to subsequent generations.

There is substantial theoretical literature on the population genetics of modifier loci (Fisher, 1941; Holsinger, 1996; Karlin and McGregor, 1974). According to this literature, white genes that increase the frequency of self-fertilization are expected to increase in frequency to fixation within populations because the white gene should be transmitted differentially to progeny of white maternal plants via self-fertilization. This assumes that the selfing (white) gene is also transmitted to the outcross pool in proportion to its frequency in the population, where it can fertilize ovules of non-white maternal phenotypes (absence of pollen discounting). ["Pollen discounting" refers to the situation in which pollen transmission to other (nonmaternal) plants is reduced by a gene promoting self-fertilization, as might be the case when pollinators avoid particular floral displays.] The degree of advantage of self-fertilization genes, if any, clearly hinges on several additional factors, including (*i*) the extent of pollen discounting; (*ii*) differential maternal fertility associated with autopollination; and (*iii*) reduced viability of progeny derived through self-fertilization because of inbreeding depression. A number of experiments have been conducted to investigate each of these additional factors.

Rausher *et al.* (1993) found no evidence for pollen discounting in experimental populations of morning glory and instead found a nonsignificant excess of white gene fertilizations of ovules of non-white parents. In a different set of experiments, Epperson and Clegg (unpublished data) found a nonsignificant deficit of white gene fertilizations of ovules of non-white parents. Much larger experiments with greater statistical power are needed to provide a definitive answer about the role of pollen discounting, but at present there is no clear evidence that the advantage of white genes is offset by pollen discounting. Epperson and Clegg (unpublished data) have also measured the fertility of white maternal parents subjected to autopollination and find a reduction in maternal fertility that could act to moderate or eliminate the transmission advantage of white genes. Experiments to measure inbreeding depression provide evidence for a depression in fitness associated with self-fertilization, but the magnitude is not sufficient to offset the selfing advantage of white genes (Chang and Rausher, 1999). Other studies have suggested heterogeneity among families in segregation bias favoring the transmission of either white or dark alleles in different heterozygous parents (Fry and Rausher, 1997). Finally, overdominance in seed size among the progeny of white maternal parents has also been documented, but this evidently does not translate into a fitness advantage (Mojonnier and Rausher, 1997).

In summary, there is clear and convincing evidence that white phenotypes suffer some disadvantage in natural populations based on spatial autocorrelation analyses, and there is clear and convincing evidence that white genes have a transmission advantage when white maternal plants are infrequent in populations. The transmission advantage is associated with pollinator preferences and a consequent increased rate of self-fertilization among white maternal parents, but this advantage is one-sided and should diminish to zero as the frequency of white maternal types approaches 50%. Many lines of evidence argue for a compensatory disadvantage of white types to account for the global frequency in the southeastern U.S. of approximately 10%, but the precise nature of the disadvantage is unresolved. It may be that a number of separate components such as reduced maternal fertility under autopollination, inbreeding depression, and segregation bias all sum to provide the needed balance. Because small effects demand very large experiments to provide adequate statistical power, it may not prove feasible to measure each component of fitness with sufficient accuracy to identify all of the factors that comprise a balance. It is also important to note that, despite much experimental effort over two decades, no evidence has emerged for differential selection among other flower color phenotypes such as those determined by the P/p and I/i loci.

CONCLUSIONS

We began this work with the conviction that a full understanding of the mechanisms of adaptation would entail an integration of the ecological and the genetic dimensions of biology. This wisdom was by no means original but, rather, was embodied in Ledyard Stebbins' monumental contributions to plant evolutionary biology. As Stebbins emphasized, the point of contact between these two levels of biological organization is the phenotype. Whether a phenotype is adapted to a particular environment depends on a host of biotic and abiotic factors that collectively translate into survival and reproductive success. It is often difficult to identify the environmental elements that occasion phenotypic success. We began with the notion that flower color provided a simple model for the study of adaptation because one aspect of the environment seemed, *a priori*, to be predominant—the behavior of insect pollinators. To a large extent a substantial body of experimental work has validated this assumption, but along the way we have been forced to begin to confront the full complexity of plant reproductive biology. In a similar vein, the phenotypic bases of flower color variation appeared to be susceptible to analysis because much was already known about the biochemistry and genetics of flower color determination and because the tools for molecular analysis were rapidly appearing on the horizon. What we did not anticipate was the remarkable ecology of plant genomes in which most genes are redundant and in which mobile elements are a major player in the generation of phenotypic diversity.

One complexity that may be more apparent than real is the problem of genetic redundancy. As noted in this article, most of the genes of flavonoid biosynthesis occur in multiple copies, and sorting through this redundancy to find the particular genes responsible for individual flower color phenotypes appeared daunting at first sight. The actual findings are encouraging, in that particular gene copies of CHS and DFR are shown to be causally responsible for particular flower color phenotypes. In the case of CHS, we know that the other redundant copies are predominantly expressed at other stages in development (e.g., CHS-E in the floral tube) or they have diverged in catalytic properties (e.g., CHS-A, -B, and -C).

Another complexity is the pleiotropic nature of biosynthetic pathways. At first sight it appears that each mutation in a gene encoding a pathway component is likely to have many phenotypic effects. Is it possible to associate a main effect with one aspect of phenotype? Perhaps, if gene redundancy and specialization in expression patterns approximates a one-to-one correspondence between gene and phenotype. Whether the unraveling of phenotypic determination will be so simple remains to be seen, but there is some reason to be optimistic based on flower color analyses.

Another fascinating result of this work is the overwhelming importance of mobile elements as generators of phenotypic diversity. All except one of the flower color phenotypes analyzed so far in both *I. purpurea* and *I. nil* are the result of transposon insertions. This strongly implies that transposable elements are the major cause of mutations that yield an obvious phenotype in these plant species. Because the distribution of transposons appears to be heterogeneous across plant genomes, it seems reasonable to speculate that some species may experience higher mutation rates and therefore may be more flexible in adapting to environmental changes than other plant species. In the case of *I. purpurea*, a significant environmental change was the appearance of human plant domesticators who selected and propagated unusual phenotypes for esthetic and perhaps other purposes. The plant has clearly been successful, at least in the short term, by virtue of its ability to adapt to this circumstance.

A number of questions remain to be resolved before this work can be seen as complete. First, the gene that is clearly subject to selection in nature is not a structural gene determining an enzyme in the anthocyanin pathway but, instead, a regulatory gene that determines the floral distribution of pigmentation (W/w locus). To date, this gene has not been characterized at the molecular level, and we do not know the nature of the mutational changes that cause the white phenotype. We do know a lot about the A/a locus, which also determines a white (albino) phenotype, and we can probably assume that this phenotype is also discriminated against by insect pollinators, but the albino phenotype is rare in populations. So major gaps exist in our present knowledge, but we have every reason to expect these gaps to fill in over time. Interestingly, we do know the molecular bases for the P/p locus phenotypes, but so far there is no evidence of selection in nature at this locus, although in a broader sense it is clear that man has selected this polymorphism for propagation.

A second limitation is that most population work has focused on the southeastern U.S., where the plant is introduced, rather than on native populations in the central highlands of Mexico. The southeastern U.S. is not the geographical region in which the species evolved, and it is not possible to make inferences about the longer term ecological circumstances that shaped the evolution of this species based on investigations in other geographical and ecological settings. Despite this reservation, the southeastern U.S. populations represent a crude series of experiments that trace to introductions within the last several hundred to one thousand years, and the ability to place temporal limits on the history of these populations is a strength of the morning glory program. Common patterns among populations are indicative of common initial conditions or common selective forces. Thus, the lack of spatial autocorrelation for the W/w locus phenotypes over populations is strongly indicative of a disadvantage that is independent of location. Similarly, the case for an isolation-by-distance model of population structure for the P/p locus is strongly supported by the consistency of this result over local populations.

A third consideration is the problem of statistical power. Because the structure of statistical hypothesis testing is deliberately biased against accepting the alternative hypothesis (power is usually much less than the size of the critical region, α, under the null hypothesis), very large experiments must be carried out to detect moderately large effects. Put another way, magnitudes of selection that can be quite effective from an evolutionary standpoint cannot be detected in experiments of reasonable size. This may explain the failure to detect pollen discounting and the failure to detect significant inbreeding depression in experimental populations of *I. purpurea*. Limited statistical power can in part be overcome by experiments that run over many generations in which the effects are cumulative. This is the strength of the geographical and population studies in the southeastern U.S., which represents the cumulative outcome of many generations rather than the result of single generation experiments.

Despite the limitations noted above, the study of model evolutionary systems is just as important as the study of other model systems. Model systems should reveal the important questions and provide a basis for inventing novel approaches that can then be extended to more refractory systems. As one example, Stebbins pioneered the analysis of plant development as an essential prerequisite to understanding the determination of phenotype and hence to the understanding of adaptation. This approach is just as important today as it was more than 40 years ago, when Stebbins began his classic work with barley awn development as his model system. Today we have a rich suite of molecular tools that assist in unraveling the determination of phenotype and in penetrating the complexities associated with the coordinated action of many genes. The determination of floral color development is an area of special promise because the translation between genes and phenotype is tractable. Similarly, the translation between environment and phenotype is more transparent for flower color than in most other cases, and there is still much to be learned from this research strategy.

We thank Dr. Shigeru Iida of the National Institute for Basic Biology in Okazaki, Japan for his critical review and suggestions and for sharing his unpublished data. We also thank Dr. Hiroshi Noguchi from the School of Pharmaceutical Sciences, University of Shizuoka, Shizuoka, Japan for sharing his unpublished data on the substrate specificity testing of the I. purpurea CHS genes. This work was supported in part by a grant from the Alfred P. Sloan Foundation.

REFERENCES

Abe, Y., Hoshino, A. & Iida, S. (1997) Appearance of flower variegation in the mutable *speckled* line of the Japanese morning glory is controlled by two genetic elements. *Genes Genet. Syst.* 72, 57–62.

Austin, D. F. & Huaman, Z. (1996) A synopsis of *Ipomoea* (Convolvulaceae) in the Americas. *Taxon* 45, 3–38.

Barker, E. E. (1917) Heredity studies in the morning glory (*Ipomoea purpurea*). *Bull. Cornell Univ. Agric. Exp. Sta.* 392, 121–154.

Brown, B. A. & Clegg, M. T. (1984) Influence of flower color polymorphism on genetic transmission in a natural population of the common morning glory, *Ipomoea purpurea*. *Evolution* 38, 796–803.

Chang, S. M. & Rausher, M. D. (1999) The role of inbreeding depression in maintaining the mixed mating system of the common morning glory, *Ipomoea purpurea*. *Evolution* 53, 1366–1376.

Clegg, M. T. (2000) Limits to knowledge in population genetics. In *Limits to Knowledge in Evolutionary Genetics*, ed. Clegg, M. T. (Plenum Press, New York) in press.

Durbin, M. L., Learn, G. H., Huttley, G. A. & Clegg, M. T. (1995) Evolution of the chalcone synthase gene family in the genus *Ipomoea*. *Proc. Natl. Acad. Sci. U.S.A.* 92, 3338–3342.

Durbin, M. L., McCaig, B. & Clegg, M. T. (2000) Molecular evolution of the chalcone synthase multigene family in the morning glory genome. *Pl. Molec. Biol.* 42, 79–92.

Ennos, R. A. & Clegg, M. T. (1983) Flower color variation in morning glory, *Ipomoea purpurea* Roth. *J. Hered.* 74, 247–250.

Epperson, B. K. & Clegg, M. T. (1986) Spatial-autocorrelation analysis of flower color polymorphisms within substructured populations of morning glory (*Ipomoea purpurea*). *Amer. Natur.* 128, 840–858.

Epperson, B. K. & Clegg, M. T.. (1987a) Frequency-dependent variation for outcrossing rate among flower color morphs of *Ipomoea purpurea*. *Evolution* 41, 1302–1311.

Epperson, B. K. & Clegg. M. T. (1987b) Instability at a flower color locus in morning glory, *Ipomoea purpurea*. *J. Hered.* 78, 346–352.

Epperson, B. K. & Clegg, M. T. (1988) Genetics of flower color polymorphism in the common morning glory, *Ipomoea purpurea*. *J. Hered.* 79, 64–68.

Epperson, B. K. (1990) Spatial autocorrelation of genotypes under directional selection. *Genetics* 124, 757–771.

Epperson, B. K. & Clegg, M. T. (1992) Unstable white flower color genes and their derivatives in the morning glory. *J. Hered.* 83, 405–409.

Fisher, R. (1941) Average excess and average effect of a gene substitution. *Ann. Eugen.* 11, 53–63.

Fry, J. D. & Rausher, M. D. (1997) Selection on a floral color polymorphism in the tall morning glory (*Ipomoea purpurea*): Transmission success of the alleles through pollen. *Evolution* 51, 66–78.

Fukada-Tanaka, S., Hoshino, A., Hisatomi, Y., Habu, Y., Hasebe, M. & Iida, S. (1997) Identification of new chalcone synthase genes for flower pigmentation in the Japanese and common morning glories. *Plant and Cell Physiol.* 38, 754–758.

Glover, D. E., Durbin, M.L., Huttley, G. & Clegg, M. T. (1996) Genetic diversity in the common morning glory. *Plant Species Biol.* 11, 41–50.

Habu, Y., Fukada-Tanaka, S., Hisatomi, Y. & Iida, S. (1997) Amplified restriction fragment length polymorphism-based mRNA fingerprinting using a single restriction enzyme that recognizes a 4-bp sequence. *Biochem. Biophys. Res. Comm.* 234, 516–521.

Habu, Y., Hisatomi, Y. & Iida, S. (1998) Molecular characterization of the mutable flaked allele for flower variegation in the common morning glory. *Plant Journal* 16, 371–376.

Hamrick, J. L. & Godt, M.J. (1989) Allozyme diversity in plant species. In *Plant Population Genetics, Breeding, and Genetic Resources*, eds. Brown, A. D. H, Clegg, M. T., Kahler, A. L. & Weir, B. S. (Sinauer Associates, Sunderland, MA), pp. 43–63.

Hisatomi, Y., Yoneda, Y., Kasahara, K. Inagaki, Y. & Iida, S. (1997a) DNA rearrangements at the region of the dihydroflavonol 4-reductase gene for flower pigmentation and incomplete dominance in morning glory carrying the mutable flaked mutation. *Theor. Appl. Genet.* 95, 509–515.

Hisatomi, Y., Hanada, K. & Iida, S. (1997b) The retrotransposon RTip1 is integrated into a novel type of minisatellite, MiniSip1, in the genome of the common morning glory and carries another new type of minisatellite, MiniSip2. *Theor. Appl. Genet.* 95, 1049–1056.

Holsinger, K. E. (1996) Pollination biology and the evolution of mating systems in flowering plants. In *Evolutionary Biology*, eds. Hecht, M. K., MacIntyre, R. J. & Clegg, M. T. (Plenum, New York), pp. 107–149.

Hoshino, A., Abe, Y., Saito, N., Inagaki, Y. & Iida, S. (1997a) The gene encoding flavanone 3-hydroxylase is expressed normally in the pale yellow flowers of the Japanese morning glory carrying the speckled mutation which produce neither flavonol nor anthocyanin but accumulate chalcone, aurone and flavanone. *Plant Cell Physiol.* 38, 970–974.

Hoshino, A. & Iida, S. (1997b) Molecular analysis of the mutable *speckled* allele of the Japanese morning glory. ABSTRACT. *Genes Genet. Syst.* 72, 422.

Hoshino, A. & Iida, S. (1997c) Identification and characterization of the mutable speckled allele in the Japanese morning glory. Collected abstracts of the 5th Internation Congress of Plant Molecular Biology (Singapore). *Plant Mol. Biol. Rep. Suppl.* 15, 3.

Hoshino, A. & Iida, S. (1999) Molecular characterization of the En/Spm related transposable element found within the *CHS* Gene in the r-1 mutant of the Japanese morning glory. ABSTRACT *Plant Cell Physiol.* 40, s26.

Huttley, G. A., Durbin, M. L., Glover, D. E. & Clegg, M. T. (1997) Nucleotide polymorphism in the chalcone synthase-A locus and evolution of the chalcone synthase multigene family of common morning glory *Ipomoea purpurea*. *Molec. Ecol.* 6, 549–558.

Imai, Y. (1927) The vegetative and seminal variations observed in the Japanese morning glory, with special reference to its evolution under cultivation. *J. Coll. Agric. Imp. Univ. Tokyo* 9, 199–222.

Inagaki, Y., Hisatomi, Y., Suzuki, T., Kasahara, K. & Iida, S. (1994) Isolation of a suppressor-mutator enhancer-like transposable element, TPN1, from Japanese morning glory bearing variegated flowers. *Plant Cell* 6, 375–383.

Inagaki, Y., Johzuka-Hisatomi, Y., Mori, T., Takahashi, S., Hayakawa, Y., Peyachoknagul, S, Ozeki, Y. & Iida, S. (1999) Genomic organization of the genes encoding dihydroflavonol 4-reductase for flower pigmentation in the Japanese and common morning glories. *Gene* 226, 181–188.

Johzuka-Hisatomi, Y., Hoshino, A., Mori, T., Habu, Y. & Iida, S. (1999) Characterization of the chalcone synthase genes expressed in flowers of the common and Japanese morning glories. *Genes Genet. Syst.* 74, 141–147.

Karlin, S. & McGregor, J. (1974) Towards a theory of the evolution of modifier genes. *Theor. Pop. Biol* 5, 59–103.

Koes, R. E., Spelt, C. E., Mol, J. N. M. & Gerats, A. G. M. (1987) The chalcone synthase multigene family of *Petunia hybrida* (V30): sequence homology, chromosomal localization and evolutionary aspects. *Plant Mol. Biol.* 10, 159–169.

Koes, R. E., Quattrocchio, F. & Mol, J. N. M. (1994) The flavonoid biosynthetic pathway in plants—function and evolution. *Bioessays* 16, 123–132.

Kreuzaler, F. & Hahlbrock, K. (1975) Enzymatic synthesis of aromatic compounds in higher plants. *Arch. Biochem. Biophys.* 169, 84–90.

Lönnig, W.-E. & Saedler, H. (1997) Plant transposons: contributors to evolution? *Gene* 205, 245–253.

McCaig, B. (1998) The molecular evolution and expression of anthocyanin multigene families in *Ipomoea purpurea* (common morning glory): pathway evolution. Ph.D. Dissertation, University of California, Riverside.

Miles, C. O. & Main, L. (1985) Kinetics and mechanism of the cyclisation of 2', 6'-Dihydroxy-4,4,-dimethoxychalcone: influence of the 6'-hydroxy group on the rate of cyclisation under neutral conditions. *J. Chem. Soc. Perkin Trans.* 2, 1639–1642.

Mojonnier, L. E. & Rausher, M. D. (1997) A floral color polymorphism in the common morning glory (*Ipomoea purpurea*): The effects of overdominance in seed size. *Evolution* 51, 608–614.

Morita, Y., Hoshino, A., Tanaka, Y., Kusumi, T., Saito, N. & Iida, S. (1999) Identification of the *magenta* mutations for flower pigmentation in the Japanese and common morning glories. ABSTRACT *Plant Cell Physiol.* 40, Suppl., 124.

Nei, M. (1972) Genetic distance between populations. *Am. Nat.* 106, 283–292.

Nitasaka, E. (1997) Molecular cloning of a floral homeotic gene *duplicated* in *Ipomoea nil.* ABSTRACT *Genes Genet. Syst.* 72, 421.

O'Neill, S. D., Tong, Y., Sporlein, B., Forkmann, G. & Yoder, Y. I. (1990) Molecular genetic analysis of chalcone synthase in *Lycopersicon esculentum* and an anthocyanin-deficient mutant. *Mol. Gen. Genet.* 224, 279–288.

Rausher, M. D., Augustine, D., & Vanderkooi, S. (1993) Absence of pollen discounting in a genotype of *Ipomoea purpurea* exhibiting increased selfing. *Evolution* 47, 1688–1695.

Reinhold, U., Krüger, M., Kreuzaler, M. & Hahlbrock, K. (1983) Coding and 3' non-coding nucleotide sequence of chalcone synthase mRNA and assignment of amino acid sequence of the enzyme. *EMBO J.* 2, 1801–1805.

Schoen, D. J., Giannasi, D. E., Ennos, R. A. & Clegg, M. T. (1984) Stem color and pleiotropy of genes determining flower color in the common morning glory. *J. Hered.* 75, 113–116.

Schoen, D. J. & Clegg, M. T. (1985) The influence of flower color on outcrossing rate and male reproductive success in *Ipomoea purpurea. Evolution* 29, 1242–1249.

Shiokawa, K., Inagaki, Y., Morita, H., Hsu, T., Iida, S. & Noguchi, H. (2000) The functional expression of the CHS-D and CHS-E genes of the common morning glory (*Ipomoea purpurea*) in *Escherichia coli* and characterization of their gene products. *Plant Biotechnol.* 17, in press.

Sokal, R. R. & Wartenberg, D. E. (1983) A test of spatial autocorrelation analysis using an isolation-by-distance model. *Genetics* 105, 219–237.

Sokal, R. R. & Oden, N. L. (1978) Spatial autocorrelation in biology. *J. Linn. Soc.* 10, 199–228.

Takahashi, S., Inagaki, Y., Satoh, H., Hoshino, A. & Iida, S. (1999) Capture of a genomic HMG domain sequence by the En/Spm-related transposable element Tpn1 in the Japanese morning glory. *Molecular and General Genetics* 261, 447–451.

Turner, M. E., Stephens, J. C. & Anderson, W. W. (1982) Homozygosity and patch structure in plant populations as a result of nearest-neighbor pollination. *Proc. Natl. Acad. Sci. USA* 79, 203–207.

Wessler, S. R., Bureau. T. E. & White, S. E. (1995) LTR-retrotransposons and mites—important players in the evolution of plant genomes. *Current Opinion in Genetics & Development* 5, 814–821.

13

Gene Genealogies and Population Variation in Plants

BARBARA A. SCHAAL AND KENNETH M. OLSEN

Early in the development of plant evolutionary biology, genetic drift, fluctuations in population size, and isolation were identified as critical processes that affect the course of evolution in plant species. Attempts to assess these processes in natural populations became possible only with the development of neutral genetic markers in the 1960s. More recently, the application of historically ordered neutral molecular variation (within the conceptual framework of coalescent theory) has allowed a reevaluation of these microevolutionary processes. Gene genealogies trace the evolutionary relationships among haplotypes (alleles) within populations. Processes such as selection, fluctuation in population size, and population substructuring affect the geographical and genealogical relationships among these alleles. Therefore, examination of these genealogical data can provide insights into the evolutionary history of a species. For example, studies of *Arabidopsis thaliana* have suggested that this species underwent rapid expansion, with populations showing little genetic differentiation. The new discipline of phylogeography examines the distribution of allele genealogies in an explicit geographical context. Phylogeographic studies of plants have documented the recolo-

---

Department of Biology, Washington University, St. Louis, MO 63130

This paper was presented at the National Academy of Sciences colloquium "Variation and Evolution in Plants and Microorganisms: Toward a New Synthesis 50 Years After Stebbins," held January 27–29, 2000, at the Arnold and Mabel Beckman Center in Irvine, CA.

nization of European tree species from refugia subsequent to Pleistocene glaciation, and such studies have been instructive in understanding the origin and domestication of the crop cassava. Currently, several technical limitations hinder the widespread application of a genealogical approach to plant evolutionary studies. However, as these technical issues are solved, a genealogical approach holds great promise for understanding these previously elusive processes in plant evolution.

In the following succinct statements, G. L. Stebbins presents what would become the framework for the study of plant evolutionary mechanisms for the next 50 years (Stebbins, 1950).

> Individual variation, in the form of mutation and gene recombination, exists in all populations; . . . the molding of this raw material . . . into variation on the level of populations by means of natural selection, fluctuation in population size, random fixation and isolation is sufficient to account for all of the differences, both adaptive and non-adaptive, which exist between related races and species . . .
>
> The problem of the evolutionist is . . . evaluating on the basis of all available evidence the role which each of these known forces has played in any particular evolutionary line . . .

A central thesis of Stebbins' seminal book, *Variation and Evolution in Plants* (Stebbins, 1950), is the notion that to understand evolution we must examine its action at the level of populations within species. This reasoning may seem obvious to contemporary readers, but at the time of Stebbins' writing, the importance of population-level processes for evolution was far from apparent. Stebbins' elucidation of this connection is one of his most enduring contributions to plant evolutionary biology.

Fifty years ago, the study of plant evolution was necessarily concerned with the phenotype, much of which is subject to selection. Morphology, karyotypes, and fitness components are central traits for understanding evolution and adaptation, but they limit which evolutionary processes can be studied. In his book, Stebbins discusses such events as fluctuations in population size, random fixation (genetic drift), and isolation as all affecting the process of evolution (see passage quoted above; Stebbins, 1950). However, the study of these mechanisms requires markers that are not under selection.

In the years after the publication of Stebbins' book, among the first major technical advances in evolutionary biology were the development of protein electrophoresis and the identification of allozyme variation in natural populations. Many allozymes are selectively neutral, and thus, for the first time, evolutionary biologists could attempt to assess the amount of neutral genetic variation within species as well as its spatial distribution. Plant species were found to vary widely, both in levels of genetic variation and in the apportionment of this variation within and among populations. These observations spurred researchers to examine the mechanisms underlying the process of genetic differentiation in plants. One of the most common approaches for doing this analysis has been to look for correlations between the life history characteristics of a species (e.g., mechanisms of pollen and seed dispersal, system of mating, and generation length) and patterns of population genetic differentiation (Hamrick and Godt, 1989). Neutral allelic variation from allozymes also can be used to estimate levels of gene flow among populations. The population genetic theory developed by Wright, Fisher, and Malecot (among others) established that, for a group of populations at equilibrium, the level of genetic differentiation is roughly inversely proportional to the level of interpopulation gene flow per generation. This relationship is expressed by Wright's (1951) familiar equation for estimating gene flow under an island model: $F_{ST} \approx 1/(4N_e m + 1)$, where F_{ST} is the standardized variance in allele frequencies among populations, N_e is the effective population size, and m is the migration rate.

The use of allozymes has led to more than 30 years of insight into how plant populations evolve. However, inferring population structure solely from allele frequencies has its limitations. Allozyme alleles (or their DNA analogs, restriction fragment length polymorphisms, amplified fragment length polymorphisms, and microsatellites) are unordered, meaning that the genealogical pattern of relationships among alleles cannot be easily inferred. As a result, these data cannot be used in directly assessing genetic change over time but rather require indirect approaches based on models that often assume equilibrium conditions. For example, Wright's equation (above) for quantifying gene flow under the island model assumes that populations have reached an equilibrium between gene flow and random genetic drift. This equilibrium perspective can be biologically misleading, particularly for species in which recent history is a major determinant of population structure. We speculate that very few plant species have reached, or will ever reach, a gene flow-drift equilibrium. Many plants, both temperate and tropical, have altered their range subsequent to glaciations in the last 20,000 years, a recent event on an evolutionary time scale. Likewise, many plant species have a metapopulation structure with subpopulations continually being colonized, dispersing migrants, and going extinct. Such metapopulations may reach a system-wide equilibrium in which probabilities of extinction and recolonization are constant given enough time, but such a situation is unlikely considering the relatively short time frame of global climatic changes in the past.

Equilibrium in plants is for the most part a theoretical construct with little relation to reality.

We would argue that the second major development since the publication of Stebbins' work has been the application of ordered, genealogical data to the study of population-level processes. This development, which has begun to reach its potential only in the last decade, was predicated on two major advances, one technical and the other conceptual. The technical advance has been the widespread availability of ordered genetic variation at the intraspecific level, typically in the form of DNA sequence variation or mapped restriction-site data. For such data, mutational differences among genetic variants indicate the patterns of relationship among variants. These data therefore provide the raw information needed to reconstruct genealogical relationships among alleles (i.e., gene trees).

The conceptual advance has been the application of coalescent theory to the study of microevolutionary processes (for a recent review, see Fu and Li, 1999). For a population of constant size, new alleles are continually arising through mutation, and others are going extinct over successive generations (assuming neutrality). Therefore, the extant alleles of a gene in a population are all derived from (i.e., coalesce to) a single common ancestral allele that existed at some point in the past (Fig. 1). Coalescent theory provides a framework for studying the effects of population-level processes (e.g., population size fluctuations, selection, and gene flow) on the expected time to common ancestry of alleles within a gene tree. The application of gene genealogies to population genetics has allowed the study of population-level processes within a temporal, nonequilibrium framework. Thus, microevolution can be studied as a dynamic, historical process, changing over time within a species.

At the foundation of all genealogical analyses is the gene tree, which represents the inferred genealogical relationships among alleles observed in a species. Most intraspecific gene trees are unrooted, because one often cannot determine the temporal polarity of mutations, even with an outgroup (Castelloe and Templeton, 1994). A common means of representing the inferred genealogy is with a "minimum spanning tree" (e.g., Smouse, 1998), for which the number of mutational changes among alleles is minimized (Fig. 1). If homoplasy (mutational convergence or reversal) is infrequent, then a single most parsimonious minimum spanning tree often can be inferred by using maximum parsimony search algorithms (Swofford, 1993). For data showing high levels of homoplasy, more complicated tree estimation algorithms may be required (e.g., Templeton *et al.*, 1992). Extant alleles on a gene tree are often separated by more than one mutational step, and thus, the gene tree typically contains a number of inferred intermediate alleles; these unobserved alleles may be extinct, may have been missed during population sampling, or may never have existed at all (if mutations did not accumulate in single steps).

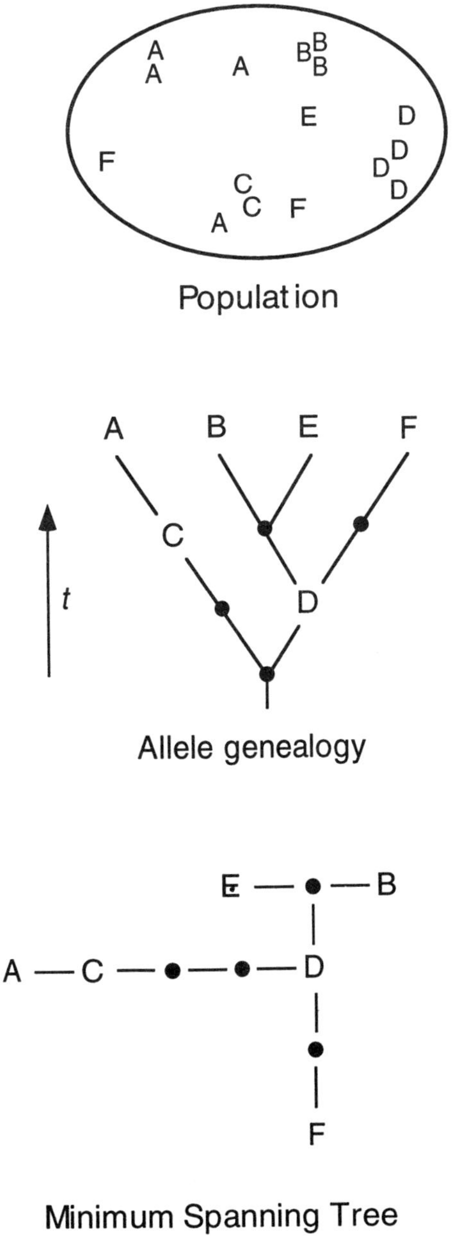

FIGURE 1. Hypothetical population (Top) showing geographical distribution of alleles; allele genealogy (Middle) indicating true history of allelic divergence over time; unrooted minimum spanning tree (Bottom) showing inferred genealogical relationships among alleles.

Allele genealogies can inform us about the effects of microevolutionary forces on organismal and population lineages. However, as has become well established in the last decade, a gene tree is far from equivalent to the population lineages through which it is transmitted (see review in Avise, 2000). Therefore, caution must be used in drawing inferences about population-level processes from genealogical data. The hypothetical allele genealogy in Fig. 2 illustrates the potential incongruity that can exist between a gene tree and the populations in which it exists. After two populations have become isolated from each other (and barring subsequent gene flow), the populations will diverge genetically until eventually all of the alleles within each population are more closely related to each other than to those from the other population (Fig. 2). At this point, the alleles show reciprocal monophyly with respect to the two populations and accurately reflect the history of population divergence. Before reaching reciprocal monophyly, however, alleles are expected to be poly-

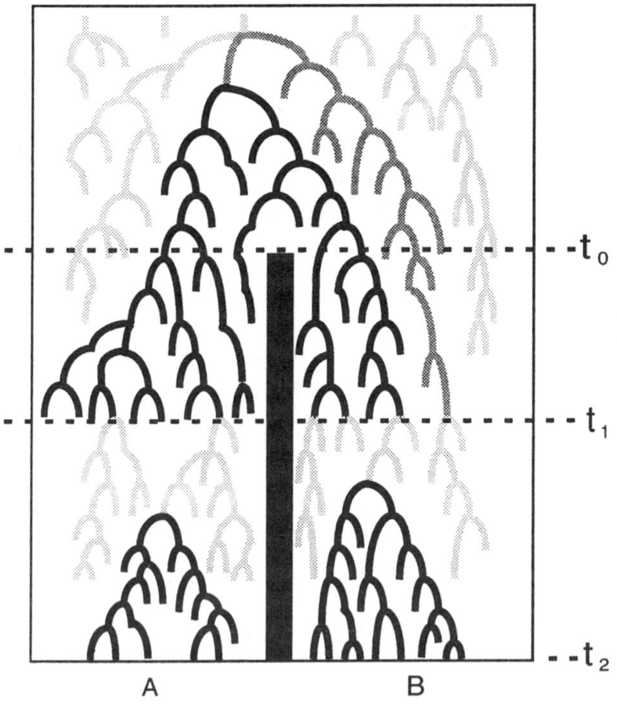

FIGURE 2. Hypothetical allele genealogy in populations A and B that became isolated from each other at time t_0. At time t_1, genealogical relationships show paraphyly with respect to the two populations. At time t_2, alleles show reciprocal monophyly and are congruent with the history of population divergence.

phyletic, then paraphyletic, with respect to the populations (Neigel and Avise, 1986). In these cases, genealogical relationships among alleles are not expected to correspond to population identity (Fig. 2). Thus, for recently diverged populations, inferences about the history of population divergence based on the gene tree may be misleading or erroneous. In some cases, ancestral allelic variation may actually persist in populations after population divergence. These shared ancestral polymorphisms can easily be misinterpreted as evidence of interpopulation gene flow.

Below, we present several studies that exemplify the usefulness of gene genealogies for studying population-level processes. We begin with several examples from *Arabidopsis thaliana*, including the homeotic loci *APETALA3* and *PISTILLATA* and the disease-resistance locus *RPS2*, all of which are subject to selection. Then, we illustrate the utility of genealogies for tracing the postglacial range expansion in a variety of plant species. Finally, the usefulness of a genealogical approach for documenting crop origin is shown for cassava, a staple crop of the tropics.

GENE GENEALOGY: AN EXAMPLE FROM *ARABIDOPSIS*

The model plant, *A. thaliana*, is being used increasingly often for evolutionary studies. *Arabidopsis* offers many advantages as a study system, including its small size, simple genome, and rapid generation time. Molecular biologists have elucidated the function of many genes in *Arabidopsis*; mechanisms of development have been detailed, and the sequence of its genome is nearly complete. All of this work provides fertile ground for evolutionary biologists. There are an increasing number of excellent studies that use this information. For example, the role of homeotic genes in the development of floral structures has furthered understanding of tissue differentiation in plants; this work has provided the background for studies that investigate the evolutionary diversification of morphogenic pathways. Purugganan and Suddith (1999) have examined the molecular evolution of the homeotic loci *APETALA3* and *PISTILLATA*, which affect petal and stamen development in *Arabidopsis* flowers. They have compared the gene genealogies of these sequences with variation at five other nuclear loci of *Arabidopsis*. Based on an excess of low-frequency nucleotide polymorphisms and elevated within-species replacement polymorphisms, the authors conclude that *A. thaliana* has undergone rapid expansion in population numbers and size. Likewise, patterns of variation in restriction fragment length polymorphisms of several nuclear loci and the construction of a multilocus haplotype network have indicated that *A. thaliana* populations exhibit little to no geographical structuring (Bergelson *et al.*, 1998), a conclusion that is consistent with rapid population expansion.

In the above examples from *Arabidopsis*, population history has strongly affected patterns of genealogy and molecular evolution. Other loci within the *Arabidopsis* genome may reflect different evolutionary processes, in particular selection. *Arabidopsis* has served as a model system for unraveling disease-resistance response, a trait presumed under strong selection. In *Arabidopsis*, the *RPS2* gene is involved in the recognition of the plant pathogen *Pseudomonas syringae* pv. *tomato*. *RPS2* interacts with an avirulence gene, *avrRpt2*, of the pathogen to initiate the cascade of events that led to disease resistance. Both the avirulence gene in the pathogen and the resistance gene in the host must be functional to elicit resistance. These genes interact in a specific "gene-for-gene" manner. The close relationship between avirulence genes and resistance genes as well as the obvious fitness consequences of resistance for a plant have led to speculation on the evolutionary dynamics of resistance genes.

RPS2 encodes a 909-amino acid gene product. The gene contains several motifs that suggest it is part of a signaling pathway, including a leucine zipper, leucine-rich repeats, a hydrophobic region, and a nucleotide-binding site. A gene genealogy for the RPS2 locus has been constructed to investigate the molecular evolution of the gene (Caicedo *et al.*, 1999); 17 accessions of *A. thaliana*, representing a diversity of ecotypes, were sequenced for *RPS2*, and their resistance to *Pseudomonas* was determined. The resulting genealogy reveals an intriguing pattern (Fig. 3). Disease-resistance haplotypes (alleles) are clustered on the gene tree, in-

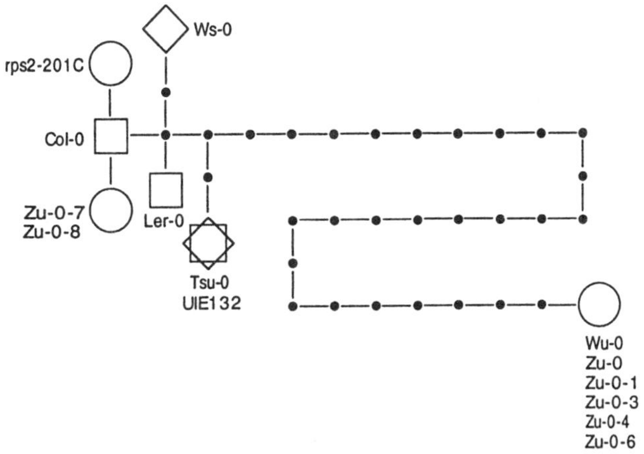

FIGURE 3. Gene tree for the RPS2 locus of *A. thaliana*. Open circles represent susceptible haplotypes; open squares are resistant haplotypes; and open diamonds are haplotypes intermediate in resistance. Closed circles are haplotypes not present in the sample but inferred from single-step mutations. The figure is modified from Caicedo et al. (1999).

dicating that resistance haplotypes are closely related. In contrast, a susceptible ecotype is 23 mutational steps from this cluster (the other two susceptible ecotypes represent a mutation to a stop codon and a strain created by mutagenesis). Silent mutations are distributed more often on the long branch of the tree, whereas nonsilent mutations occur more frequently on the short branches of the genealogy. Such genealogies have potential for inferring gene function as well as unraveling the dynamics of molecular evolution.

PHYLOGEOGRAPHY

One of the most successful applications of genealogical methods in natural populations has been in field of phylogeography. The conceptual approach of phylogeography was pioneered by John Avise and colleagues (Avise *et al.*, 1987). Avise (2000) defines phylogeography as "a field of study concerned with the principles and processes governing the geographic distribution of genealogical lineages, especially those within and among closely related species." Phylogeographic studies draw inferences about the history of population divergence based on associations between the geographical distribution of the alleles and their genealogical relationships. Because these studies are not based on equilibrium assumptions of gene flow and genetic drift, they have proved insightful in studying historical changes in patterns of gene flow, isolation, and secondary contact among divergent populations. The vast majority of phylogeography studies have focused on animal systems, and most of these have relied on the rapidly evolving regions of the mitochondrial genome as a source of genetic variation. Phylogeographic studies in plants have lagged behind those of animal studies, primarily because of difficulties in finding ordered, neutral intraspecific variation required for constructing gene trees (see *Conclusions* below).

POSTGLACIAL MIGRATION

Some of the most elegant studies of phylogeography in plants have examined the postglacial migration of species from Pleistocene refugia. A series of studies, using polymorphism in the chloroplast genome, on European trees, such as oaks (*Quercus* spp.), beech (*Fagus sylvatica*), and black alder (*Alnus glutinosa*), have shown similar patterns of variability; these species show a strong east–west cline in variation. Investigators interpret this cline as a result of postglacial migration from the same glacial refugia, leading to the concordance of variation patterns among species (Newton *et al.*, 1999). Phylogeographic studies of eight oak species have demonstrated that recolonization of Europe subsequent to the last glaciation was from several refugia in the peninsulas of Iberia, Italy, and the Balkans (Dumolin-Lapègue *et al.*, 1997). In this case, each refugium was represented by a distinct haplotype lineage. Fine-scale phylogeographic analysis further indicated that chloroplast DNA polymorphisms are shared between several oak species and, in this case, are attributed to hybridization and introgression subsequent to the recolonization of Europe. Similarly, chloroplast DNA analysis has shown concordance between beech (Demesure *et al.*, 1996) and black alder (King and Ferris, 1998) phylogeography. Both species are believed to have colonized Europe after glaciation from a refugium in the Carpathian Mountains. Moreover, the data indicate that an additional refugium for these species in Italy did not contribute to the recolonization of Europe. Similar concordance in phylogeographic patterns associated with postglacial spread is observed between plant species in the Pacific Northwest of North America. Soltis *et al.* (1997) have shown, via chloroplast DNA phylogenies, similar patterns in the structuring of variation among several different types of plants, including ferns, trees, and several members of the Saxifragaceae, suggesting that the present genetic structure of these species is strongly affected by their postglacial pattern of colonization.

PHYLOGEOGRAPHY AND PLANT DOMESTICATION: *MANIHOT ESCULENTA*

We have used a genealogical approach in examining two questions involving the species *M. esculenta* (Euphorbiaceae): the origin of the staple root crop cassava (*M. esculenta* subsp. *esculenta*) and the phylogeography of cassava's closest wild relative (*M. esculenta* subsp. *flabellifolia*). Cassava (manioc) is the sixth most important crop in the world (Mann, 1997). It is the primary source of calories in sub-Saharan Africa and serves as the main carbohydrate source for over 500 million people in the tropics worldwide (Best and Henry, 1992; Cock, 1985). Cassava is mostly grown by subsistence farmers, and despite its global importance as a food crop, it has traditionally received less attention by researchers than have temperate cereal crops. One fundamental question that has remained unresolved concerns the crop's geographical and evolutionary origins. Cassava was traditionally proposed to be a "compilospecies" derived from multiple hybridizing progenitor species in the genus *Manihot* (Jennings, 1995; Sauer, 1993). *Manihot* includes ≈98 species occurring in both northern South America (≈80 spp.) and in Mexico/Central America (≈17 spp.); sites of domestication were proposed from much of this vast geographical area.

Traditional phylogenetic approaches were only partially successful in determining cassava's origin. Species of *Manihot* show low levels of divergence in both morphological and molecular characters, probably reflecting a recent diversification of the genus (Rogers and Appan, 1973; Olsen and Schaal, 1999). A phylogeny of the genus based on DNA sequences in the nuclear ribosomal ITS region (B.A.S., unpublished data) is not highly resolved but does place cassava in a clade of South American species. This finding was consistent with the proposition, based on morphological characters (Allem, 1994), that cassava is derived from a single wild South American progenitor (referred to as *M. esculenta* subsp. *flabellifolia* under present taxonomy).

To test the hypothesis that cassava is derived from *flabellifolia*, we examined DNA sequence variation in 20 crop accessions, 27 populations of *flabellifolia*, and 6 populations of a closely related species, *Manihot pruinosa*, which has been proposed to hybridize with *flabellifolia* (Allem, 1992, 1999). Populations of *flabellifolia* occur in mesic transitional forest patches in the ecotone between the lowland rainforest of the Amazon basin and the seasonally dry cerrado (savanna–scrub) found to the south and east on the Brazilian Shield plateau (Fig. 4). Populations were sampled in two transects, one along the southern border of the Amazon and the other along the eastern border. *M. pruinosa* is a cerrado species that occurs within the eastern range of *flabellifolia*. We included this species to test whether it has contributed to the genetic diversity of cassava through hybridization with *flabellifolia*.

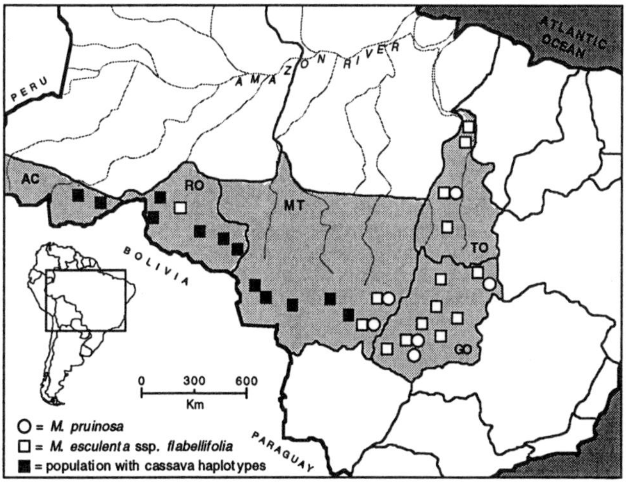

FIGURE 4. Locations of populations of *M. esculenta* subsp. *flabellifolia* (squares) and *M. pruinosa* (circles) sampled for the G3pdh phylogeography study. Shaded squares indicate populations containing one or more haplotypes found in domesticated cassava accessions. The figure is modified from Olsen and Schaal (1999).

The study was based on sequence variation within a portion of the low-copy nuclear gene *G3pdh*, which encodes glyceraldehyde 3-phosphate dehydrogenase (Olsen and Schaal, 1999). Using primers designed by Strand *et al.* (1997), we PCR amplified and sequenced a 962-bp region that spanned three exons, four introns, and parts of two flanking exons. From the 424 alleles (212 individuals) that were sequenced, we observed a total of 63 nucleotide polymorphisms, which characterized 28 different haplotypes. Maximum parsimony analysis yielded two negligibly different gene tree topologies, one of which is shown in Fig. 5.

Because the domestication of cassava is an extremely recent event evolutionarily speaking, one would not expect the divergence between cassava and *flabellifolia* to be reflected in the *G3pdh* gene tree. However, by looking at the haplotypes that are shared between cassava and the wild taxa and by examining the geographical locations of these haplotypes in the wild populations, we were able to draw several insights into the ori-

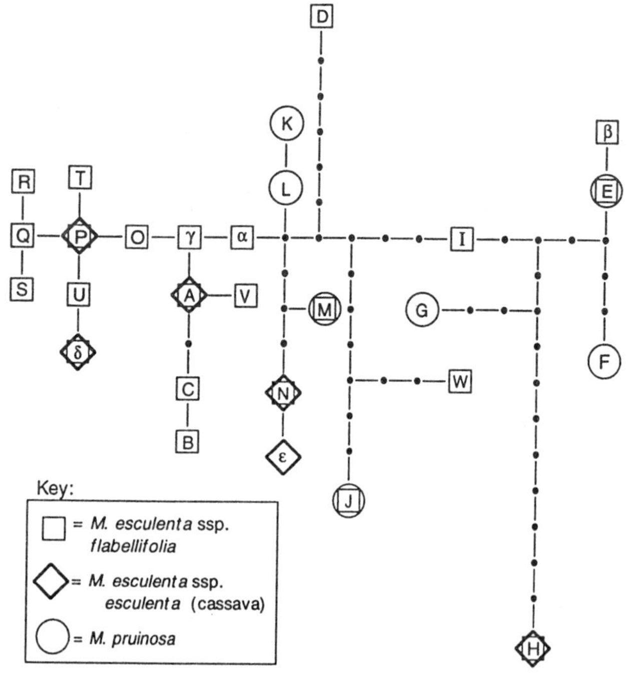

FIGURE 5. *G3pdh* gene tree for *M. esculenta* and *M. pruinosa*. Letters correspond to haplotype designations in GenBank accession numbers (AF136119–AF136149). Shapes around letters indicate the taxon or taxa in which a haplotype was found, as indicated in the key. The figure is modified from Olsen and Schaal (1999).

gin of the crop. First, we found that genetic variation in the crop is almost entirely a subset of that found in *flabellifolia*. *Flabellifolia* contains 24 haplotypes, of which 6 are found in cassava; cassava's haplotype diversity, therefore, represents 25% of that found in *M. esculenta* overall. Thus, the crop is most likely derived directly from *flabellifolia*, rather than from several hybridizing progenitor species as traditionally thought. In addition, we found that the cassava haplotypes occur in *flabellifolia* populations along the southern border of the Amazon basin and not along the eastern border. This finding points to the southern Amazonian region as the likely site of domestication of cassava. Interestingly, paleobotanical and other anthropological data indicate this region as a probable zone of domestication shared with peanut, two species of chili pepper, and jack bean (Piperno and Pearsall, 1998). Finally, we found that none of the cassava haplotypes occur in *M. pruinosa*, suggesting that this species is not a progenitor of the crop. All of these conclusions are corroborated by an analysis of this same study system with microsatellite markers (K.M.O. and B.A.S., unpublished work).

The phylogeographic aspect of the study has focused on historical patterns of population divergence in *flabellifolia* and between *flabellifolia* and *pruinosa*. The distribution of the rainforest–cerrado ecotone where these species occur is likely to have shifted during the climatic changes of the Pleistocene (Behling, 1998; Burnham and Graham, 1999). Although there is not yet a consensus on the pattern or extent of habitat shifts, cooler/drier periods (associated with glaciations in temperate latitudes) are expected to have favored the expansion of cerrado and transitional forest; during warmer, humid periods (including the present), these habitats would be expected to be more restricted and fragmented as rainforest expanded. The repeated climate fluctuations of the Pleistocene are therefore predicted to have led to cycles of population fragmentation followed by range expansions and secondary contact in populations of *flabellifolia* and *pruinosa*. If these events have occurred, they should be reflected in the present phylogeographic structure of these taxa.

These hypotheses are being tested currently through a nested cladistic analysis (Templeton *et al.*, 1995) of the *G3pdh* data set (K.M.O., unpublished data). Although the statistical analyses are not complete, some preliminary insights are possible by visual inspection of the *G3pdh* gene tree. One interesting finding is that three haplotypes are shared between *flabellifolia* and *pruinosa*, suggesting interspecific introgression and/or shared ancestral polymorphisms that predate the divergence of these species. Two of the shared haplotypes (E and J) are common in eastern *flabellifolia* populations, and each is found in a single *pruinosa* individual from a population in close proximity to *flabellifolia* populations. This pattern suggests introgression from *flabellifolia* into *M. pruinosa*. The position of these haplotypes near the tips of the gene tree also favors this explanation over shared ancestral polymorphisms; tip haplotypes are likely to be younger than interior haplotypes (Castelloe and Templeton, 1994) and therefore would be less likely to represent ancestral variation. The third shared haplotype (M) is also a tip haplotype. However, this haplotype is common in *M. pruinosa* and is found in a single *flabellifolia* population approximately 1,000 km west of the current range of *M. pruinosa*. Although clearly not the result of contemporary gene flow, this pattern could possibly have arisen through hybridization in the recent past. Palynological data indicate that during the last glacial maximum (<18,000 years B.P.), cerrado vegetation expanded into areas along the southern border of the Amazon basin that are presently rainforest (reviewed in Burnham and Graham, 1999). Thus, hybridizing *pruinosa* populations could have existed in this region as recently as 11,000 years B.P.

Haplotypes on the *G3pdh* tree are not clustered by species (Fig. 4). Because *flabellifolia* and *pruinosa* are closely related taxa within a recently radiated genus, they would not necessarily be expected to have reached a pattern of reciprocal monophyly with respect to *G3pdh* haplotypes (Fig. 2). The phylogeographic structure within each species is also complex. However, although there is no simple concordance between the geographical distributions of haplotypes and their genealogical relationships, contingency analyses (Posada *et al.*, 1999) reveal that nested clades within the gene tree are geographically structured. Thus, the phylogeographic structure reflects more than just the random sorting of ancestral polymorphisms among populations. Detailed phylogeographic analysis (Templeton *et al.*, 1995) and the analysis of DNA sequence data from two additional nuclear genes (K.M.O., unpublished data) will be useful in elucidating the historical processes that have led to the current phylogeographic structure in this study system.

CONCLUSIONS

Gene genealogies have lead to several important insights into plant evolution and have the potential for far greater contributions. Many of the processes that affect the evolution of plant populations, such as selection, isolation, size fluctuations, and gene flow, are amenable to genealogical analysis. In particular, the use of genealogies within the framework of coalescence theory will allow us to understand in greater detail the role of historical fluctuations in population size, colonization, and range expansion. Although the large-scale metapopulation structure of many plants is clearly documented, there are relatively few studies of the genetic dynamics of this structure: colonization and establishment of subpopulations, gene migration, and extinction. The genetic aspects of such processes largely remain to be explored, and a historical genealogical approach will be particularly instructive.

The major impediment to wide application of gene genealogies for phylogeographic studies in plants is identifying DNA sequences with appropriate levels of ordered variation within chloroplast, mitochondrial, or nuclear genomes (Schaal *et al.*, 1998). In many cases, the chloroplast spacer regions that have been informative for some species (see above) show little to no intraspecific variation in other plant species. Moreover, chloroplast restriction fragment length polymorphism genealogies based on length variation alone can be confounded by homoplasy. The nuclear genome remains problematic because of the difficulty in finding regions that have sufficient levels of neutral variation and that are not involved in intragenic recombination. Moreover, the effective population size of a nuclear gene is four times that of an organelle gene, because it is diploid and biparentally inherited. The larger effective population size results in increased coalescent times, which in turn, increases the likelihood of encountering ancestral polymorphisms. High-resolution nuclear markers such as random amplified polymorphic DNAs and amplified fragment length polymorphisms are historically unordered, and variants cannot be related easily in a genealogical manner. Because of the difficulty in finding genealogically informative markers, many plant studies have been phylogeographic only in the broad sense, meaning that they detect an association between patterns of genetic variation and geography. Such studies do not incorporate a genealogical perspective.

The search for appropriate markers has turned to nuclear genes that are increasingly the focus for genealogical studies. Nuclear genes often contain multiple introns, and many of the introns contain high levels of neutral variation. This approach has been applied successfully in several animal species: e.g., oysters (Hare and Avise, 1998), fish (Bagley and Gall, 1998), and birds (Degnan, 1993). Nuclear sequences of plants have been used to understand the genetic relationships of wild populations of *A. thaliana* (Bergelson *et al.*, 1998), selection, and evolution of homeotic genes (Purugganan and Suddith, 1999), as well as in the example from cassava above. Numerous studies of nuclear gene genealogies are currently under way and promise to provide new insights into the processes identified by Stebbins a half century ago as central for the evolution of plants.

This work was supported in part by a grant from the Explorer's Club, by National Science Foundation Doctoral Dissertation Improvement Grant DEB 9801213 to K.M.O., and by grants from the Rockefeller and Guggenheim Foundations to B.A.S.

REFERENCES

Allem, A. (1992) *Manihot* germplasm collecting priorities. In *Report of the First Meeting of the International Network for Cassava Genetic Resources,* eds. Roca, W. M. & Thro, A. M. (Cent, Int. Agric. Trop., Cali, Colombia), pp. 87–110.

Allem, A. (1994) The origin of *Manihot esculenta* Crantz (Euphorbiaceae) *Genet. Res. Crop Evol.* 41, 133–150.

Allem, A. (1999) The closest wild relatives of cassava (*Manihot esculenta* Crantz). *Euphytica* 107, 123–133.

Avise, J. (2000) *Phylogeography: the History and Formation of Species* (Harvard, Cambridge, MA).

Avise, J., Arnold, J., Ball, R., Bermingham, E., Lamb, T., Neigel, J. E., Reeb, C. A. & Saunders, N. C. (1987) Intraspecific phylogeography: the mitochondrial DNA bridge between population genetics and systematics. *Annu. Rev. Ecol. Syst.* 18, 489–522.

Bagley, J. & Gall, G. (1998) Mitochondrial and nuclear DNA sequence variability among populations of rainbow trout (*Oncorhynchus mykiss*). *Mol. Ecol.* 7, 945–961.

Behling, H. (1998) Late Quaternary vegetational and climatic changes in Brazil. *Rev. Palaeobot. Palynol.* 99, 143–156.

Bergelson, J., Stahl, E., Dudek, S. & Kreitman, M. (1998) Genetic variation within and among populations of *Arabidopsis thaliana. Genetics* 148, 1289–1323.

Best, R. & Henry, G. (1992) Cassava: towards the year 2000. In *Report of the First Meeting of the International Network for Cassava Genetic Resources,* eds. Roca, W. M. & Thro, A. M. (Cent. Int. Agric. Trop., Cali, Colombia), pp. 3–11.

Burnham, R. & Graham, A. (1999) The history of neotropical vegetation: new developments and status. *Ann. Mo. Bot. Gar.* 86, 546–589.

Caicedo, A. L., Schaal, B. A. & Kunkel, B. N. (1999) Diversity and molecular evolution of the *RPS2* resistance gene in *Arabidopsis thaliana. Proc. Natl. Acad. Sci. USA* 96, 302–306.

Castelloe, J. & Templeton, A. R. (1994) Root probabilities for intraspecific gene trees under neutral coalescent theory. *Mol. Phylo. Evol.* 3, 102–113.

Cock, J. (1985) *Cassava: new potential for a neglected crop* (Westfield, London, UK).

Degnan, S. (1993) The perils of single gene trees—mitochondrial versus single-copy nuclear DNA variation in white-eyes (Aves: Zosteropidae). *Mol. Ecol.* 2, 219–225.

Demesure, B., Comps, B. & Petit, R. (1996) Chloroplast DNA phylogeography of the common beech (*Fagus sylvatica* L.) in Europe. *Evolution* 50, 2515–2520.

Dumolin-Lapègue, S., Demesure, B., Fineschi, S., Le Corre, V. & Petit, R. (1997) Phylogeographic structure of white oaks throughout the European continent. *Genetics* 146, 1475–1487.

Fu, Y. & Li, W.-H. (1999) Coalescing into the 21st century: an overview and prospects of coalescent theory. *Theor. Pop. Bio.* 56, 1–10.

Hamrick, J. & Godt, M. (1989) Allozyme diversity in plant species. In *Plant Population Genetics, Breeding, and Genetic Resources,* eds. Brown, A., Clegg, M., Kahler, A. & Weir, B. (Sinauer, Sunderland, MA), pp. 43–63.

Hare, M. & Avise, J. (1998) Population structure in the American oyster as inferred by nuclear gene genealogies. *Mol. Biol. Evol.* 15, 119–128.

Jennings, D. (1995) Cassava: *Manihot esculenta* (Euphorbiaceae). In *Evolution of Crop Plants,* eds. Smartt, J. & Simmonds, N. (Wiley, New York), pp. 128–132.

King, R. & Ferris, C. (1998) Chloroplast DNA phylogeography of *Alnus glutinosa* (L.) Gaertn. *Mol. Ecol.* 7, 1151–1163.

Mann, C. (1997) Reseeding the green revolution. *Science* 277, 1038–1043.

Neigel, J. & Avise, J. (1986) Phylogenetic relationships to mitochondrial DNA under various demographic models of speciation In *Evolutionary Processes and Theory,* eds. Karlin, S.& Nevo, E. (Academic, New York), pp. 515–534.

Newton, A., Allnutt, T., Gilles, A., Lowe A. & Ennos, R. (1999) Molecular phylogeography, intraspecific variation and the conservation of tree species *Trends Ecol. Evol.* 14, 140–145.

Olsen, K. M. & Schaal, B. A. (1999) Evidence on the origin of cassava: phylogeography of *Manihot esculenta. Proc. Natl. Acad. Sci. USA* 96, 5586–5591.

Piperno, D. & Pearsall, D. (1998) *The Origins of Agriculture in the Lowland Neotropics,* (Academic, New York).

Posada, D., Crandall, K. & Templeton, A. (1999) *Geodis, version 2.0* (Brigham Young Univ., Provo, UT).

Purugganan, M. & Suddith, J. (1999) Molecular population genetics of floral homeotic loci: departures from the equilibrium–neutral model at the *APETALA3* and *PISTILLATA* genes of *Arabidopsis thaliana. Genetics* 151, 839–848.

Rogers, D. & Appan, S. (1973) *Manihot and Manihotoides (Euphorbiaceae): a computer assisted study* (Hafner, New York).

Sauer, J. (1993) *Historical Geography of Crop Plants* (CRC, Boca Raton, FL).

Schaal, B. A., Hayworth, D. A., Olsen, K. M., Rauscher J. T. & Smith, W. A. (1998) Phylogeographic studies in plants: problems and prospects. *Mol. Ecol.* 7, 465–474.

Smouse, P. (1998) To tree or not to tree. *Mol. Ecol.* 7, 399–412.

Soltis, D., Gitzendanner, M., Strenge, D. & Soltis, P. (1997) Chloroplast DNA intraspecific phylogeography of plants from the Pacific Northwest of North America. *Plant Syst. Evol.* 206, 353–373.

Stebbins, G. L. (1950) *Variation and Evolution in Plants* (Columbia, New York).

Strand, A., Leebens-Mack, J. & Milligan, B. (1997) Nuclear DNA-based markers for plant evolutionary biology. *Mol. Ecol.* 6, 113–118.

Swofford, D. (1993) *PAUP: Phylogenetic Inference Using Parsimony*, version 3.1, Ill. Nat. Hist. Survey (Champaign, IL).

Templeton, A., Crandall, K. & Sing, C. (1992) A cladistic analysis of phenotypic associations with haplotypes inferred from restriction endonuclease and DNA sequence data. III. Cladogram estimation. *Genetics* 132, 619–633.

Templeton, A., Routman, E. & Phillips, C. (1995) Separating population structure from population history: A cladistic analysis of the geographical distribution of mitochondrial DNA haplotypes in the tiger salamander, *Ambystoma tigrinum. Genetics* 140, 767–782.

Wright, S. (1951) The genetical structure of populations. *Ann. Eugen.* 15, 322–354.

Part V

TRENDS AND PATTERNS IN PLANT EVOLUTION

The study of angiosperm fossils has experienced a "paradigm shift" during the last three decades. In 1950, when *Variation and Evolution in Plants* was published, angiosperm paleobotany consisted of matching fossils, mostly leaves, to extant genera, contributing but little towards understanding patterns and rates of plant evolution. Angiosperms from the Cretaceous and early Tertiary are now known that have become extinct or are only distantly related to living genera. The evolutionary biology of angiosperms is nowadays largely addressed on the basis of detailed character-based analyses that follow cladistic methodologies. According to David Dilcher ("Toward a New Synthesis: Major Evolutionary Trends in the Angiosperm Fossil Record," Chapter 14), three basic radiation nodes have been identified: the closed carpel and radially symmetrical flower, the bilateral flower, and fleshy fruits with nutritious nuts and seeds. The genetic systems of the angiosperms promoted their evolution towards outcrossing reproduction, with the strongest selection directed towards flowers, fruits, and seeds.

There is a variety of reproductive systems among the 250,000 known species of vascular plants. Evolutionary explanations of this variety have in the past been based on population-level differences. Thus, selfing or asexual plants are said to be more highly adapted to immediate circumstances but less able to adapt to changing environments than sexual outcrossers. Kent E. Holsinger argues in "Reproductive Systems and Evolution in Vascular Plants" (Chapter 15) that for understanding the origin and persistence of particular reproductive styles we must relate them to differences expressed among individuals within populations. Holsinger points out that selfers have fewer genotypes within populations, but greater genetic diversity among populations, than sexual outcrossers. Selfers and asexuals may thereby be less able to respond adaptively to changing environments; they also accumulate deletion mutations more rapidly. Sexual outcrossers suffer from a cost of outcrossing and may be impacted by circumstances that handicap the union of gametes produced by different individuals. These costs of outcrossing and reduced reproductive assurance lead to an over-representation of selfers and asexuals in newly-formed progeny, which may displace sexual outcrossers unless these enjoy compensating advantages in survival and reproduction.

The damage wrought by invasive species costs $122 billion per year in the United States. Successful plant and animal invasions impact ecologically and demographically the endemic flora and fauna and may have considerable evolutionary import. Norman C. Ellstrand and Kristina A. Schierenbeck ("Hybridization as a Stimulus for the Evolution of Invasiveness in Plants?" Chapter 16) note that invasions typically involve long lag eriods before they become successful, and require multiple introductions. Their explanation is that hybridization between the invaders and resident populations is a stimulus often required for successful invasion. Hybrid progenies may enjoy genetic advantages over their progenitors. Ellstrand and Schierenbeck show that, as predicted by their model, invasiveness can evolve.

Stebbins devoted two chapters (nearly 100 pages) to polyploidy. Pamela S. Solits and Douglas E. Soltis ("The Role of Genetic and Genomic Attributes in the Success of Polyploids," Chapter 17) set forth the genetic attributes that account for the great success of polyploid plants: about 50% of all angiosperm species and nearly 95% of all ferns. Polyploids maintain higher levels of genetic variation and heterozygosity, and exhibit lesser inbreeding depression, than diploids. These may be the case because most polyploid species have arisen more than once, from genetically different diploid parents, in addition to the presence of more than two homologues. Genome rearrangement seems to be a common attribute of polyploids; and many plant species may be ancient polyploids (see the case of maize in ref. 24). Soltis and Soltis conclude that the advances of the last 50 years notwithstanding, there remains much to be known about polyploid plant species, including their general mode of formation.

14
Toward a New Synthesis: Major Evolutionary Trends in the Angiosperm Fossil Record

DAVID DILCHER

Angiosperm paleobotany has widened its horizons, incorporated new techniques, developed new databases, and accepted new questions that can now focus on the evolution of the group. The fossil record of early flowering plants is now playing an active role in addressing questions of angiosperm phylogeny, angiosperm origins, and angiosperm radiations. Three basic nodes of angiosperm radiations are identified: (i) the closed carpel and showy radially symmetrical flower, (ii) the bilateral flower, and (iii) fleshy fruits and nutritious nuts and seeds. These are all coevolutionary events and spread out through time during angiosperm evolution. The proposal is made that the genetics of the angiosperms pressured the evolution of the group toward reproductive systems that favored outcrossing. This resulted in the strongest selection in the angiosperms being directed toward the flower, fruits, and seeds. That is why these organs often provide the best systematic characters for the group.

Here I focus on the fossil record of the same plants that Stebbins did in his book, *Variation and Evolution of Plants* (1950), the angiosperms. This contribution has the advantage of being written more than 50 years after Stebbins wrote about his view of the fossil record

Florida Museum of Natural History, University of Florida, Gainesville, FL 32611-7800

This paper was presented at the National Academy of Sciences colloquium "Variation and Evolution in Plants and Microorganisms: Toward a New Synthesis 50 Years After Stebbins," held January 27–29, 2000, at the Arnold and Mabel Beckman Center in Irvine, CA.

of the angiosperms. His use of the fossil record of angiosperms as a model in his evolutionary synthesis was hampered because in the 1940s the paradigm in angiosperm paleobotany was to match fossils, especially leaves, to extant genera (Dilcher, 1974). The successes of the angiosperm paleobotanists (e.g., D. Axelrod, H. Becker, E. W. Berry, R. Brown, R. Chaney, and H. MacGinitie, and many others for 100 years before them) were judged by their ability to match a high percentage of fossils to living genera (Fig. 1). Once the identifications were made to living genera, their focus was on questions of phytogeography and paleoclimate. This meant that almost no fossil angiosperms were recognized as extinct; it was quite impossible to focus questions of plant evolution on the fossil record of the angiosperms in 1950 as George Gaylord Simpson had done with the fossil vertebrate record in his classic *Tempo and Mode in Evolution* in 1944 (Simpson, 1944).

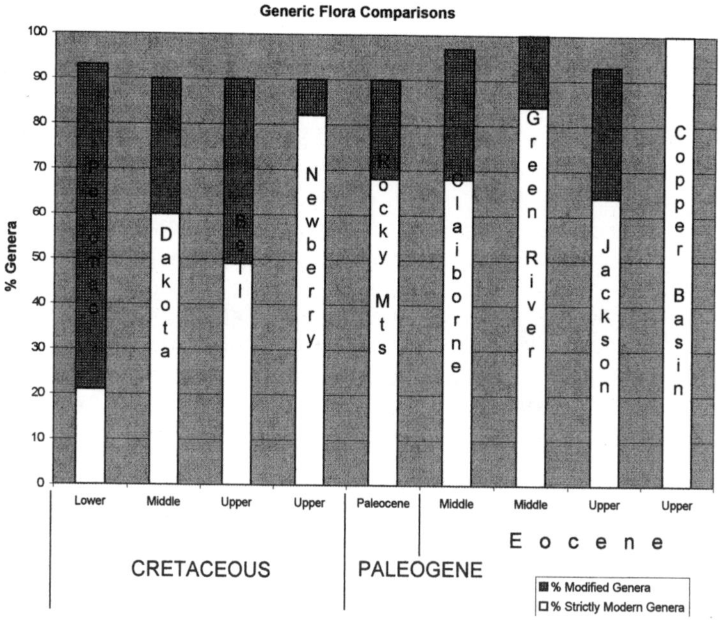

FIGURE 1. Selected floras published from the late 1800s to the 1960s, ranging in time from the Lower Cretaceous to the Upper Eocene (Fontaine, 1889; Lesquereux, 1891; Newberry, 1898; Berry, 1924; Bell, 1957; Brown, 1962; Axelrod, 1966; MacGinitie, 1969). The open area represents percent of the species in the flora that were given extant generic names. The shaded area represents the percent of species in the flora that were given fossil generic names based on a modern genus to which they were perceived to be similar. The short fall, less than 100% for each flora, represent genera perceived to be truly extinct.

Stebbins wrote in "Fossils, Modern Distribution Patterns and Rates of Evolution," chapter 14 of *Variation and Evolution of Plants* (Stebbins, 1950), about the disjunct distribution of modern genera of fossil plants. His rates of evolution were based on the various modern genera described in the fossil record of North America and currently living in southeastern Asia or South America. His arguments about the rates of evolution from the fossil record may have some validity when based on fossils from the Miocene (about 25 million years) and younger. However, many of the fossils from the Paleocene, Eocene, and Oligocene reported as living genera have been subject to revisions (Manchester, 1994) as shown in Fig. 2. This trend that had dominated angiosperm paleobotany for more than 100 years continued into the early 1970s. The supposed failure of the fossil record to contribute to understanding the evolution of the early angiosperms was still evident in 1974 when Stebbins published *Flowering Plants: Evolution Above the Species Level* (Stebbins, 1974). In chapter 10, "The Nature and Origin of Primitive Angiosperms," there is no substantive use of the fossil record to address this question. The theories and hypothesis

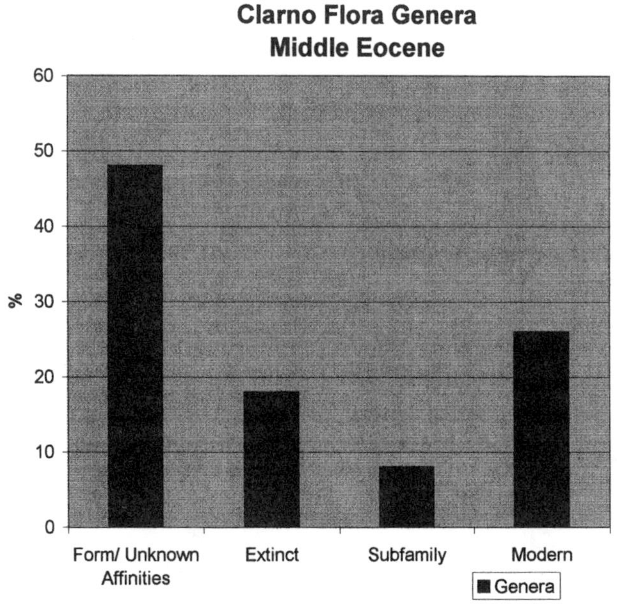

FIGURE 2. Representation of the Middle Eocene Clarno Flora from eastern Oregon (Manchester, 1994) based on several thousands of fruits and seeds collected over 60 years. The bars represent the percent of the genera identified to angiosperm genera of various degrees of similarity to living genera. Note that less than 30% of the fruits and seeds can be identified with living genera.

presented by Stebbins are based on the comparative morphology and anatomy of living angiosperms considered primitive at that time rather than the fossil record of early angiosperms.

However, at the same time, the early 1970s, special attention was being focused on the fine features of the morphology of angiosperm leaf venation and the cuticular anatomy of living and fossil angiosperms (Hickey, 1973; Dilcher, 1974; Hickey and Wolfe, 1975; Doyle and Hickey, 1976). Most of the early angiosperms from the Cretaceous and early Tertiary were being found to be extinct or only distantly related to living genera (Fig. 3). Grades and clades of relationships were being established upon the basis of careful character analysis (Dilcher *et al.*, 1976b; Roth and Dilcher, 1979). During this time, it became scientifically acceptable to be unable to identify a fossil to a modern genus. Fossil angiosperms were analyzed on the basis of multiple detailed objective characters, and degrees of relationships could be established based on the extent to which these same combinations of characters were found in living families, subfamilies, or genera (Jones and Dilcher, 1980). Analyses of the fossil angiosperm record were being constructed that included vast amounts of data based on careful anatomical and morphological analysis of the diversity of characters found in living genera and modern families. Large collections of cleared leaves and cuticular preparations were developed, and whole families were surveyed to establish their range of venation and

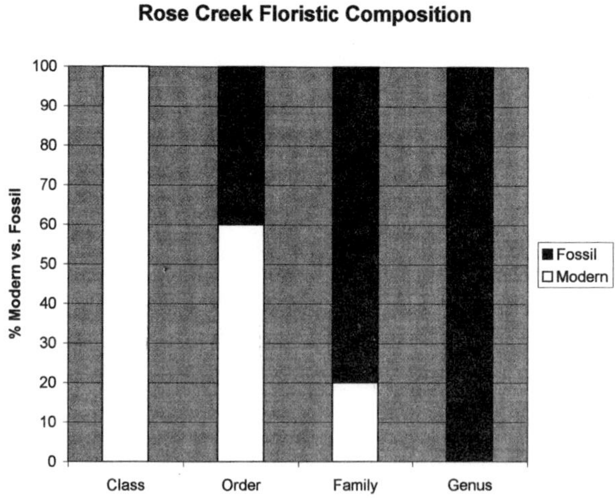

FIGURE 3. Representation of modern vs. fossil taxonomic groups published for the mid-Cretaceous, Dakota Formation, Rose Creek Flora (Upchurch and Dilcher, 1990). This flora is based on leaves. Note no modern genera identified as opposed to the 60% identified for the same flora illustrated in Fig. 1.

cuticular characters, along with fruit and seed anatomy and morphology in order to research the fossil history of a family (Sheffy, 1972; Dolph, 1974, 1975; Dilcher *et al.*, 1976b; Jones and Dilcher, 1980; Manchester, 1981, 1994; Roth, 1981; Jones, 1984; Schwarzwalder, 1986; Herendeen, 1990; Herendeen and Dilcher, 1990a, b., 1991; Herendeen *et al.*, 1992; LAWG, 1999). The anatomical/morphological style of systematic-based angiosperm paleobotany was a distinct change from the floristic approaches that focused on paleogeographic and paleoclimatic questions and dominated the field before 1970.

This new paradigm shift opened the door for a new synthesis of the fossil record of angiosperms. New questions about the evolutionary biology of the fossil record of angiosperms could now be addressed based on detailed character-based data of living and fossil angiosperms often organized with the help of cladistic analysis (Crane 1985; Doyle and Donoghue 1986, 1993; Doyle *et al.* 1994). At this same time there was renewed interest in exploring the fossil plant record to determine the origin and early evolutionary history of the angiosperms (Doyle, 1969; Doyle and Hickey, 1976; Dilcher, 1979). The techniques of careful analysis and the concerted effort to open up a new fossil record of early angiosperms by the use of small, often charcoalified plant remains (Kovach and Dilcher, 1988), or fragments of cuticle sieved from sediment (Huang, 1992) from newly collected material of the Jurassic to the Upper Cretaceous were very successful. A whole new area of the study of intermediate-sized fossil plants, often termed mesofossils as opposed to microfossils or megafossils, expanded to occupy the majority of angiosperm research in some laboratories with good success (Nixon and Crepet, 1993; Crepet and Nixon 1998a, b; Gandolfo *et al.*, 1998a, b, c; Friis *et al.*, 1999; Crepet *et al.*, 2000 and references cited therein). It is the success of these new techniques applied to the fossil record of angiosperms that now provides a new database from which to analyze some of the major trends in angiosperm evolution and allows us to ask new questions.

WHAT IS KNOWN ABOUT EARLY ANGIOSPERM DIVERSITY DURING THE CRETACEOUS?

New Tools of Analysis

The study of angiosperm fossils has undergone rapid and profound changes during the past 30 years as discussed in the introduction to this paper. Although the study of angiosperm fossils is only as reliable as the individual investigator, resources are now available, such as cleared leaf collections and cuticular reference slide collections from vast herbarium holdings. This allows angiosperm paleobotanists to survey the nature of characters circumscribed by a particular living family or genus before reaching a conclusion about the relationship(s) of a fossil angiosperm organ. It is now understood that some organs may contain more useful characters for determining relationships than others. It is not only acceptable, but desirable, to list the available characters and how these are distributed among several living genera in a family rather than to select only a single living genus that has such characters and on this basis refer the fossil to that living genus.

New Paradigm Applied

We are moving toward a well-defined and repeatable objective character-based analysis of the angiosperm fossil record. Much of this analysis is based on the study of the anatomy and morphology of fossil plant organs. Stebbins (1950) recognized the need for this type of study when he wrote that "The method of identification is simple comparison between the fossil and the leaves of living species, but various approaches have greatly increased its accuracy." He then cites Bandulska (1924), Edwards (1935), and Florin (1931), all early pioneers in the use of anatomy and morphology in the study of the systematics of fossil plants. However, using the new tools, it became apparent that there were many fossils that could not be related to living taxa even when such careful analyses were applied (Figs. 2 and 3). There came a time when it was necessary to give names to the various organs of fossil angiosperms that reflected their extinct nature, recognizing them as separate from any living genus (Dilcher and Crane, 1984). With few exceptions (Sun *et al.*, 1998), workers have not yet taken such bold steps as defining and naming extinct angiosperm plant families, orders, or classes.

Rapid Changes in the Data

The great strides in developing techniques of investigation for understanding Devonian fossil plants, on the basis of seemingly nondescript structurally-preserved compressed remains (Leclercq and Andrews, 1960; Leclercq and Banks, 1962), and the excellent application of anatomy and morphology to the study of Pennsylvanian age plants (i.e., Delevoryas, 1955, and examples cited in Taylor and Taylor, 1993) influenced me to apply similar techniques to fossil angiosperms to extract as much information as possible from compressed leaves and flowers. The amount of information that can be determined about a fossil leaf, fruit, flower, pollen grain, or wood by using these techniques allows character-based comparisons to be made. These data have become available at the same time that cladistic-based (i.e., character-based) data were being assembled for the living angiosperms. Now, with the study of megafossils, mesofossils, and microfossils all yielding new information about the characters of the early angiosperms, there are huge amounts of new data available each year. In particular, because they had not been studied before, mesofossils are adding a new set of valuable information that is changing our concept of early angiosperm diversity. The Lower Cretaceous sediments from Portugal have yielded 105 different kinds of flowers with 13 associated pollen types (Friis *et al.*, 1999). The lower Upper Cretaceous sediments from New Jersey are yielding a large number of new taxa (Nixon and Crepet, 1993; Crepet and Nixon, 1998a, b; Gandolfo *et al.*, 1998a, b, c; Crepet *et al.*, 2000). The application of character-based analyses of fossil angiosperm remains has been used (Magallón *et al.*, 1999) to demonstrate the presence of systematic groups of angiosperms through time (Fig. 4).

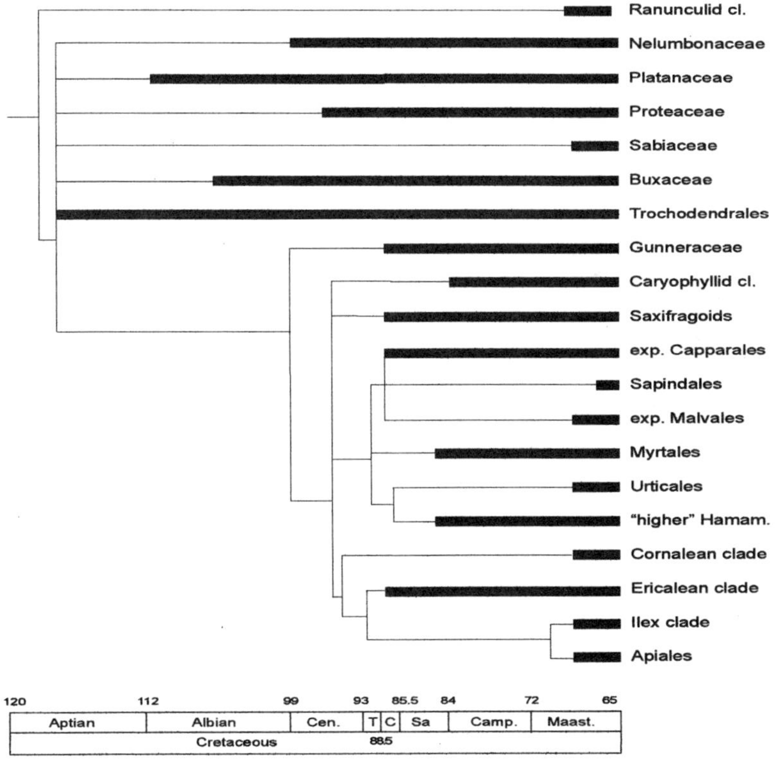

FIGURE 4. Angiosperm diversity as recognized through Cretaceous time. The solid bars extend from the earliest identified fossils of the clades listed on the right side. Modified from Haq and Eysinga, 1998 and Magallón *et al.*, 1999.

HOW HAS ANGIOSPERM REPRODUCTIVE BIOLOGY CHANGED THROUGH TIME?

Evolution of the Closed Carpel

The closed carpel is the one major feature that separates the angiosperms from other vascular seed plants. The closure most often is complete and entirely seals off the unfertilized ovules from the outside environment. Suggestions that this provided protection for the vulnerable ovules from beetles or other herbivores have been proposed as a reason for the closure of the carpel. However, I think that the closure of the carpel may be more directly related to the evolution of the bisexual flower (Dilcher, 1995). During the evolution of the flower, as the male and female organs of the flower were brought into proximity, the need for protection against self-fertilization was so important that biochemical and mechanical barriers were developed very early in flowering plant ancestors. The mechanical barrier is the closed carpel and the biochemical barrier is the incompatibility systems that developed to prevent the successful growth of pollen tubes. Some living angiosperms have loosely closed carpels or lack any firm closure at all. It has been suggested that these have sufficient exudates to fill the carpel opening so that the carpel has a biochemical barrier against self-fertilization (Endress, 1994).

Although the closed carpel is the fundamental strategy for preventing self-pollination, the addition or loss of sepals, petals, and stamens must have been important events ensuring outcrossing. It is reasonable to assume that the development of attractive colored organs and nectaries, the clustering together of female (ovule-bearing) organs and male (pollen-bearing) organs, and, finally, the association of the female and male organs together on the same axis were all changes designed to increase the effectiveness of insect pollination. The closed carpel and biochemical incompatibility are natural early steps that followed or took place at the same time as the evolution of the floral features just mentioned. The closed carpel in a showy flower ensured outcrossing by animal pollinators while increasing pollen exchange with bisexual flowers. The closed carpel serves as a plant's control mechanism to guarantee that outcrossing happens. Any mechanical protection it offered probably always has been of secondary importance and can be easily overcome by insects.

Evolution of Floral Form and Patterns

Radial symmetry. The floral organs of all early angiosperms are radially symmetrical, a symmetry exhibited by all of the floral organs and flowers whether they are small or large, unisexual or bisexual. The earliest known angiosperm flowers suggest that individual carpels were borne helically on an elongated axis with pollen organs if present, subtending and helically arranged on the same axis (Sun *et al.*, 1998). Similarly, the early small flowers (Friis *et al.*, 1999), unisexual or bisexual, have axes with radially arranged organs. In small flowers the elongation of many early flowering axes is compressed so that the organs appear radially arranged. This organization is clearly seen in larger flowers such as *Archaeanthus* (Dilcher and Crane, 1984) and the Rose Creek flower (Basinger and Dilcher, 1984). This radial arrangement of organs dominated floral form until late into the Late Cretaceous or the Paleocene and still persists in many flowers today.

Bilateral symmetry. By Paleocene and Eocene time, there are several evidences in the fossil record of bilateral flowers. This evolution probably began during the Upper Cretaceous. The evolution of bilateral flowers is associated with the presence of social insects in the Upper Cretaceous (Michener and Grimaldi, 1988a, b) and the coevolution of bilateral flowers occurred at different stages in the evolution of several living families. In some angiosperm families, bilateral symmetry may be present in only a part of the family, while in other families the entire family, is characterized by bilateral symmetry. As discussed below, this must relate to the time at which different groups evolved in relation to these coevolutionary events.

Evolution of Small and Large Flowers

Flower size in living angiosperms is quite variable. Only during the past 25 years have numerous new fossil flowers been discovered from the Cretaceous. The record that has been developed demonstrates that both medium- and small-sized flowers are present very early. Certainly, flower size must relate to pollinator size. The variability in size of the early flowers suggests that a variety of pollinators were involved in their pollination biology (Grimaldi, 1999). In addition to insect pollinators, both wind and water were important in the pollination of early angiosperms. Because the wind and the water have changed very little since the Cretaceous, there has been little change in the floral anatomy and morphology of these plants. Therefore, they are examples of some of the most ancient lines of living flowering plants. Those angiosperms that have modified their pollination biology to accommodate new or different animal pollinators are plants that probably have undergone the most extensive changes and whose fossil ancestors should be the most different from their modern descendants. These would include bird and bat pollinated flowers.

Evolution of Floral Presentation

In flowers that are insect pollinated, the display of the flower is critical. There seem to be clear distinctions between the presentation of the large *Archaeanthus* flower and the small fossil dichasial (Dilcher and Muller, 2000) flowers. The large *Archaeanthus* flower appears to have been terminal on a moderately large axis similar to the flowers of *Liriodendron* or *Magnolia* today. This allows for sturdy support and a colorful display to attract a pollinator. The dichasial flower, in contrast, is small and clustered into an umbel-like arrangement. This allows for a showy display of flowers in different stages of maturity and a broad area of clustered flowers upon which a pollinator can land and move about. However, small unisexual florets such as those of wind pollinated platanoid-like inflorescences and water pollinated ceratophylloid-like plants have been little affected by animal pollinators. For this reason, they persist today only slightly changed from their forms in the Early Cretaceous.

Unisexual vs. Bisexual Flowers

The earliest flowers now known appear to be gynodioecious. One axis has only carpels with a clear indication that no other organs subtended them, while an attached axis has both carpels and stamens (Sun *et al.*, 2000). So, was the first flower unisexual or bisexual? It appears to have had the potential to be both. Some early flowers, such as the platanoids and ceratophylloids, appear to be unisexual and never to have had a bisexual ancestry. Others such as *Archaefructus*, many of the small flowers from Portugal and the larger flowers from the Dakota Formation, are certainly bisexual. I suggest that the ancestral lineage of the angiosperms was most likely unisexual, and that with the availability of insect pollinators the efficiency of bisexual flowers won the day.

WHAT ARE THE SIGNIFICANT NODES OF ANGIOSPERM EVOLUTION?

There are three major nodes or events through time that resulted in major radiations of the angiosperms. These nodes include the evolution of showy flowers with a closed carpel, the evolution of bilateral flowers, and the evolution of nuts and fleshy fruits. At each of these events, there is a burst of adaptive radiation within the angiosperms that can be interpreted as an attempt to maximize the event for all of the diversity possible and to use the event for increased reproductive potential.

The evolution of the closed carpel and the evolution of the showy radial flower must have occurred at nearly the same time. This was the first adaptive node marking a distinct coevolution of early flowering plants and animal (insect) pollinators. The success of this involvement of insects in the reproductive biology of plants was not new. Dating back into the Paleozoic, insects most probably were involved in pollination of some of the seed ferns such as *Medullosa* (Dilcher, 1979; Retallack and Dilcher, 1988). During the Mesozoic, several non-angiospermous plants were certainly using animals for pollination as part of their reproductive biology. These include plants such as the Cycadoidea, (Delevoryas, 1968; Crepet, 1974) *Williamsonia, Williamsoniella,* and, perhaps, some seed ferns such as *Caytonia*. Insect diversity increased parallel to the increasing diversity of the angiosperms during the Mesozoic (Labandeira *et al.*, 1994; Labandeira, 1998; Magallón *et al.*, 1999). This node of evolution corresponds to the initial coevolution of animals and flowering plants in gamete transport. These early showy flowers came in many sizes, were displayed on the plant in many different ways, and were uniform in the types of organs they contained and the radial symmetry of these organs. They must have accommodated many different types of pollinators as evidenced by the variety of their anthers, stigmatic surfaces, nectaries, and the sizes and positions of the floral organs (Dilcher *et al.*, 1976a; Dilcher, 1979; Crepet and Nixon, 1998a, b; Gandolfo *et al.*, 1998a, b, c; Friis *et al.*, 1999; Crepet *et al.*, 2000). It was through the success of this coevolution that the angiosperms became the dominant vegetation during the early Late Cretaceous. Ordinal and family clades began to become identifiable during the later Early Cretaceous and the early Late Cretaceous (Crepet and Nixon, 1998a, b; Gandolfo *et al.*, 1998a, b, c; Magallón *et al.*, 1999; Crepet *et al.*, 2000). However, at the same time, some of the angiosperms never developed showy flowers and used other means of gamete transport for cross-pollination such as wind (early platinoids) and water (early ceratophylloids).

The evolution of bilateral flowers happened about 60 million years after the origin of the angiosperms. This node in coevolution never affected the water- or wind-pollinated groups that were already established. The evolution of the bees late in the Late Cretaceous (Michener and Grimaldi, 1988a, b) was a coevolutionary event with the evolution of bilateral flowers. This occurred independently in many different clades of flowering plants that were already established by the mid-Late Cretaceous. The potential for flowers to further direct the behavior of insects to benefit their pollination had a profound influence on those clades that evolved during the late Upper Cretaceous and early Tertiary. Flowers not only presented their sex organs surrounded by sterile floral organs with attractive patterns and colors, exuding attractive fragrances and filled with nectar and pollen for food, but the bilateral flowers could show the animals which way to approach them and how to enter and exit them. This allowed flowers to maximize the potential for precise gamete exchange that was impossible with radially symmetrical flowers. Such clades as the Papilionoideae (legume subfamily), Polygalaceae, and Orchidaceae, among others, demonstrate this coevolution. The success of these clades and especially the Orchidaceae, with its vast number of species, demonstrates the potential of this coevolutionary event.

The evolution of large stony and fleshy fruits and seeds is the last major coevolutionary node of the angiosperms. This is not to say that there were not the occasional attractive fruits produced earlier, but a large radiation of fruit and seed types of the angiosperms occurred during the Paleocene and Eocene. The change in angiosperm fruit size was noted by Tiffney (1984) who associated this change with the radiation of rodents and birds. This coevolutionary node allowed for both the further radiation of the angiosperms and the radiation of the mammals and birds. Stone (1973) noted that there was a tendency to develop animal-dispersed fruit types in the Juglandaceae several times in different clades of this family. Many angiosperm families took advantage of the potential to disperse their fruits and seeds by bird and mammal vectors during the early Tertiary as evidenced by the bursts of the evolution of fruits and seeds during this time (Reid and Chandler, 1933; Manchester, 1994). It is interesting to note that at this same time the angiosperms also were experiencing a radiation of wind-dispersed fruits and seeds (Call and Dilcher, 1992). This radiation of fruit and seed dispersal strategies in the angiosperms, late in their evolution (early Tertiary), is yet one more example of a means to promote outcrossing for the group.

WHY DID ANGIOSPERMS EVOLVE?

Coevolutionary events are largely responsible for the origin and subsequent nodes of evolution and radiation of the angiosperms. As we begin to find reproductive material of very early angiosperms (Taylor and Hickey, 1990; Sun *et al.*, 1998; Friis *et al.*, 1999), it becomes clear that some or most angiosperms developed bisexual insect-pollinated flowers very early, while some lines also maintained unisexual flowers with abiotic means of pollination (Dilcher, 1979). The coevolution with insects sparked a tremendous potential for plants to outcross by co-opting animals to carry their male gametes (pollen) to other individuals and other populations of the same species.

Each node of angiosperm evolution established genetic systems that favor outcrossing. The showy bisexual flower, the more specialized bilateral flower, and the nutritious nuts and fleshy fruits all are means by which the flowering plants increase their potential for outcrossing. The majority of angiosperm evolution is centered on this increased potential for outcrossing through coevolution with a wide variety of animals. In most cases the animals benefited as well from this coevolutionary association. Wind and water pollination syndromes also allowed for outcrossing and have continued to exist since the Early Cretaceous. However, they have never developed the diversity of those angiosperms pollinated by animals. Also several abiotically pollinated angiosperms, for example the Fagaceae (*Quercus* or oaks) and the Juglandaceae (*Carya* or pecans), later accommodated themselves for animal dispersal of their fruits or seeds. The importance of outcrossing cannot be underestimated as a driving force in the evolution of the angiosperms (Dilcher, 1995, 1996).

The ability of the angiosperms to accommodate and maximize benefits from animal behavior has been responsible for the evolutionary success of the group. As individual clades made use of particular coevolutionary strategies the diversity of both the angiosperms and animal groups increased. The benefits to the angiosperms were the benefits of the genetics of outcrossing. Because this is a sexual process, it was accomplished by means of evolutionary changes to flowers and fruits and seeds. This is why these particular organs have been centers of angiosperm evolution and why they are so useful in angiosperm systematics today.

I acknowledge with thanks the help of Terry Lott and Katherine Dilcher in the preparation of this manuscript. Thanks to Peter Raven who read and commented on this paper and also to the many students and J. William Schopf and colleagues who shared perspectives of angiosperm evolution with me. I thank the organizers of the symposium at which this paper was presented: Francisco Ayala, Walter Fitch, and Michael Clegg.

REFERENCES

Axelrod, D. I. (1966) The Eocene Copper Basin Flora of Northeastern Nevada. *Univ. Calif. Pub. Geol. Sci.* 59, 1–83.

Bandulska, H. (1924) On the cuticles of some recent and fossil Fagaceae. *Linn. Soc. Bot. (London)* 46, 427–441.

Basinger, J. F. & Dilcher, D. L. (1984) Ancient bisexual flowers. *Science* 224, 511–513.

Bell, W. A. (1957) Flora of the Upper Cretaceous Nanaimo Group of Vancouver Island, British Columbia. *Geol. Surv. Canada Memoir* 293, 1–84.

Berry, E. W. (1924) The Middle and Upper Eocene Floras of Southeastern North America. *U.S. Geol. Sur. Prof. Paper* 92, 1–206.

Brown, R. W. (1962) Paleocene Flora of the Rocky Mountains and Great Plains. *U.S. Geol. Surv. Prof. Paper* 375, 1–119.

Call, V. & Dilcher, D. L. (1992) Early angiosperm reproduction: survey of wind-dispersed disseminules from the Cretaceous-Paleogene of North America. In Organisation Internationale de Paleobotanique 4eme Conference (Organisation Francaise de Paleobotanique Information N: Special 16B, Paris), p. 36.

Crane, P. R. (1985) Phylogenetic analysis of seed plants and the origin of angiosperms. *Ann. Missouri Bot. Gard.* 72, 716–793.

Crepet, W. L. (1974) Investigations of North American cycadeoids: the reproductive biology of Cycadeoidea. *Paleontographica B* 148, 144–169.

Crepet, W. L. & Nixon K. C. (1998a) Fossil Clusiaceae from the Late Cretaceous (Turonian) of New Jersey and implications regarding the history of bee pollination. *Am. J. Bot.* 85, 1122–1133.

Crepet, W. L. & Nixon. K. C. (1998b) Two new fossil flowers of magnoliid affinity from the Late Cretaceous of New Jersey. *Am. J. Bot.* 85, 1273–1288.

Crepet, W. L., Nixon, K. C. & Gandolfo, M. A. (2000) Turonian flora of New Jersery. In Asociacion Paleontologica Argentina, Publicacion Especial. VII International Symposium on Mesozoic Terrestrial Ecosystem (Buenos Aires, Argentina).

Delevoryas, T. (1955) The Medullosae-structure and relationships. *Palaeontogr. Abt. B* 97, 114–167.

Delevoryas, T. (1968) Investigations of North American cycadeoids: structure, ontogeny, and phylogenetic considerations of cones of Cycadeoidea. *Palaeontographica B* 121, 122–133.

Dilcher, D. L. (1974) Approaches to the identification of fossil leaf remains. *Bot. Rev.* 40, 1–157.

Dilcher, D. L. (1979) Early angiosperm reproduction: an introductory report. *Rev. Palaeob. Palynol.* 27, 291–328.

Dilcher, D. L. (1995) Plant reproductive strategies: using the fossil record to unravel current issues in plant reproduction. In *Experimental and Molecular Approaches to Plant Biosystematics,* eds. Hoch, P. C. & Stephenson, A. G. (Missouri Botanical Gardens, St. Louis), pp. 187–198.

Dilcher, D. L. (1996) La importancia del origen de las angiospermas y como formaron el mundo alrededor de ellas. In *Conferencias VI Congreso Latinoamericano De Botanica, Mar Del Plata—Argentina 1994* (Royal Botanic Gardens, Kew), pp. 29–48.

Dilcher, D. L. & Crane, P. R. (1984) *Archaeanthus*: an early angiosperm from the Cenomanian of the Western Interior of North America. *Ann. Missouri Bot. Gard.* 71, 351–383.

Dilcher, D. L., Crepet, W. L., Beeker, C. D. & Reynolds, H. C. (1976a) Reproductive and vegetative morphology of a Cretaceous angiosperm. *Science* 191, 854–856.

Dilcher, D. L., Potter, F. W. & Crepet, W. L. (1976b) Investigations of angiosperms from the Eocene of North America: Juglandaceous winged fruits. *Am. J. Bot.* 63, 532–544.

Dilcher, D. L. & Muller, M. (2000) *Rev. Palaeob. Palynol.* (in press).

Dolph, G. F. (1974) Studies in the Apocynaceae, Part 1: a statistical analysis of *Apocynophyllum mississippiensis*, Ph.D. dissertation (Indiana University, Bloomington), 139p.

Dolph, G. F. (1975) A statistical analysis of *Apocynophyllum mississippiensis. Palaeontogr. Abt. B* 151, 1–51.

Doyle, J. A. (1969) Cretaceous angiosperm pollen of the Atlantic Plain and its evolutionary significance. *J. Arnold Arboretum* 50, 1–35.

Doyle, J. A. & Donoghue, M. J. (1986) Seed plant phylogeny and the origin of angiosperms: An experimental cladistic approach. *Bot. Rev.* 52, 321–431.

Doyle, J. A. & Donoghue, M. J. (1993) Phylogenies and angiosperm diversification. *Paleobiology* 19, 141–167.

Doyle, J. A., Donoghue, M. J. & Zimmer, E. A. (1994) Integration of morphological and ribosomal RNA data on the origin of angiosperms. *Ann. Missouri Bot. Gard.* 81, 419–450.

Doyle, J. A. & Hickey, L. J. (1976) Pollen and leaves from the mid-Cretaceous Potomac Group and their bearing on early angiosperm evolution. In *Origin and Early Evolution of Angiosperms,* ed. Beck, C. B. (Columbia Univ. Press, New York), pp. 139–206.

Edwards, W. N. (1935) The systematic value of cuticular characters in recent and fossil angiosperms. *Biol. Rev.* 1: 442–459.

Endress, P. K. (1994) Diversity and Evolutionary Biology of Tropical Flowers (Cambridge Univ. Press, Cambridge), 511p.

Florin, R. (1931) Untersuchungen zur stammesge-schichte der coniferales und cordaitales. *K. Svenska Vetensk. Akad. Handl.* 10, 1–588.

Fontaine, W. M. (1889) The Potomac or younger Mesozoic flora. *U.S. Geol. Survey, Monogr.* 15, 1–377.

Friis, E. M., Pedersen, K. R. & Crane, P. R. (1999) Early angiosperm diversification: the diversity of pollen associated with angiosperm reproductive structures in Early Cretaceous floras from Portugal. *Ann. Missouri Bot. Gard.* 86, 259–296.

Gandolfo, M. A., Nixon, K. C. & Crepet, W. L. (1998a) *Tylerianthus crossmanensis* gen. et sp. nov. (Aff. Hydrangeaceae) from the Upper Cretaceous of New Jersey. *Am. J. Bot.* 85, 376–386.

Gandolfo, M. A., Nixon, K. C. & Crepet, W. L. (1998b) A new fossil flower from the Turonian of New Jersey: *Dressiantha bicarpellata* gen. et sp. nov. (Capparales). *Am. J. Bot.* 85, 964–974.

Gandolfo, M. A., Nixon, K. C., Crepet, W. L., Stevenson, D. W. & Friis, E. M. (1998c) Oldest known fossils of monocotyledons. *Nature (London)* 394, 532–533.

Grimaldi, D. (1999) The co-radiations of pollinating insects and angiosperms in the Cretaceous. *Ann. Missouri Bot. Gard.* 86, 373–406.

Haq, B. U. & van Eysinga, F. W. B. (1998) Geological Time Table (Elsevier Science, Amsterdam), 5th Ed.

Herendeen, P. S. (1990) Fossil History of the Leguminosae from the southeastern North America, Ph. D. dissertation (Indiana University, Bloomington), 282p.

Herendeen, P. S. & Dilcher, D. L. (1990a) *Diplotropis* (Leguminosae, Papilionoideae) from the Middle Eocene of southeastern North America. *Syst. Bot.* 15, 526–533.

Herendeen, P. S. & Dilcher, D. L. (1990b) Fossil mimisoid legumes from the Eocene and Oligocene of southeastern North America. *Rev. Palaeobot. Palynol.* 62, 339–361.

Herendeen, P. S. & Dilcher, D. L. (1991) *Caesalpinia* subgenus Mezoneuron (Leguminosae, Caesalpinioideae) from the Tertiary of North America. *Am. J. Bot.* 78, 1–12.

Herendeen, P. S., Crepet, W. L. & Dilcher, D. L. (1992) The fossil history of the Leguminosae: Phylogenetic and Biogeographic implications. *In Advances in Legume Systematics: Part 4, The Fossil Record,* eds. Herendeen, P. S. & Dilcher, D. L. (Royal Botanic Gardens, Kew), pp. 303–316.

Hickey, L. J. (1973) Classification of the architecture of dicotyledonous leaves. *Am. J. Bot.* 60, 17–33.

Hickey, L. J. & Wolfe, J. A. (1975) The bases of angiosperm morphology. *Ann. Missouri Bot. Gard.* 62, 538–589.

Huang, Q. C. (1992) The paleoecological and stratigraphic implications of dispersed cuticle from the Mid-Cretaceous Dakota Formation of Kansas and Nebraska, Ph. D. dissertation (Indiana University, Bloomington), 221p.

Jones, J. H. & Dilcher, D. L. (1980) Investigations of angiosperms from the Eocene of North America: *Rhamnus marginatus* (Rhamnaceae) reexamined. *Am. J. Bot.* 67, 959–967.

Jones, J. H. (1984) Leaf architectural and cuticular analysis of extant Fagaceae and "Fagaceous" leaves from the Paleocene of southeastern North America, Ph.D. dissertation (Indiana University, Bloomington), 328p.

Kovach, W. L. & Dilcher, D. L. (1988) Megaspores and other dispersed plant remains from the Dakota Formation (Cenomanian) of Kansas. *Palynology* 12, 89–119.

Labandeira, C. C., Dilcher, D. L., Davis, D. R. & Wagner, D. L. (1994) Ninety–seven million years of angiosperm-insect association: Paleobiological insights into the meaning of coevolution. *Proc. Natl. Acad. Sci. USA* 91, 12278–12282.

Labandeira, C. C. (1998) How old is the flower and the fly? *Science* 28: 57–59.

LAWG (Leaf Architecture Working Group). (1999) Manual of Leaf Architecture: Morphological Description and Categorization of Dicotyledonous and Net-Veined Monocotyledonous Angiosperms (Smithsonian, Washington, DC.), 65p.

Leclercq, S. & Andrews, H. N. (1960) *Calamophyton bicephalum*, a new species from the Middle Devonian of Belgium. *Ann. Missouri Bot. Gard.* 47, 1–23.
Leclercq, S. & Banks, H. P. (1962) *Pseudosporochnus nodosus* sp. nov., a Middle Devonian plant with cladoxylalean affinities. *Palaeontogr. Abt. B* 110, 1–34.
Lesquereux, L. (1891) The flora of the Dakota Group. *U.S. Geol. Surv. Monogr.* 17, 1–400.
MacGinitie, H. D. (1969) The Eocene Green River flora of Northwestern Colorado and Northeastern Utah. *Univ. Calif. Pub. Geol. Sci.* 83, 1–140.
Magallón, S., Crane, P. R. & Herendeen, P. S. (1999) Phylogenetic pattern, diversity, and diversification of eudicots. *Ann. Missouri Bot. Gard.* 86, 297–372.
Manchester, S. R. (1981) Fossil history of the Juglandaceae, Ph.D. dissertation (Indiana University, Bloomington), 209p.
Manchester, S. R. (1994) Fruits and seeds of the Middle Eocene Nut Beds Flora, Clarno Formation, Oregon. *Paleontogr. Am.* 58, 1–205.
Michener, C. D. & Grimaldi, D. (1988a) A *Trigona* from Late Cretaceous amber of New Jersey (Hymenoptera: Apidae: Meliponinae). *Amer. Mus. Novit.* 2917, 1–10.
Michener, C. D. & Grimaldi, D. (1988b) The oldest fossil bee: apoid history, stasis, and the evolution of social behavior. *Proc. Natl. Acad. Sci. USA* 85, 6424–6426.
Newberry, J. S. (1898) The later extinct floras of North America. *U.S. Geol. Surv. Monogr.* 35, 1–295.
Nixon, K. C. & Crepet, W. L. (1993) Late Cretaceous fossil flowers of ericalean affinity. *Am. J. Bot.* 80, 616–623.
Reid, E. M. & Chandler, M. E. J. (1933) *The London Clay Flora* (British Museum of Natural History, London), 561p.
Retallack, G. J. & Dilcher, D. L. (1988) Reconstructions of selected seed ferns. *Ann. Missouri Bot. Gard.* 75, 1010–1057.
Roth, J. L. & Dilcher, D. L. (1979) Investigations of angiosperms from the Eocene of North America: stipulate leaves of the Rubiaceae. *Am. J. Bot.* 66, 1194–1207.
Roth, J. L. (1981) Epidermal studies in the Annonaceae and related families, Ph.D. dissertation (Indiana University, Bloomington), 218p.
Schwarzwalder, R. N. (1986) Systematics and early evolution of the Platanaceae, Ph.D. dissertation (Indiana University, Bloomington), 198p.
Sheffy, M. (1972) A Study of the Myricaceae from Eocene sediments of southeastern North America, Ph.D. dissertation (Indiana University, Bloomington), 134p.
Simpson, G. G. (1944) *Tempo and Mode in Evolution* (Columbia Univ. Press, New York), 237p.
Stebbins, G. L. Jr. (1950) *Variation and Evolution in Plants* (Columbia Univ. Press, New York), 643p.
Stebbins, G. L. Jr. (1974) *Flowering Plants: Evolution above the Species Level* (Harvard Univ. Press, Cambridge), 399p.
Stone, D. E. (1973) Patterns in the evolution of amentiferous fruits. *Brittonia* 25, 371–384.
Sun G., Dilcher, D. L., Zheng, S. L. & Zhou, Z. K. (1998) In search of the first flower: a Jurassic angiosperm, *Archaefructus*, from Northeast China. *Science* 282, 1692–1695.
Sun, G., Dilcher, D. L., Zheng, S. & Wang, X. (2000) *Rev. Palaeob. Palynol.* (in press).
Taylor, D. W. & Hickey, L. J. (1990) An Aptian plant with attached leaves and flowers: implications for angiosperm origin. *Science* 247, 702–704.
Taylor, T. N. & Taylor, E. L. (1993) *The Biology and Evolution of Fossil Plants* (Prentice Hall, Englewood Cliffs), 982p.
Tiffney, B. H. (1984) [1985] Seed size, dispersal syndromes, and the rise of the angiosperms: evidence and hypothesis. *Ann. Missouri Bot. Gard.* 71, 551–576.
Upchurch, G. R. & Dilcher, D. L. (1990) Cenomanian angiosperm leaf megafossils from the Rose Creek locality of the Dakota Formation, southeastern Nebraska. *U.S. Geological Survey Bulletin* 1915, 1–55.

15

Reproductive Systems and Evolution in Vascular Plants

KENT E. HOLSINGER

Differences in the frequency with which offspring are produced asexually, through self-fertilization and through sexual outcrossing, are a predominant influence on the genetic structure of plant populations. Selfers and asexuals have fewer genotypes within populations than outcrossers with similar allele frequencies, and more genetic diversity in selfers and asexuals is a result of differences among populations than in sexual outcrossers. As a result of reduced levels of diversity, selfers and asexuals may be less able to respond adaptively to changing environments, and because genotypes are not mixed across family lineages, their populations may accumulate deleterious mutations more rapidly. Such differences suggest that selfing and asexual lineages may be evolutionarily short-lived and could explain why they often seem to be of recent origin. Nonetheless, the origin and maintenance of different reproductive modes must be linked to individual-level properties of survival and reproduction. Sexual outcrossers suffer from a cost of outcrossing that arises because they do not contribute to selfed or asexual progeny, whereas selfers and asexuals may contribute to outcrossed progeny. Selfing and

Department of Ecology and Evolutionary Biology, U-3043, University of Connecticut, Storrs, CT 06269-3043

This paper was presented at the National Academy of Sciences colloquium "Variation and Evolution in Plants and Microorganisms: Toward a New Synthesis 50 Years After Stebbins," held January 27–29, 2000, at the Arnold and Mabel Beckman Center in Irvine, CA.

asexual reproduction also may allow reproduction when circumstances reduce opportunities for a union of gametes produced by different individuals, a phenomenon known as reproductive assurance. Both the cost of outcrossing and reproductive assurance lead to an over-representation of selfers and asexuals in newly formed progeny, and unless sexual outcrossers are more likely to survive and reproduce, they eventually will be displaced from populations in which a selfing or asexual variant arises.

The world's quarter of a million vascular plant species (Heywood and Watson, 1995) display an incredible diversity of life histories, growth forms, and physiologies, but the diversity of their reproductive systems is at least as great. In some ferns, individual haploid gametophytes produce both eggs and sperm. In others, individual gametophytes produce only one or the other. In seed plants, pollen- and ovule-producing structures may be borne together within a single flower, borne separately in different structures on the same plant, or borne on entirely different plants. In both groups of plants, the pattern in which reproductive structures are borne influences the frequency with which gametes from unrelated individuals unite in zygotes, and it is a predominant influence on the amount and distribution of genetic diversity found in a species.

Evolutionary explanations for the diversity in mating systems once focused on differences in population-level properties associated with the different reproductive modes. Selfing or asexual plants were, for example, presumed both to be more highly adapted to immediate circumstances and to be less able to adapt to a changing environment than sexual outcrossers, and these differences were used to explain the association of different reproductive modes with particular life histories, habitats, or both (Mather, 1943; Stebbins, 1957). We now realize that to explain the origin and the maintenance of particular reproductive modes within species we must relate differences in reproductive mode to differences that are expressed among individuals within populations (Lloyd, 1965). Nonetheless, differences in rates of speciation and extinction may be related to differences in reproductive modes. As a result, understanding broad-scale phylogenetic trends in the evolution of plant reproductive systems will require us to learn more about the patterns and causes of those relationships.

MODES OF REPRODUCTION

In higher animals, meiosis produces eggs and sperm directly. The sexual life cycle of vascular plants is more complex. Multicellular haploid and diploid generations alternate. Diploid sporophytes produce haploid spores through meiosis, and those spores develop into multicellular haploid gametophytes. In pteridophytes (ferns, club mosses, and horestails) the gametophyte is free-living. In seed plants (gymnosperms and angiosperms) the female gametophyte is borne within the ovule and only the male gametophyte (pollen) leaves the structure in which it was produced. Gametophytes produce haploid egg and sperm through mitosis, and these unite to form diploid zygotes from which new sporophytes develop. Asexual reproduction in plants, as in animals, occurs when offspring are produced through modifications of the sexual life cycle that do not include meiosis and syngamy (see Fig. 1).

When vascular plants reproduce asexually, they may do so either by budding, branching, or tillering (vegetative reproduction) or by producing spores or seed genetically identical to the sporophytes that produced them (agamospermy in seed plants, apogamy in pteridophytes). Vegetative reproduction is extremely common in perennial plants, especially in grasses and aquatic plants, and it can have dramatic consequences. The water-weed *Elodea canadensis*, for example, was introduced into Britain in about 1840 and spread throughout Europe by 1880 entirely by vegetative reproduction (Gustafsson, 1947). Exclusive reliance on vegetative reproduction is, however, the exception rather than the rule. More commonly,

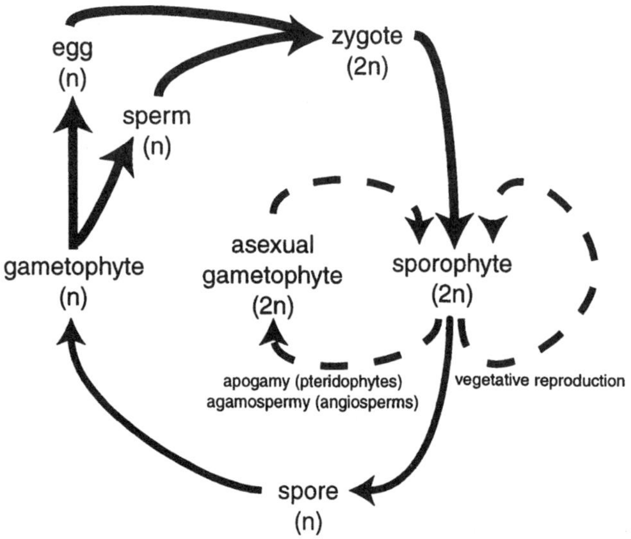

FIGURE 1 Diagram of basic vascular plant life cycles. Asexual life cycles are indicated with dashed lines. The sexual life cycle is indicated with solid lines. See Mogie (1992) for a more complete description of asexual life cycles.

species like white clover (*Trifolium repens*), reproduce both through vegetative reproduction and through sexually produced seed (Burdon, 1980).

Agamospermy is less widespread than vegetative reproduction, although it has been reported from at least 30 families of flowering plants (Gustafsson, 1947; Grant, 1981), and it is especially common in grasses and roses. Agamospermous species are often polyploids derived from hybridization between reproductively incompatible progenitors. When they have arisen many times, as in the hawk's beards (*Crepis*) of western North America (Babcock and Stebbins, 1938) or European blackberries (*Rubus*; Gustafsson, 1943), the pattern of variation makes it difficult to identify distinct lineages that can be called species. The taxonomic distribution of apogamy in pteridophytes is not well known because of the technical difficulties associated with studying spore development. Nonetheless, Manton (1950) cites examples from at least seven genera of ferns and points out that it has been known for more than a century that apogamy can be experimentally induced in many other groups (Lang, 1898).

When vascular plants reproduce sexually, the reproductive structures may be borne in many different ways. In some pteridophytes, like the club moss *Selaginella*, and in all seed plants, eggs and sperm are produced by different gametophytes. In other pteridophytes a single gametophyte may produce both eggs and sperm, as in most ferns. Even when eggs and sperm are produced on the same gametophyte, however, zygotes most frequently are formed through union of eggs and sperm from different gametophytes (Soltis and Soltis, 1992). Differences in the time at which male and female reproductive structures form often are reinforced by antheridiogens released by gametophytes in female phase that induce nearby gametophytes to remain in male phase (Döpp, 1959). The antheridiogen system of *Cryptogramma crispa*, for example, appears to enforce outcrossed reproduction even though individual gametophytes are developmentally capable of producing both eggs and sperm (Parajón *et al.*, 1999).

Eggs and sperm are produced by different gametophytes in flowering plants, but anthers and stigmas most often are borne in a single flower. Despite the apparent opportunity for self-fertilization, zygotes most frequently are formed through the union of eggs and sperm derived from different plants (Barrett and Eckert, 1990). Genetically determined self-incompatibility mechanisms appear to have evolved several times in flowering plants (Holsinger and Steinbachs, 1997), but differences in the time at which pollen is released and stigmas are receptive within a flower and spatial separation between anthers and stigmas promote outcrossing, even in many self-compatible plants (Bertin, 1993; Chang and Rausher, 1998). In some species of flowering plants a polymorphism in stigma height is associated with a complementary polymorphism in anther height, a condition known as heterostyly. Short anthers are found in flowers with long stigmas and vice versa. Darwin (1877) described the classic example of this system in *Primula veris*. In that species as in many others, morphological differences are associated with compatibility differences that allow pollen derived from short anthers to germinate only on short stigmas and pollen derived from long anthers to germinate only on long stigmas (see also Barrett, 1992). In other species of flowering plants and in all gymnosperms sexual functions are separate from one another. Either pollen and ovules are produced in different structures on the same plant (monoecy) or they are produced on different plants (dioecy). The separation of sexual functions also may be associated with physiological and ecological differences between the sexes, as in *Siparuna grandiflora* in which females and males have different patterns of distribution within populations (Nicotra, 1998).

CONSEQUENCES OF REPRODUCTIVE SYSTEMS

The genetic structure of a species comprises the identity and frequency of genotypes found within populations and the distribution of genotypes across populations. The reproductive system has long been recognized as a predominant influence on the genetic structure of plant species. Asexual progeny are genetically identical to the individuals that produced them, except for differences caused by somatic mutation. Selfed progeny may differ from their parent as a result of segregation at heterozygous loci, but selfing usually produces far fewer genotypes among offspring than outcrossing. As a result, fewer genotypes usually are found in populations in which either form of uniparental reproduction is common than in those in which outcrossing is the norm.

Among sexually reproducing species, selfers have populations with a smaller and more variable effective sizes (Schoen and Brown, 1991) and with less exchange of alleles among individuals within and among populations. As a result, selfing species are usually more homozygous than close relatives and have fewer genotypes per population than outcrossers. They also typically have fewer polymorphic loci and fewer alleles per polymorphic locus than closely related outcrossers (Brown, 1979; Gottlieb, 1981). In addition, the diversity found within selfing species is more a result of differences among populations than of differences among individuals within populations. Over 50% of the allozyme diversity found in selfers is attributable to differences among populations, whereas only 12% is attributable to differences among populations in outcrossers (Hambrick and Godt, 1989).

Allozyme and restriction site analyses of chloroplast DNA (cpDNA) in *Mimulus* (Scrophulariaceae) and nucleotide sequence analyses of introns associated with two nuclear genes in *Leavenworthia* (Brassicaceae) illustrate the dramatic impact mating systems can have on the genetic structure of plant species. In *Mimulus* both allozyme and nucleotide sequence diversity in a selfing species are only one-fourth that of a closely related outcrossing species (Fenster and Ritland, 1992). The lower nucleotide diversity in cpDNA might not be expected, because it is maternally inherited, but in highly selfing species background selection against deleterious alleles at nuclear loci can substantially reduce diversity in both nuclear and cytosolic genomes (Charlesworth *et al.*, 1995; Charlesworth *et al.*, 1997). In *Leavenworthia* populations of selfers are composed almost entirely of a single haplotype at each of the two loci studied, and each population is characterized by a different haplotype. In outcrossers, on the other hand, individuals belonging to the same population are only a little more similar to one another than were individuals belonging to different populations (Liu *et al.*, 1998, 1999). In addition, balancing selection appears to be responsible for maintaining an electrophoretic polymorphism at the locus encoding phosphoglucose isomerase in outcrossing species of *Leavenworthia* (Filatov and Charlesworth, 1999). Thus, selfers may have lower individual fitness than outcrossers, because they are genetically uniform at this locus.

The consequences of asexual reproduction are in some ways similar to those of selfing. In a strictly asexual population there is no exchange of genes among family lines, just as there is none within a completely selfing population. In contrast to selfers, however, asexual genotypes reproduce themselves exactly, except for differences caused by somatic mutation. Thus, the frequency of heterozygotes can be large in asexual populations even if the number of genotypes found is quite small, especially because many apogamous or agamospermous plants are derived from products of hybridization (Manton, 1950; Stebbins, 1950; Grant, 1981). Agamospermous *Crepis* in western North America, for example, are polyploids derived from hybridization between different pairs of seven narrowly distributed diploid progenitors (Babcock and Stebbins, 1938), and local populations are composed of relatively few genotypes. Moreover, most of the genetic diversity in the entire set of agamospermous species, which are facultatively sexual, is attributable to multiple origins rather than sexual recombination (Whitton, 1994; Holsinger *et al.*, 1999).

Although asexual populations are virtually guaranteed to have many fewer genotypes than sexual populations with similar allele frequencies, the number of genotypes within a population can still be quite large. Allozyme studies revealed between 15 and 47 clones in populations of the salt-marsh grass *Spartina patens* on the east coast of North America (Silander, 1984) and 13–15 clones of the daisy *Erigeron annuus* (Hancock and Wilson, 1976). When the number of genotypes per population is large, however, most genotypes are found in only one population and only a few are found in more than two or three populations (Ellstrand and Roose, 1987). When the number of genotypes per population is small, as it often is in agricultural weeds, each one may be quite widespread. More than 300 distinct forms of skeleton wire-weed, *Chondrilla juncea*, are found in Eurasia and the Mediterranean, but none is widespread. In Australia, however, only three forms are found, but each is widespread and the species is a serious agricultural pest (Hull and Groves, 1973; Burdon *et al.*, 1980).

In addition to effects on variation at individual loci, both selfing and asexuality may reduce the ability of populations to respond to a changing environment via natural selection, because they reduce the amount of genetic variability in populations. By exposing recessive alleles to selection, selfing may promote the loss of currently deleterious alleles that would be adaptively advantageous in other environments. In fact, selfers may maintain as little as one-fourth of the heritable variation outcrossers would maintain in a population of comparable size (Charlesworth and Charlesworth, 1995). In addition, by preventing gene exchange among family lineages, selfing and asexual populations reduce the diversity of genotypes on which natural selection can act. In fact, the proportion of variation caused by differences among individuals within a family is expected to decline almost linearly as a function of the selfing rate in populations. In completely selfing populations, virtually all genetic differences are differences among maternal families. In outcrossing populations, genetic differences within maternal families are expected to be almost as great as those among maternal families (Holsinger and Steinbachs, 1997). In both selfing and asexual species, therefore, the genetic structure of their populations may limit their ability to respond adaptively to natural selection. Because the causes of this constraint are the same for both types of uniparental reproduction, it is convenient to refer to it as the uniparental constraint.

Empirical analyses of phenotypic variation are consistent with the patterns of variation predicted by uniparental constraint in asexual populations (Ellstrand and Roose, 1987), but less so in populations of selfers. Comparisons of closely related outcrossers and selfers in *Phlox*, for example, found that outcrossers had more among family variation in 11 of 20 morphological traits than selfers (Clay and Levin, 1989), although the analysis did not distinguish between genetic and environmental effects on morphological differences. A similar study in *Collinsia heterophylla* (Scrophulariaceae), however, estimated genetic components of variance and found no relationship between genetically estimated rates of selfing in populations and the partitioning of genetic variance within and among families (Charlesworth and Mayer, 1995).

In addition to reducing the ability of populations to respond to a changing environment via natural selection, both selfing and asexuality also reduce a population's effective size (Pollack, 1987). As a result, deleterious alleles that would have little chance of drifting to fixation in an outcrossing population may be effectively neutral in one that is mostly or completely self-fertilizing. If such an allele is fixed and if it also reduces the reproductive potential of the population, the population will become smaller, making it possible for alleles that are even more deleterious to drift to fixation. This autocatalytic process, the "mutational meltdown," can lead to population extinction and can do so much more easily in completely selfing or asexual populations than in outcrossing ones (Lynch and Gabriel, 1990; Gabriel *et al.*, 1993). It appears, however, that a small amount of outcrossing in primarily selfing species can greatly retard the rate at which the process occurs.

Both the uniparental constraint and mutational meltdown hypotheses make an important prediction: an obligate selfing or obligate asexual lineage will be more short-lived than an otherwise comparable sexual outcrossing lineage. Unfortunately, phylogenetic analyses of the distribution of selfing and asexuality in plants are too few to allow us to assess this prediction directly. It is, however, a botanical commonplace that selfing species are often derivatives of outcrossing progenitors (see the discussion of selfers in *Arenaria* and *Linanthus* below, for example). Because derivatives must be younger than their progenitors, the average age of selfing species is probably less than that of outcrossing species, suggesting that selfers are also more short-lived.

EVOLUTION OF REPRODUCTIVE SYSTEMS

Explanations for the diversity of reproductive systems in flowering plants in terms of a tradeoff between short-term adaptive benefit and long-term flexibility were attractive, in part, because they emphasized the synergistic role of reproductive systems in plant evolution. Because of their impact on the amount and distribution of genetic variation within and among populations, reproductive systems can play an important role in determining the pattern and extent of population responses to natural selection on many other traits. As we have just seen, both obligate selfers and obligate asexuals are expected to harbor less genetic variability and to accumulate deleterious mutations more rapidly than sexual outcrossers, which could limit their ability to respond adaptively to environmental change. As a result, we would expect selfing and asexual lineages of plants to be relatively short-lived. Nonetheless, both selfers and asexuals have evolved repeatedly, and it is vital that we understand the circumstances under which they have evolved.

Failure to recognize the frequency with which self-fertilization has evolved, in particular, has led to many taxonomic mistakes. In *Arenaria* (Caryophyllaceae), for example, *A. alabamensis* is a self-pollinating derivative of *A. uniflora*, and populations of *A. alabamensis* in Georgia are independently derived from those in North and South Carolina. Because the floral features that distinguish *A. alabamensis* from *A. uniflora* are convergently derived, both sets of populations are now included within *A. uniflora* (Wyatt, 1988). *Linanthus* sect. *Leptosiphon* provides an even more striking example (Fig. 2). Self-fertilization evolved independently at least three times in this group of 10 taxa, and one selfing taxon (*Linanthus bicolor*) appears to have three separate origins (Goodwillie, 1999).

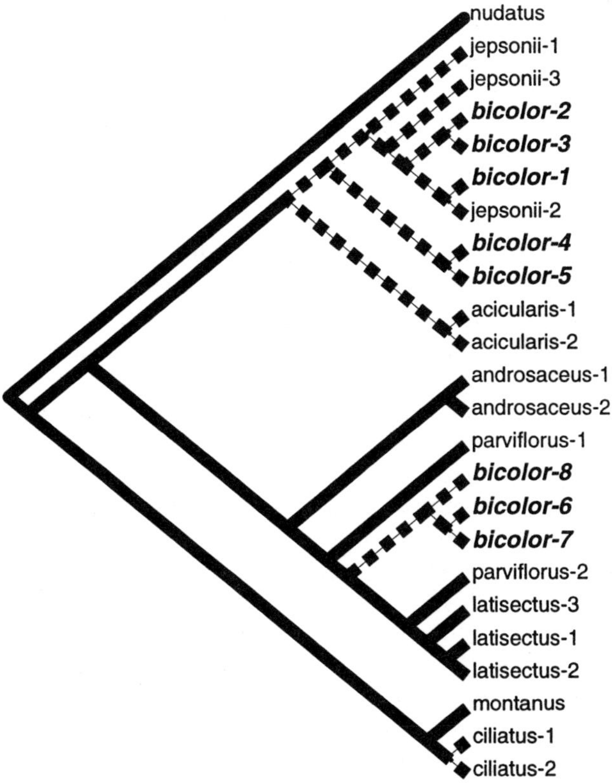

FIGURE 2 Phylogeny of *Linanthus* sect. *Leptosiphon* based on 450 bp from the internal transcribed spacer of the nuclear rDNA (redrawn from Goodwillie, 1999). Hatched branches show lineages where self-fertilization evolved. *Linanthus bicolor* appears in bold, italic print.

Despite the possible impacts that self-fertilization might have on long-term persistence of populations, we have known for more than 35 years that self-fertilization "can evolve only because of a selective advantage before fertilization" (Lloyd, 1965). Moreover, there are only two types of advantage that self-fertilization can provide. It can increase reproductive success when lack of pollinators or inefficient pollen transfer limits reproductive success (reproductive assurance), or it can increase success as a pollen parent when pollen devoted to selfing is more likely to accomplish fertilization than pollen devoted to outcrossing (automatic selection) (Holsinger, 1996). Unfortunately, we are not yet able to predict when selfing will provide either reproductive assurance or an automatic selection advantage.

In *Aquilegia formosa* (Ranunculaceae), for example, pollinator exclusion does not affect seed set, suggesting that individuals are able to self-pollinate to ensure seed set. Hand pollinations increase seed set relative to open-pollinated controls, suggesting that pollen transfer limits reproductive success. Nonetheless, open-pollinated, emasculated flowers set as much seed as open-pollinated, unemasculated flowers, demonstrating that self-pollination provides little reproductive assurance in a species where both the capacity for autonomous self-pollination and pollen-limited seed set exist (Eckert and Schaefer, 1998).

The automatic selection advantage of self-fertilization, first pointed out by Fisher (1941), can arise because in a stable population an outcrossing individual will, on average, serve as ovule parent to one member of the next generation and as pollen parent to one other. A selfing individual in the same population will, however, serve as both ovule and pollen parent to its own selfed progeny and as pollen parent to one outcrossed progeny of another individual in the population. Thus, an allele promoting self-fertilization has a 3:2 transmission advantage relative to one promoting outcrossing. So alleles promoting self-fertilization are expected to spread, unless selfed progeny suffer a compensating disadvantage in survival or reproduction. Fisher's argument assumes that morphological changes promoting self-fertilization do not diminish the selfer's ability to serve as an outcross pollen parent. The extent to which selfing reduces an individual's contribution to the outcross pollen pool is referred to as pollen discounting (Holsinger *et al.*, 1984).

Relatively few attempts have been made to measure the extent of pollen discounting in plant populations. In one experiment in *Eichhornia paniculata* (Pontederiaceae) (Kohn and Barrett, 1994) and another in *Ipomoea purpurea* (Convolvulaceae) (Rausher *et al.*, 1993) selfers were actually more successful as outcross pollen parents than outcrossers. In observations derived from natural populations of *Mimulus* (Scrophulariaceae) (Ritland, 1991), selfers appeared not to contribute any pollen to the outcross pollen pool. Because the selfers were morphologically quite different from the outcrossers in *Mimulus* and much less so in *E. paniculata* and *I. purpurea*, it may be that differences in the extent of pollen discounting are related to differences in floral morphology. This would be consistent with the observation that pollinator movement within multiple-flowered influorescences led to observable differences in the degree of pollen discounting in other experiments on *E. paniculata* (Harder and Barrett, 1995).

Just as the forces favoring evolution of self-fertilization, reproductive assurance, and automatic selection are well-known, so also is the primary force opposing its spread, inbreeding depression. Thomas Knight pointed out more than 200 years ago that the selfed progeny of garden peas are less vigorous and fertile than are outcrossed progeny (Knight, 1799). The impact that inbreeding depression has on the evolution of self-fertilization is, however, more complex than might be expected. The fate of a variant causing an increase in the rate of selfing depends not only on the magnitude of inbreeding depression, but also on the genetic basis of inbreeding depression, and on the magnitude of the difference in selfing rates that the variant induces (see Uyenoyama *et al.*, 1993 for a detailed review). The complexity arises because different family lineages within a population may exhibit different degrees of inbreeding depression. Because selection among family lines is an important component of natural selection in partially self-fertilizing populations (see above), inbred families (those with a high frequency of alleles promoting self-fertilization) may show less inbreeding depression than less inbred families. If the extent of the association between family inbreeding depression and mating system is strong enough, selfing variants may spread even in the face of high population inbreeding depression (Holsinger, 1988).

The extent of associations between genetic variants affecting the mating system and levels of inbreeding depression in natural populations is not known. If the genomic rate of mutations to recessive or nearly recessive lethals is sufficiently high, levels of inbreeding depression are relatively insensitive to selfing rates (Lande *et al.*, 1994), which will cause even families that differ substantially in their selfing rate to have similar levels of inbreeding depression. The relationship between within-population mating system differences and inbreeding depression also may be weak if mating system differences are polygenically controlled (Schultz and Willis, 1995). Experimental results are mixed. In *Lobelia siphilitica* (Campanulaceae) no differences in inbreeding depression could be found between females, which must outcross, and hermaphrodites, which self to some extent (Mutikainen and Delph, 1998), whereas in *Gilia achilleifolia* (Polemoniaceae) individuals with anthers and stigmas well separated (more outcrossing) have greater amounts of inbreeding depression than those in which anthers and stigmas are not well separated (more selfing) (Takebayashi and Delph, 2000).

Ideas about the evolution of agamospermy and apogamy in plants tend to focus on the genetic consequences of agamospermy and the mechanisms by which it might arise (Mogie, 1992). In flowering plants, for example, agamospermous reproduction resulting from asexual development of gametophytic tissue is almost invariably associated with polyploidy. Whitton (1994) suggests that this correlation arises because the same process, formation of unreduced female gametophytes, contributes both to agamospermous reproduction and to the origin of polyploids. Although these arguments may shed light on the evolutionary correlates of agamospermy, they shed no light on the process by which a genetic variant promoting agamospermy is able to establish itself within populations. Fortunately, it is easy to construct arguments parallel to those for the automatic selection advantage of self-fertilization to show why a similar advantage might accrue to asexual plants in a population of hermaphroditic outcrossers.

In a stable population of hermaphrodites, each outcrosser will replace itself, serving once as a seed parent and once as a pollen parent to the outcrossed progeny of another individual. Suppose a genetic variant that causes complete agamospermy is introduced into this population and that this variant has no effect on the pollen production of individuals carrying it. Then an agamospermous individual will replace itself with agamospermous seed, but it also will serve as pollen parent to the outcrossed progeny of sexual individuals. In short, some of the seed progeny of sexuals will carry the genetic variant causing agamospermy and will be agamospermous themselves, whereas all of the seed progeny of agamosperms also will be agamospermous. Thus, agamospermy has an automatic selection advantage over outcrossing, and it will tend to spread through populations, unless agamosperms have a compensating disadvantage in survival and reproduction relative to outcrossers. I am not aware of studies that investigate the extent of the automatic selection advantage agamosperms might have in natural populations.

THE COST OF SEX

Mathematical analyses of models for the evolution and maintenance of sexual reproduction suggest that asexuals can be favored either because they avoid the "cost of males" or because they avoid the "cost of meiosis" (Williams, 1975; Maynard Smith, 1978). The cost of males arises because the number of females in a population more often limits its rate of population growth than the number of males, a consequence of Bateman's principle (Bateman, 1948). As a result, an asexual population composed entirely of females may have a higher intrinsic rate of increase and therefore displace an otherwise equivalent sexual population with separate sexes. In hermaphrodites, however, the cost of males will exist only when vegetative reproduction allows individuals to produce more offspring per unit of resource than reproduction through seed (Lively and Lloyd, 1990) or when selfers or agamosperms are able to divert resources from pollen to seed production (Schoen and Lloyd, 1984).

The automatic selection advantage of selfers and agamosperms often is attributed to the cost of meiosis. More careful analysis of the similarities between the evolution of selfing and the evolution of agamospermy suggests that cost of meiosis is not an apt description for the forces governing either process. The phrase cost of meiosis refers to the idea that the genetic coefficient of relatedness between individuals and their outcrossed offspring is smaller than the coefficient of relatedness would be between those same individuals and their selfed or asexual offspring. Notice, however, that in a population of complete selfers an agamosperm would not have an automatic selection advantage over selfers because the pollen it produces would not fertilize any ovules. The asexual progeny of an agamosperm are genetically identical to their parent (barring rare somatic mutation) and the selfed progeny of a selfer are genetically variable to the extent that there is segregation at heterozygous loci. Thus, the asexual progeny of an agamosperm are more closely related to their parent than the sexual progeny of a selfer. Nonetheless, the relative fitness of selfed and agamospermous offspring will determine the outcome of natural selection, not the extent to which selfed or agamospermous progeny resemble their parents.

These observations suggest that cost of outcrossing is a better phrase to describe the automatic selection advantage of selfers and agamosperms relative to sexual outcrossers. The cost of outcrossing arises because selfers and agamosperms can serve as pollen parents of progeny produced by sexual outcrossers, but outcrossers are prevented from serving as pollen parents to the selfed progeny of selfers and the asexual progeny of agamosperms. So long as pollen devoted to selfing is more likely to accomplish fertilization than pollen devoted to outcrossing (Holsinger, 1996) and so long as agamosperms are able to serve as pollen parents to the outcrossed progeny of other individuals, selfing and agamospermy will be overrepresented in newly formed progeny of the next generation. Unless, natural selection against selfed or agamospermous progeny is sufficiently strong, the cost of outcrossing will cause the frequency of outcrossers to decline.

CONCLUSIONS

The direct genetic consequences of self-fertilization and asexual reproduction are quite different. Self-fertilization causes progeny to be heterozygous at only half as many loci, on average, as their parents, and highly selfing populations are composed primarily of homozyous genotypes. Asexual reproduction, however, results in progeny that are genetically identical to their parents, barring somatic mutation, and asexual populations therefore may be highly heterozygous. Because self-fertilization and asexual reproduction both prevent exchange of genetic material among family lineages, however, the number of genotypes found in populations of predominant selfers or obligate asexuals is usually much smaller than would be found in a population of sexual outcrossers with the same allele frequencies. For similar reasons, surveys have repeatedly shown that a greater proportion of the genetic diversity found in selfing or asexual species is a result of differences among populations than in sexual outcrossers (Brown, 1979; Gottlieb, 1981; Hamrick and Godt, 1989, 1996). In short, the great diversity of reproductive systems exhibited by vascular plants is matched by a similar diversity of genetic structures within and among their populations. Indeed, the diversity of reproductive systems may be the predominant cause of the diversity in genetic structure.

Differences in genetic structure associated with differences in reproductive systems were once commonly invoked as evolutionary forces governing their origin (Stebbins, 1957; Grant, 1958; Baker, 1959). Although such differences may help us to understand why some lineages persist and diversify and others do not, we now realize that to understand the origin of alternative reproductive systems we must look for benefits and costs associated with individual survival and reproduction. Moreover, the comparison of self-fertilization and agamospermy shows that the most important distinction for hermaphroditic organisms is between uniparental and biparental forms of reproduction. Uniparental reproduction, whether through selfing, agamospermy, or apogamy, excludes sexual outcrossers from contributing to some offspring that will form the next generation. If selfers, agamosperms, or apogams are able to contribute to some sexual offspring, genotypes promoting that mode of reproduction will be over-represented in the next generation, reflecting the cost of outcrossing.

To the segregational cost of outcrossing can be added another: selfers, agamosperms, and apogams are able to produce offspring even under conditions that prevent the union of gametes produced by different individuals. The benefit of this reproductive assurance seems so apparent that it is surprising how few experimental studies have been done to document it and how equivocal their results are (Eckert and Schaefer, 1998). When plants reproducing uniparentally benefit from reproductive assurance or are over-represented in the next generation as a result of donating gametes to the outcrossed progeny of other individuals, they eventually will replace sexual outcrossers in the population, unless the progeny of sexual outcrossers are substantially more likely to survive and reproduce. Thus, the relative fitness of different types of individuals competing in a population and the frequency with which different types are formed will determine whether outcrossers persist in the short term or are replaced by selfers or asexuals.

In the long term, however, differences in the ability of outcrossers, selfers, and asexuals to respond to environmental change and resist the accumulation of deleterious alleles may cause lineages with different reproductive systems to persist for different lengths of time. The smaller number of genotypes in highly selfing and asexual populations may reduce the efficiency with which natural selection can operate, limiting the ability of their populations to respond adaptively to a changing environment—the uniparental constraint. In addition, highly selfing populations have an effective size only about half that of an outcrosser with the same number of individuals (Pollak, 1987), and their size also tends to be more variable (Schoen and Brown, 1991). Both selfers and asexuals, therefore, are more likely to accumulate deleterious mutations than sexual outcrossers, and these mutations could decrease their reproductive capacity and contribute to their early extinction through a mutational meltdown (Lynch and Gabriel, 1990; Gabriel *et al.*, 1993).

In this sense, therefore, Stebbins (1957), Grant (1958), and Baker (1959) were right to contend that selfers and asexuals lack the long-term flexibility characteristic of sexual outcrossers. Indeed, this lack of flexibility may explain why selfers and asexuals originate frequently from outcrossing ancestors, but often seem to be evolutionarily short-lived. If we are to have a comprehensive understanding of broad-scale evolutionary patterns in plants, therefore, we must begin to investigate the relationships among reproductive systems, rates of speciation, and rates of extinction, and we must begin to understand the causes of the relationships we find.

Two anonymous reviewers made suggestions on an earlier version of this paper that led to substantial improvements. Greg Anderson, Janine Caira, Cindi Jones, and members of the plant mating systems discussion group at the University of Connecticut also read an earlier draft of this paper, and I am indebted to them for their helpful suggestions.

REFERENCES

Babcock, E. B. & Stebbins, G. L. (1938) The American species of *Crepis*. *Carnegie Inst. Wash.* Publ. No. 504, 1–199.

Baker, H. G. (1959) Reproductive methods as factors in speciation in flowering plants. *Cold Spring Harbor Symposia on Quantitative Biology* 24, 177–191.

Barrett, S. C. H. (Ed.) (1992) *Evolution and Function of Heterostyly* (Springer-Verlag, Berlin).

Barrett, S. C. H. & Eckert, C. G. (1990) Variation and evolution of mating systems in seed plants. In *Biological Approaches and Evolutionary Trends in Plants*, ed. Kawano, S. (Academic Press, London) pp. 229–254.

Bateman, A. J. (1948) Intra-sexual selection in *Drosophila. Heredity* 2, 349–368.

Bertin, R. I. (1993) Incidence of monoecy and dichogamy in relation to self-fertilization in angiosperms. *American Journal of Botany* 80(5), 556–560.

Brown, A. H. D. (1979) Enzyme polymorphism in plant populations. *Theoretical Population Biology* 15(1), 1–42.

Burdon, J. J. (1980) Intra-specific diversity in a natural population of *Trifolium repens. Journal of Ecology* 68, 717–735.

Burdon, J. J., Marshall, D. R. & Groves, R. H. (1980) Isozyme variation in *Chondrilla juncea* l. in Australia. *Australian Journal of Botany* 38, 193–198.

Chang, S.-M. & Rausher, M. D. (1998) Frequency-dependent pollen discounting contributes to maintenance of a mixed mating system in the common morning glory *Ipomoea purpurea. American Naturalist* 152(5), 671–683.

Charlesworth, B., Nordborg, M. & Charlesworth, D. (1997) The effects of local selection, balanced polymorphism and background selection on equilibrium patterns of genetic diversity in subdivided populations. *Genetical Research* 70(2), 155–174.

Charlesworth, D. & Charlesworth, B. (1995) Quantitative genetics in plants: the effect of the breeding system on genetic variability. *Evolution* 49(5), 911–920.

Charlesworth, D., Charlesworth, B. & Morgan, M. T. (1995) The pattern of neutral molecular variation under the background selection model. *Genetics* 141(4), 1619–1632.

Charlesworth, D. & Mayer, S. (1995) Genetic variability of plant characters in the partial inbreeder *Collinsia heterophylla (Scrophulariaceae). American Journal of Botany* 82(1), 112–120.

Clay, K. & Levin, D. A. (1989) Quantitative variation in *Phlox*: comparison of selfing and outcrossing species. *American Journal of Botany* 76(4), 577–588.

Darwin, C. (1877) *The Different Forms of Flowers on Plants of the Same Species* (Appleton, New York).

Döpp, W. (1959) Über eine hernmende und eine fördernde Substanz bei der Antherdienbildung in den Prothallien von *Pteridium aquilinum. Berichte der Deutschen Botanischen Gessellschaft* 72, 11–24.

Eckert, C. G. & Schaefer, A. (1998) Does self-pollination provide reproductive assurance in *Aquilegia canadensis* (Ranunculaceae). *American Journal of Botany* 85(7), 919–924.

Ellstrand, N. C. & Roose, M. L. (1987) Patterns of genotypic diversity in clonal plant species. *American Journal of Botany* 74(1), 123–131.

Fenster, C. B. & Ritland, K. (1992) Chloroplast {DNA} and isozyme diversity in two *Mimulus* species (Scrophulariaceae) with contrasting mating systems. *American Journal of Botany* 79, 1440–1447.

Filatov, D. A. & Charlesworth D. (1999) DNA polymorphism, haplotype structure and balancing selection in the leavenworthia PgiC locus. *Genetics* 153(3), 1423–1434.

Fisher, R. A. (1941) Average excess and average effect of a gene substitution. *Annals of Eugenics* 11, 53–63.

Gabriel, W., Lynch, M. & Bürger, R. (1993) Muller's ratchet and mutational meltdowns. *Evolution* 47(6), 1744–1757.

Goodwillie, C. (1999) Multiple origins of self-compatibility in *Linanthus* section *Leptosiphon* (Polemoniaceae): phylogenetic evidence from internal-transcribed-spacer sequence data. *Evolution* 53(5), 1387–1395.

Gottlieb, L. D. (1981) Electrophoretic evidence and plant populations. In *Progress in Phytochemistry*, Volume 7, eds. Reinhold, L., Harborne, J. B. & Swain T. (Pergamon Press, Oxford), pp. 1–46.

Grant, V. (1958) The regulation of recombination in plants. *Cold Spring Harbor Symposia on Quantitative Biology* 23, 337–363.

Grant, V. (1981) *Plant Speciation*, 2nd ed. (Columbia Univ. Press, New York, NY).

Gustafsson, A. (1943) The genesis of the European blackberry flora. *Lunds Universitets Årsskrift* 39(6), 1–200.

Gustafsson, A. (1946–1947) Apomixis in higher plants. *Lunds Universitets Årsskrift* 42–43, 1–370.

Hamrick, J. L. & Godt, M. J. W. (1989) Allozyme diversity in plant species. *In Plant Population Genetics, Breeding, and Genetic Resources* eds. Brown, A. H. D., Clegg, M. T., Kahler, A. T. & Weir, B. S. (Sinauer Associates, Sunderland, MA), pp. 43–63.

Hamrick, J. L. & Godt, M. J. W. (1996) Effects of life history traits on genetic diversity in plant species. *Philosophical Transactions of the Royal Society of London, Series B* 351, 1291–1298.

Hancock, J. F. & Wilson R. E. (1976) Biotype selection in erigeron annuus during old field succession. *Bulletin of the Torrey Botanical Club* 103, 122–125.

Harder, L. D. & Barrett, S. C. H. (1995) Mating cost of large floral displays in hermaphrodite plants. *Nature* 373, 512–515.

Heywood, V. H. & Watson, R. T. (Eds.) (1995) *Global Biodiversity Assessment* (Cambridge University Press, Cambridge).

Holsinger, K. E. (1988) Inbreeding depression doesn't matter: the genetic basis of mating system evolution. *Evolution* 42, 1235–1244.

Holsinger, K. E. (1996) Pollination biology and the evolution of mating systems in flowering plants. *Evolutionary Biology* 29, 107–149.

Holsinger, K. E., Feldman, M. W. & Chrisiansen F. B. (1984) The evolution of self-fertilization in plants. *American Naturalist* 124, 446–453.

Holsinger, K. E., Mason-Gamer, R. J. & Whitton, J. (1999) Genes, demes, and plant conservation. In *Genes, Species, and the Threat of Extinction: DNA and Genetics in the Conservation of Endangered Species*, eds. Landweber, L. F. & Dobson, A. P. (Princeton University Press, Princeton, NJ), pp. 23–46.

Holsinger, K. E. & Steinbachs, J. E. (1997) Mating systems and evolution in flowering plants. In *Evolution and Diversification of Land Plants*, eds. Iwatsuki, K. & Raven, P. H. (Springer-Verlag, Tokyo), pp. 223–248.

Hull, V. J. & Groves, R. H. (1973) Variation in *Chondrilla juncea* in south-eastern Australia. *Australian Journal of Botany* 12, 112–135.

Knight, T. (1799) Experiments on the fecundation of vegetables. *Philosophical Transactions of the Royal Society of London* 89, 195–204.

Kohn, J. R. & Barrett, S. C. H. (1994) Pollen discounting and the spread of a selfing variant in tristylous *Eichhornia paniculata*: evidence from experimental populations. *Evolution* 48(5), 1576–1594.

Lande, R., Schemske, D. W. & Schultz, S. T. (1994) High inbreeding depression, selective interference among loci, and the threshold selfing rate for purging recessive lethal mutation. *Evolution* 48(4), 965–978.

Lang, W. H. (1898) Apogamy and the development of sporangia on prothalli. *Philosophical Transactions of the Royal Society of London, Series B* 190, 187–238.

Liu, F., Charlesworth, D. & Kreitman, M. (1999) The effect of mating system differences on nucleotide diversity at the phosphoglucose isomerase locus in the plant genus *Leavenworthia*. *Genetics* 151(1), 343–357.

Liu, F. L., Zhang, L. & Charlesworth, D. (1998) Genetic diversity in *Leavenworthia* populations with different inbreeding levels. *Proceedings of the Royal Society of London, Series B* 265, 293–301.

Lively, C. M. & Lloyd, D. G. (1990) The cost of biparental sex under individual selection. *American Naturalist* 135(4), 489–500.

Lloyd, D. G. (1965) Evolution of self-compatibility and racial differentiation in *Leavenworthia* (Cruciferae). *Contributions from the Gray Herbarium* 195, 3–133.

Lynch, M. & Gabriel, W. (1990) Mutation load and the survival of small populations. *Evolution* 44, 1725–1737.

Manton, I. (1950) *Problems of Cytology and Evolution in the Pteridophyta* (Cambridge University Press, Cambridge).

Mather, K. (1943) Polygenic inheritance and natural selection. *Biological Reviews* 18, 32–64.

Maynard Smith, J. (1978) *The Evolution of Sex* (Cambridge Univ. Press, Cambridge).

Mogie, M. (1992) *The Evolution of Asexual Reproduction in Plants* (Chapman & Hall, London).

Mutikainen, P. & Delph, L. F. (1998) Inbreeding depression in gynodioecious *Lobelia siphilitica*: among-family differences override between-morph differences. *Evolution* 52(6), 1572–1582.

Nicotra, A. B. (1998) Sex ratio variation and spatial distribution of *Siparuna grandiflora*, a tropical dioecious shrub. *Oecologia* 115(1/2), 102–113.

Parajón, S., Pangua, E. & Garcia-Álvarez, L. (1999) Sexual expression and genetic diversity in populations of *Cryptogramma crispa* (Pteridaceae). *American Journal of Botany* 86(7), 964–973.

Pollak, E. (1987) On the theory of partially inbreeding finite populations. I. partial selfing. *Genetics* 117, 353–360.

Rausher, M. D., Augustine, D. & VanderKooi, A. (1993) Absence of pollen discounting in a genotype of *Ipomoea purpurea* exhibiting increased selfing. *Evolution* 47(6), 1688–1695.

Ritland, K. (1991) A genetic approach to measuring pollen discounting in natural plant populations. *American Naturalist* 138(4), 1049–1057.

Schoen, D. J. & Brown, A. H. D. (1991) Intraspecific variation in population gene diversity and effective population size correlates with the mating system in plants. *Proceedings of the National Academy of Sciences USA* 88, 4494–4497.

Schoen, D. J. & Lloyd. D. G. (1984) The selection of cleistogamy and heteromorphic diaspores. *Biological Journal of the Linnean Society* 23, 303–322.

Schultz, S. T. & Willis, J. H. (1995) Individual variation in inbreeding depression: the roles of inbreeding history and mutation. *Genetics* 141, 1209–1223.

Silander, J. A. (1984) The genetic basis of the ecological amplitude of *Spartina patens*. iii. allozyme variation. *Botanical Gazette* 145(4), 569–577.

Soltis, D. E. & Soltis, P. S. (1992) The distribution of selfing rates in homosporous ferns. *American Journal of Botany* 79, 97–100.

Stebbins, G. L. (1950) *Variation and Evolution in Higher Plants* (Columbia Univ. Press, New York, NY).

Stebbins, G. L. (1957) Self-fertilization and population variability in the higher plants. *American Naturalist* 91, 337–354.

Takebayashi, N. & Delph L. F. (2000) Association between floral traits and inbreeding depression. *Evolution* 54, 840-846.

Uyenoyama, M. K., Holsinger, K. E. & Waller, D. M. (1993) Ecological and genetic factors directing the evolution of self-fertilization. *Oxford Surveys in Evolutionary Biology* 9, 327–381.

Whitton, J. (1994) Systematic and evolutionary investigation of the North American *Crepis* agamic complex. Ph.D. dissertation, University of Connecticut.

Williams, G. C. (1975) *Sex and Evolution* (Princeton Univ. Press, Princeton, NJ).

Wyatt, R. (1988) Phylogenetic aspects of the evolution of self-pollination. In *Plant Evolutionary Biology*, eds. Gottlieb, L. D. & Jain, S. K. (Chapman & Hall, London), pp. 109–131.

16

Hybridization as a Stimulus for the Evolution of Invasiveness in Plants?

NORMAN C. ELLSTRAND* AND
KRISTINA A. SCHIERENBECK‡

Invasive species are of great interest to evolutionary biologists and ecologists because they represent historical examples of dramatic evolutionary and ecological change. Likewise, they are increasingly important economically and environmentally as pests. Obtaining generalizations about the tiny fraction of immigrant taxa that become successful invaders has been frustrated by two enigmatic phenomena. Many of those species that become successful only do so (i) after an unusually long lag time after initial arrival, and/or (ii) after multiple introductions. We propose an evolutionary mechanism that may account for these observations. Hybridization between species or between disparate source populations may serve as a stimulus for the evolution of invasiveness. We present and review a remarkable number of cases in which hybridization preceded the emergence of successful invasive populations. Progeny with a history of hybridization may enjoy one or more potential genetic benefits relative to their progenitors. The observed lag times and multiple introductions that seem a prerequisite for certain species to evolve invasiveness may be a correlate of the time

*Department of Botany and Plant Sciences and Center for Conservation Biology, University of California, Riverside, CA 92521-0124; and ‡Department of Biology, California State University, Chico, CA 93740

This paper was presented at the National Academy of Sciences colloquium "Variation and Evolution in Plants and Microorganisms: Toward a New Synthesis 50 Years After Stebbins," held January 27–29, 2000, at the Arnold and Mabel Beckman Center in Irvine, CA.

necessary for previously isolated populations to come into contact and for hybridization to occur. Our examples demonstrate that invasiveness can evolve. Our model does not represent the only evolutionary pathway to invasiveness, but is clearly an underappreciated mechanism worthy of more consideration in explaining the evolution of invasiveness in plants.

Invasive species have always held a special place for ecologists and evolutionary biologists. Successful invaders that have colonized new regions within historical time provide real-life examples of ecological and evolutionary change. The demographic change from a small number of colonists to a sweeping wave of invaders is a dramatic ecological event. Likewise, those demographic changes—a founder event followed by a massive increase in numbers—may have dramatic evolutionary consequences. Not surprisingly, whole books have been dedicated to the basic science of invasive species (for example, see Elton, 1958; Mooney and Drake, 1986).

Also, the applied biology of invasive species has become increasingly important as intentional and unintentional anthropogenic dispersal moves species from continent to continent at unprecedented rates. Invasive plants and animals are often thought of as agricultural pests, but they also pose a hazard for a variety of human concerns, including health, transportation, and conservation (U.S. Congress, Office of Technology Assessment, 1993). Invasive species not only directly impact human well being, but they also are recognized as agents that alter community structure and ecosystem function (for example, see Horvitz *et al.*, 1998). In the United States alone, the damage wrought by invasive species totals approximately $122 billion per year (Pimentel *et al.*, 2000).

Only a tiny fraction of introduced species become successful invasives (Williamson, 1993). Given that invasives are important for so many reasons, considerable effort has been spent trying to develop generalizations to determine which species are likely to become successful. In particular, ecological, taxonomic, and physiological correlates of invasive success have been sought to predict which introduced species might become successful (for example, see Bazzaz, 1986; Daehler, 1998; Pyšek, 1997, 1998; Rejmanek, 1996). Less frequently, possible genetic correlates have been sought (for example, see Gray, 1986). Very little attention has been given to the possibility of the evolution of invasiveness after colonization.

Are invasives "born" (that is, are they released from fitness constraints) or are they "made" (that is, do they evolve invasiveness after colonization)? The fact that certain correlates of invasive success have been identified suggests that invasives are born. Also, Darwin's (1859) observation that non-native genera are more likely to be successful invaders than are native genera supports the view that successful invasives are preadapted and do not evolve invasiveness *in situ*. Certain specific cases of invasives fit this model well. For example, the fact that invasiveness can sometimes be reversed by a biological control agent [(e.g., prickly pear in Australia (Dodd, 1959) and Klamath weed in the American Pacific Northwest (Huffaker and Kennett, 1959)] suggests that invasiveness can appear simply once an organism is released from its primary biological enemies. Also, it has been observed that "a strong predictor of invasiveness . . . is whether the organism has been invasive . . . elsewhere" (Ewel *et al.*, 1999, p. 627). Although such correlates may be statistically strong, they are typically weak in predicting invasions, leading one reviewer of the field to assert, "serendipity is often an important element in successful invasions" (Gray, 1986, p. 655) and another to lament, "It could be that invasions . . . are intrinsically unpredictable" (Williamson, 1999, p. 10).

But for some successful invasive species, it may well be that a series of events *after* colonization is more important than intrinsic "colonizing ability." In fact, two enigmatic phenomena associated with successful invasives suggest that many species are not preadapted to become successful invasives and that the right circumstances must transpire for invasiveness to occur (and perhaps evolve). The first is the observation that there is often a considerable lag phase between the establishment of local populations and their aggressive spread (Ewel *et al.*, 1999; Mack, 1985). For example, Kowarik (1995) reviewed 184 invasive woody species with known dates of first cultivation in Brandenburg, Germany. The mean delay in invasion was 131 years for shrubs and 170 years for trees. Delays on the order of decades may occur for herbaceous invasives as well (Pyšek and Prach, 1993). If these species were simply preadapted, then we would expect evidence of invasiveness relatively quickly. Second, multiple introductions often are correlated with the eventual success of non-native species establishment and invasiveness (Barrett and Husband, 1990). For example, North America's most successful invasive birds, the European Starling and the House Sparrow, both became invasive only after repeated introduction (Ehrlich *et al.*, 1988). Collectively considered, these observations suggest genetic change and adaptive response play a role in the ultimate establishment of some invasive species.

We contend that hybridization may result in critical evolutionary changes that create an opportunity for increased invasiveness. As Anderson and Stebbins (1954) pointed out, "hybridization between populations having very different genetic systems of adaptation may lead to . . . new adaptive systems, adapted to new ecological niches" (Anderson and Stebbins, 1954, p. 378). Stebbins further examined what he came to call "the catalytic effects of such hybridization" (Stebbins, 1974) in subsequent articles (Stebbins 1959, 1969). Although Anderson and Stebbins did not consider the case of invasive species, they did acknowledge that human activities could be a powerful agent for bringing together cross-compatible species that had been previously isolated by ecology or geography.

Indeed, Abbott (1992) observed that interspecific hybridization involving non-native plant species has often served as a stimulus for the evolution of entirely new, and sometimes invasive, species. Specifically, he noted that hybridization involving a non-native species and another (either native or non-native) has led to a number of new sexually reproducing plant species. The 10 examples he gives are either stabilized introgressants or allopolyploids. Some of these species have remained localized, but most have spread successfully far beyond their sites of origin. The latter group of his examples, plus many more we have accumulated, are listed in Table 1.

Abbott, Anderson, and Stebbins focused on interspecific hybridization. But their ideas should work equally well for hybridization among previously isolated populations of the same species. Therefore, we proceed below with a broad perspective.

We extend the ideas of Stebbins, Anderson, and Abbott to specifically address hybridization as a stimulus for the evolution of invasiveness. We restrict our examples to plants, but the model we develop may apply to other organisms as well. Below, we first provide many examples in which hybridization seems to have served as a stimulus for the evolution of a new invasive line. Second, we explain why plants with a history of hybridization may have a fitness advantage relative to those without such a history. Third, we discuss some scenarios that might lead to such hybridization. Finally, we examine how our model for interspecific hybridization could work equally well for hybridization between previously isolated populations of the same species.

MATERIALS AND METHODS

We sought at least 25 well documented examples of the evolution of invasiveness in plants after a spontaneous hybridization event. We did not intend our review to be exhaustive, but instead concentrated on finding the most convincing examples.

We used four criteria for choosing our examples:

(*i*) More evidence than intermediate morphology must be available to support the hybrid origin of the invasive lineage. Intermediate morphology does not necessarily support the hypothesis of hybridity (Rieseberg and Ellstrand, 1993). Species-specific genetically based traits such as chromosomes, isozymes, and/or DNA-based markers provide more reliable evidence for hybrid parentage. The hypothesis also can receive support from comparison of artificially synthesized hybrids with the putative spontaneous hybrids and from the relative sterility of the putative hybrids compared with that of the parental species.

(*ii*) The hybridization event preceding the evolution of invasiveness must be spontaneous. Many artificial hybrids have escaped from cultivation to become naturalized invasives (e.g., certain mints, comfrey, poplars, and watercress; *cf.* Stace, 1975).

(*iii*) The hybrid derivatives must be established as a novel, stabilized lineage and not simply as transient, localized hybrid swarms. In some cases, genetic or reproductive mechanisms may stabilize hybridity (e.g., allopolyploidy, permanent translocation heterozygosity, agamospermy, and clonal spread; *cf.* Grant, 1981). Some have become new, reproductively isolated, recombinant species. In other cases, introgression may be so extensive that the hybrid lineage swamps out one or both of its parents, becoming a coalescent complex.

(*iv*) The new lineage must exhibit some degree of invasiveness. We define invasive populations as those that are capable of colonizing and persisting in one or more ecosystems in which they were previously absent. The minimal criterion of invasiveness for our hybrid derivative is that it must replace at least one of its parental taxa or invade a habitat in which neither parent is present. We hold to this criterion for those few cases in which one parent is itself invasive.

We did not restrict ourselves to examples of hybridization involving one or more non-natives, because the evolution of invasiveness by hybridization should be independent of the geographical source of the parental material.

RESULTS AND DISCUSSION

We found 28 examples representing 12 families where invasiveness was preceded by hybridization; these examples are detailed in Tables 1 and 2. We encountered another 2 dozen or so examples of invasive lineages thought to have a hybrid origin (e.g., *Lonicera* × *bella*, *Oenothera wolfii* × *Oenothera glazioviana*, and *Platanus racemosa* × *Platanus acerifolia*). The latter did not sufficiently meet our criteria, mostly because only morphology was offered to support their putative hybrid origin.

In some of our examples, the hybrid-derived lineages have already achieved a taxonomic epithet (detailed in Table 1). In other cases, a new invasive lineage has been identified and studied but not yet named, to our knowledge (detailed in Table 2). In each case, we give the parental species, plant family, habit of the hybrid derivative, its site of origin, and the evidence supporting a history of hybridization for the new lineage. We

TABLE 1. Invasive taxa that evolved after intertaxon hybridization

Derived taxon	Parent taxa	Family	Habit of hybrid lineage	Site of taxon's origin
Amelanchier erecta	*A. humulis* × A. *"clade B"*	Rosaceae	Shrub	N. America
Bromus hordeaceus	*B. arvensis* and *B. scoparius*	Poaceae	Annual herb	Europe
Cardamine insueta	*C. rivularis* × *C. amara*	Brassicaceae	Perennial herb	Europe
Cardamine schulzii	*C. rivularis* × *C. amara*	Brassicaceae	Perennial herb	Europe
Circaea × *intermedia*	*C. alpina* × *C. lutetiana*	Onagraceae	Perennial herb	Europe
Fallopia × *bohemica*	*F. japonica* * × *F. sachalinensis**	Polygonaceae	Shrub	Europe
Glyceria × *pedicillata*	*G. fluitans* × *G. notata*	Poaceae	Perennial herb	Europe
Helianthus annuus spp. *texanus*	*H. annuus** × *H. debilis* spp. *cucumerifolius*	Asteraceae	Annual herb	N. America
Mentha × *verticillata*	*M. aquatica* × *M. arvensis*	Lamiaceae	Perennial herb	Europe
Nasturtium sterile	*N. microphyllum* × *N. officinale*	Brassicaceae	Perennial herb	Europe
Oenothera glazioviana (O. erythrosepala, O. lamarckiana)	*O. hookeri** × *O. biennis**	Onagraceae	Biennial herb	Europe
Senecio squalidus	*S. aethensis** × *S. chrysanthemumifolius**	Asteraceae	Perennial herb	Europe
Senecio vulgaris var. hibernicus	*S. v. var. vulgaris* × *S. squalidus**	Asteraceae	Annual herb	Europe
Sorghum almum	*S. propinquum** × *S. bicolor**	Poaceae	Perennial herb	S. America
Spartina anglica	*S. alterniflora** × *S. maritima*	Poaceae	Perennial herb	Europe
Stachys × *ambigua*	*S. palustris* × *S. sylvatica*	Lamiaceae	Perennial herb	Europe
Tragopogon mirus	*T. dubius** × *T. porrifolius**	Asteraceae	Biennial herb	N. America
Tragopogon miscellus	*T. dubius* * × *T. pratensis**	Asteraceae	Biennial herb	N. America

as, Artificial synthesis; c, cytological; i, isozymes; n, nuclear DNA; o, organelle DNA; s, full or partial sterility.
*Signifies non-natives.

Evidence beyond morphology	Reference	How stabilized?	Invasiveness	Occurs in human disturbed areas?
n	Campbell *et al.* (1997)	Agamospermy	Highly invasive relative to congeners	Yes
c, i, n	Ainouche and Bayer (1996)	Allopolyploid	Aggressive ruderal	Yes
c, n, o	Urbanska *et al.* (1997)	Allopolyploid	Successfully colonizing disturbed sites	Yes
c, n, o	Urbanska *et al.* (1997)	Allopolyploid	Successfully colonizing disturbed sites	Yes
as, s	Stace (1975)	Clonal growth	Sometimes a weed, often occurs in absence of one or both parents	Yes
c, n, s	Bailey *et al.* (1995)	Clonal growth	Noxious weed	Yes
s	Stace (1975), (1991)	Clonal growth	"Example of a successful ... sterile hybrid"	Yes
c, n, o	Rieseberg (1990)	Recombinant	Weed of disturbed areas	Yes
s	Stace (1991)	Clonal growth	Often in the absence of either parent	Yes
c	Bleeker *et al.* (1997)	Recombinant	Disturbed area weeds	Yes
as, c	Cleland (1972)	Permanent translocation heterozygosity	Weed	Yes
i, o	Abbott and Milne (1995); Abbott *et al.* (2000)	Recombinant	Rapidly spreading	Yes
as, c, i	Abbott (1992)	Recombinant	Rapidly becoming ubiquitous	Yes
c, n	Paterson *et al.* (1995)	Allopolyploid	Weed	Yes
C, i	Gray *et al.* (1991)	Allopolyploid, clonal growth	Noxious weed	Yes
C, s	Stace (1975)	Clonal growth	Weed	Yes
c, i, n, o	Novak *et al.* (1991)	Allopolyploid	Substantial increase in range and numbers	Yes
c, i, n, o	Novak *et al.* (1991)	Allopolyploid	Substantial increase in range and numbers	Yes

TABLE 2. Invasive lineages that evolved after intertaxon hybridization

Parent taxa	Family	Habit of hybrid lineage	Site of new lineage's origin	Evidence beyond morphology
*Avena barbata** × *A. strigosa**	Poaceae	Annual herb	North America	i
Beta vulgaris spp. *vulgaris** × *B. v.* spp. *maritima*	Chenopodiaceae	Annual herb	Europe	n, o, b
*Carpobrotus edulis** × *C. chilense*	Aizoaceae	Perennial herb	North America	as, i
*Lythrum salicaria** × *L. alatum*	Lythraceae	Perennial herb	North America	d, s
*Onopordum acanthium** × *O. illyricum*	Asteraceae	Perennial herb	Australia	n
*Raphanus raphanistrum** × *R. sativus**	Brassicaceae	Annual herb	North America	as, c, s
*Rhododendron ponticum** × *R. catawbiense**	Ericaceae	Shrub	Europe	n, o
*Secale cereale** × *S. montanum**	Poaceae	Perennial herb	North America	i, s
*Spartina alterniflora** × *S. foliosa*	Poaceae	Perennial herb	North America	as, n, s
Viola riviniana × *V. reichenbachiana*	Violaceae	Perennial herb	Europe	c, n

as, Artificial synthesis; c, cytological; i, isozymes; n, nuclear DNA; o, organelle DNA; s, full or partial sterility.
*Signifies non-natives.

cite one or two good comprehensive references for each example. In many cases, the best reference is an article or review that cites many supporting sources of empirical research. To list each of those is beyond the scope of this paper. Finally, we present how the novel lineage is maintained and indicate the scope of its invasiveness, including whether the lineage is known to grow, at least in some instances, in human-disturbed areas.

Some characteristics of our sample seem to be quite broad; many diverse families are represented. Hybridity is stabilized by a variety of mechanisms, from cytological (polyploidy and permanent translocation heterozygosity) to apomictic (agamospermy and clonal growth). In many cases, the new hybrid lineage is a coalescent complex that absorbs one or both parental types, especially among the unnamed cases in Table 2. Likewise, invasiveness runs the gamut from cases in which the new hybrid lineage is displacing a parent or spreading into a new community to cases in which the hybrid lineage is an established noxious weed.

But we also note some interesting trends in our sample. Life history

Reference	How stabilized?	Invasiveness	Occurs in human-disturbed areas?
M. Blumler, personal communication	Selfing genotype	Spreading rapidly	Yes
Parker and Bartsch (1996)	Coalescent complex	Noxious weed	Yes
Gallagher *et al.* (1997); Vilà and D'Antonio (1998)	Clonal growth	Replacing one parent	Yes
Strefeler *et al.* (1996)	Clonal growth	Noxious weed	Yes
O'Hanlon *et al.* (1999)	Coalescent complex	Weed	Yes
Panetsos and Baker (1967)	Coalescent complex	Weed	Yes
Abbott and Milne (1995); Milne and Abbott (2000)	Coalescent complex	Noxious weed	Yes
Sun and Corke (1992)	Coalescent complex	Weed	Yes
Ayres *et al.* (1999); Daehler and Strong (1997)	Clonal growth	Replacing one parent	Yes
Neuffer *et al.* (1999)	Coalescent complex	Invading polluted forests	Yes

traits tend to be concentrated within a narrow subset of those traits possible. Almost all of our examples are herbaceous (24 of 28). However, the majority of the cases involve perennial species (19 of 28). Interestingly, these characteristics also are found to be frequent among cases of spontaneous hybridization. For example, Ellstrand *et al.* (1996) examined the 10 genera in the British flora with the highest number of different spontaneous hybrids. They found that most were perennial herbs.

These trends make sense. Perennial hybrids will persist longer than will annuals, giving more time for stabilization opportunities to occur, especially if clonal reproduction is available (as it is in about half of our examples). The predominance of herbaceous over woody examples in our Tables is consistent with Harper's (1977) prediction that colonizing plants allocate more resources to reproductive rather than to vegetative growth. Iteroparous perennial herbs appear to maximize fitness by investing in sexual structures and vegetative spread instead of investing in permanent structures (Crawley, 1986).

It has been suggested that Old World or temperate ecosystems may be less susceptible to invasives than are New World or tropical ecosystems, and that most successful plant invaders have Mediterranean or Central European origins (Di Castri, 1989). The rationale for this view is that Old World species have had a much longer evolutionary history with human disturbance, particularly agricultural disturbance. These views have been modified by the recognition that historical patterns of plant invasions simply may have followed paths of commerce; indeed, numbers of invasive species in the Old World have increased as New- to Old World commerce has increased (Binggeli *et al.*, 1998). Interestingly, most of our examples come from the Old World, not the New. Finally, all but two examples (*Sorghum almum* and the *Onopordum* hybrids) are Holarctic, and all are temperate. These latter patterns may have more to do with the geographic distribution of evolutionary biologists than with any biological phenomenon.

More than half the cases (18 of 28) involve at least one non-native parental taxon. This correlate may be an artifact of how difficult it is to reconstruct evolutionary events; observed changes in the distribution of non-natives provide a historical context for identifying a hybridization event. "Frequently, the history of these events is known, allowing examination of the factors which may have favoured the spread of a new taxon following its origin" (Abbott, 1992, p. 402). On the other hand, the correlate may have real evolutionary significance. Human-mediated dispersal may magnify the potential for hybridization by increasing the migration distances and the number of independent colonization events severalfold as compared with other processes.

All of our invasives grow in habitats characterized by human disturbance, at least in part of their range. Anderson and Stebbins (1954) predicted that human disturbance should both mix previously isolated floras as well as create novel niches well suited to novel hybrid-derived genotypes, that is, to create niches better suited to intermediates or segregants than to the parental species. We caution that human-disturbed habitats may be much better studied and visited more frequently than those isolated from human activity.

How Can Hybridization Stimulate the Evolution of Invasiveness?

We are well aware that not all hybridization leads to increased fitness or adaptive evolution (Arnold, 1997). But hybridization can lead to adaptive evolution in a number of ways. We examine some hypotheses that describe how hybridization can catalyze the evolution of invasiveness, gaining support from our examples in Tables 1 and 2 when appropriate. The following hypotheses are not likely to be exhaustive nor are they necessarily mutually exclusive.

Evolutionary novelty

The generation of novel genotypes is the most common hypothesis for hybridization's role in adaptive evolution (for example, see Abbott, 1992; Anderson and Stebbins, 1954; Arnold, 1997; Lewontin and Birch, 1966; Stebbins, 1959, 1969; Weiner, 1994). Stebbins (1969) explains it succinctly: ". . . recombination which inevitably takes place in . . . fertile progeny of hybrids gives rise to a large quantitative increase in . . . the gene pool. . . . Although this recombination gives rise to a great preponderance of genotypes which are not well adapted to any environment, nevertheless a minority of them may represent better adaptations to certain environments than do any of the genotypes present in the parental species populations" (Thompson, 1991, p. 26).

One of our examples seems to fit this model perfectly. When sugar beets (which are biennials) are grown for seed production near the Mediterranean Sea, some of their seed is sired by nearby populations of wild beets (which are annuals). Therefore, sugar beet seed grown for commercial purposes in northern Europe has a fraction of hybrid seed. The resulting hybrid plants are morphologically similar to the crop but are annuals, bolting, flowering, and setting seed, leaving a woody root that cannot be sold, that in fact damages harvesting and processing machinery (Parker and Bartsch, 1996). These beet hybrids have given rise to weedy lineages, whose evolutionary novelty of annuality preadapts them for invasive success in cultivated beet fields.

Additional support for this hypothesis comes from invasive hybrid lineages that colonize well defined communities that have not been colonized by either parent. Our Tables supply at least three such examples. *Viola riviniana* and *Viola reichenbachiana* hybridize occasionally throughout Europe (Stace, 1975). But in central Germany, a hybrid lineage has successfully colonized pine forests affected by calcareous pollutants (Neuffer *et al.*, 1999). Our second example involves *Rhododendron ponticum* in Britain, which colonizes areas much colder than those of its native range in Iberia. This wider ecological tolerance is correlated with its history of hybridization in Britain with the cold-tolerant *Rhododendron catawbiense* from North America (Milne and Abbott, 2000). Our final example is *Spartina anglica* of the British Isles, an allopolyploid derivative of the native *Spartina maritima* and *Spartina alterniflora,* introduced from the east coast of North America. "After initial colonization of an estuary, the species characteristically becomes a dominant component of the marsh, producing extensive and dense monospecific swards. In contrast, the progenitor species have retained a limited distribution" (Thompson, 1991, p. 393).

We have numerous examples in our Tables of invasive hybrid derivatives that either occur in the absence of either parent or are outcompeting one or both parents. It is not clear that those examples (and even the three detailed above) are necessarily cases of evolutionary novelty or just cases of superior fitness attributable to fixed heterosis (see Fixed heterosis below). It is always possible that both novelty and heterosis may occur simultaneously. Further support for the hypothesis at hand could come from experimental studies that specifically compare the fitness of hybrid-derived lines to their parental types under a variety of different environmental parameters.

Evolutionary novelty may result from the fixation of intermediate traits, from the recombination of traits from both parents, or from traits that transgress the phenotype of both parents. Although transgressive traits are well known to occur in plant hybrids and their derivatives (Grant, 1975), recently they have been found to be so frequent that it has been posited that "transgression is the rule rather than the exception" (Rieseberg *et al.*, 2000, p. 363). Of the cases mentioned above, it seems that novelty in *Beta* is caused by the recombination of traits from both parents and that novelty in *Viola* involves a trait that transgresses the niche of the parent taxa.

Genetic variation

Recombination in hybrids generates both novelty and variation. A hypothesis related to the one just discussed is that the increase in genetic variation produced in a hybrid lineage can, in itself, be responsible for the evolutionary success of that lineage (Stebbins, 1969). We recognize that this argument falls within the category of "group selection." But we also recognize that invasiveness is itself a group trait, one that is defined by the spread and persistence of groups of individuals, one that cannot be measured from a single individual.

Overall, at the population level, early successional plant species have about the same level of genetic variation as those occurring later in succession (for example, see Hamrick and Godt, 1990). Nonetheless, in our examples of *Raphanus* in California (Panetsos and Baker, 1967), of *Secale* in California (Sun and Corke, 1992 and references therein), and of *Viola* in Germany (Neuffer *et al.*, 1999), the hybrid-derived populations were found to have much more genetic variation than were those of the parental species. Not surprisingly, all of those examples involve freely recombining "coalescent complexes" as opposed to our examples in which the genotype is tightly restrained from recombination. Thus, although these examples are compatible with the genetic variation hypothesis, rigorous experimental work with such systems would be a better test of this idea.

Fixed heterosis

Genetic or reproductive mechanisms that stabilize hybridity (e.g., allopolyploidy, permanent translocation heterozygosity, agamospermy, and clonal spread) also will fix heterotic genotypes. It may well be that the fitness boost afforded by fixed heterozygosity is all that is necessary to make a hybrid lineage invasive. Given the ubiquity of heterosis in both agricultural and natural systems, we are surprised how rarely fixed heterosis is posited as a role of hybridization in adaptive evolution (but see Grant, 1981). The majority of our examples (especially in Table 1) are capable of fixing heterotic genotypes by agamospermy (e.g., *Amelanchier*), by allopolyploidy (e.g., *Bromus, Cardamine, Sorghum,* and *Tragopogon*), by permanent translocation heterozygosity (*Oenothera*), and by clonal spread (e.g., *Circaea, Fallopia, Glyceria, Mentha,* and *Stachys*).

The case of the invasive *S. anglica* in the British Isles is perhaps our most notorious example (Gray *et al.*, 1991; Thompson, 1991). This species originated by chromosome doubling of the sterile hybrid between the Old World *S. maritima* and the New World *S. alterniflora.* Genetic analysis found fixed heterozygosity at many of this species' loci, but also showed that *S. anglica* is almost totally lacking in genetic variation among individuals. Despite its relatively narrow ecological amplitude, it has invaded intertidal flats, replacing more diverse native plant communities, altering succession, and limiting the availability of food to wading birds.

But note that we also were able to use *S. anglica* as a possible example of invasive success attributable to evolutionary novelty (see *Evolutionary novelty* above). It is not clear whether invasive success in *S. anglica* and in our other examples is caused by (*i*) the fitness benefits conferred by heterosis, (*ii*) the fixation of an evolutionarily novel genotype by a mode of reproduction that frustrates recombination, or (*iii*) both. Common garden experiments could test these hypotheses by asking whether hybrids have superior fitness to one or both parental types under specific environmental conditions. We are aware of one such study among our examples, involving *Carpobrotus* and demonstrating heterosis in the hybrids (Vilà and D'Antonio, 1998).

Dumping genetic load

Populations with a history of isolation and a small population size may accumulate detrimental mutations. In such populations, mildly deleterious alleles become fixed, leading to slow erosion of average fitness (see examples in Mills and Smouse, 1994; Lande, 1995). Hybridization between such populations can afford an opportunity to escape from this mutational load, particularly if recombination permits selection to act to reduce the frequency of detrimental alleles. If recombination creates genotypes with reduced load, then they and their descendants will enjoy increased fitness relative to their progenitors, even without fixed heterozygosity. In fact, certain stabilized diploid hybrid segregates have been shown to maintain higher viability and fecundity than do their parental taxa (L. Rieseberg, unpublished data). We are not aware of prior discussions suggesting that hybridization might stimulate adaptive evolution through dumping genetic load. Nonetheless, the fitness gained might in itself be sufficient to account for invasiveness, especially if invasiveness comes at the expense of the replacement of one or both of the parental species.

Measuring genetic load is a challenging area of experimental quantitative genetics. Presently, it would be difficult to test this hypothesis without being able to assess the relative load of the hybrid derivative versus that of the parental species. We are not aware of any experimental work that has attempted such a comparison.

Human Activities and Some Hybridization Scenarios

The following anthropogenic activities could enhance both the likelihood of hybridization and the likelihood of forming new niches that favor hybrid derivatives.

Bringing together previously isolated populations

Humans have become an ecologically significant vector of dispersal, often moving species at high rates and over long distances (for example, see Sauer, 1988). Modern transportation has accelerated that process, including bringing together cross-compatible species that previously were geographically isolated. More than one-third of our invasive hybrid derivatives involves cases in which both parental species were introduced to the location where the initial hybridization event occurred. Another 25% involve cases in which one parent was introduced and the other was native. In most cases in which at least one parental species is introduced, the dispersal involved was on the order of thousands of kilometers. In fact, in all but 3 of the 18 cases, the introduced parental species were native to another continent.

Opening new "hybrid" zones

Human activities often result in ecological disturbance. Anderson (1948) noted that disturbance, human or otherwise, opens an array of niches that might be better suited for hybrids than for their parents. Furthermore, Stebbins (1959) pointed out that, with disturbance, "the initial occurrences of hybridization [will] be in many instances, much more frequent" (Stebbins, 1959, p. 248). Although all of our examples of invasive hybrid derivatives occur at least partially in disturbed sites, some of them are found almost exclusively in human-disturbed sites (*Amelanchier, Bromus, Cardamine, Helianthus, Nasturtium,* and *Viola*). It is interesting to note that these examples more frequently involve cases in which long-distance dispersal is *not* a factor.

We hypothesize, then, that human activities can encourage hybridization through (*i*) long-distance dispersal that brings together previously isolated but closely related taxa, (*ii*) disturbance that provides habitat suitable for hybrid progeny, or (*iii*) a combination of dispersal and disturbance. Once hybridization has occurred, if invasiveness evolves, it may do so instantly, for example, as a genotype fixed by a mode of reproduction that restricts recombination, or more slowly, for example, if selection works to sieve out the best adapted genotypes among an array of recombinants.

Can Hybridization *Within* Taxa Lead to Invasiveness?

There is no reason why the observations above should be restricted to interspecific hybridization. We hypothesize that a hybridization event among well differentiated populations of the same species may act in the same way as does hybridization among species to serve as a stimulus for the evolution of invasiveness. Introduction of distantly related individuals of the same species from different parts of its range may yield an evolutionary stimulus that is essentially the same as is the introduction of different species.

Just as with interspecific hybridization, we do not expect all intraspecific hybridization events to lead to invasiveness. One can posit an optimal level of relatedness yielding the genotypes most likely to become invasive (Fig. 1). Our arguments are similar to those developed to explain an optimal outcrossing distance (Waser, 1993). Hybridization among very closely related populations should not result in any evolutionary changes different from matings within a population. Likewise, very distantly related populations may have evolved cross-incompatibility or produce sterile or otherwise unfit progeny. Thus, we would expect that only a small fraction of interpopulation combinations would yield progeny with superior fitness as compared with their parents. Still, those progeny might not become invasive in an environment that was limiting abiotically (e.g., too saline or xeric) or biotically (e.g., by predators or parasites).

Nonetheless, if hybridization among populations of the same taxa played an important role in the evolution of invasiveness, then we might expect certain correlates for the appearance of invasiveness. First, we

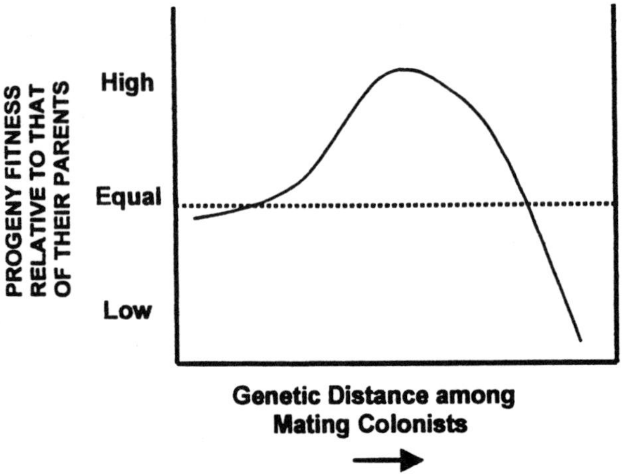

FIGURE 1. As genetic distance between mating colonists increases, so too should heterosis in their progeny—up to a point—then, progeny fitness declines as outbreeding depression becomes important.

would expect that invasiveness would occur after multiple introductions of a species, because multiple introductions would be necessary for providing genotypes from disparate sources. In fact, species that are intentionally introduced would have an advantage in this regard. Second, we would expect that invasiveness would occur after a lag time, during which hybridization and selection would act to create and increase invasive genotypes. As noted in our introduction, both of these phenomena have occurred so frequently that they have attracted the attention of students of invasive species. In fact, invasive species often originate from multiple foci, each with an independent origin (for example, see Cook *et al.*, 1998; Moody and Mack, 1988). If these foci spread and coalesce, there is an opportunity for hybridization among these independent lineages.

Finally, we might expect that if the evolution of invasiveness followed a bout of hybridization between well differentiated populations, then the resulting populations should likely be more genetically diverse than were their progenitors. This suggestion may seem surprising because of the commonly held view that invasives should be relatively genetically depauperate as a result of the bottlenecks associated with their colonization dynamics (Barrett and Husband, 1990). On the other hand, hybridization between well differentiated populations resulting from introductions from different sources ought to leave relatively high levels of within-population polymorphism as a "signature."

We have found two such examples. *Echium plantagineum* is a noxious weed of Australia. The average population there was found to be more diverse than were those genetically analyzed in its native range in Europe (Burdon and Brown, 1986). This species has been introduced more than once to Australia, both intentionally and unintentionally (Piggin and Sheppard, 1995). Similarly, North American populations of the introduced weed cheatgrass, *Bromus tectorum*, were found to have increased within-population genetic variation as compared with populations from its source range in Europe and northern Africa (Novak and Mack, 1993). Again, there is ample evidence of multiple introductions (Novak *et al.*, 1993).

CONCLUSIONS

Discussions of the population biology of invasives have focused largely on their ecology and on the evolutionary consequences of the invasive process. The evolution of invasiveness as an adaptive trait has been largely neglected. We have extended—and, indeed, hybridized—the ideas of Stebbins, Anderson, and Abbott concerning the evolutionary significance of hybridization to offer one model for the evolution of invasiveness. That is, hybridization can, through one or more mechanisms, catalyze the evolution of invasiveness. Human dispersal and human disturbance both act to accelerate the process and increase the opportunities for hybrid lineages to take hold. The process is not unique to plants. In fact, evidence recently has emerged that "a new, aggressive *Phytophthora* pathogen of alder trees in Europe" seems to have arisen through interspecific hybridization (Brashier *et al.*, 1999, p. 5878). Likewise, hybridization between different honeybee subspecies has given rise to the infamous Africanized bees of the New World (Camazine and Morse, 1988).

Certain caveats are in order. We recognize that only a fraction of hybridization events will lead to the evolution of invasiveness. We do not claim that all invasive species have evolved invasiveness. As we note in our introduction, sometimes certain ecological explanations appear to be the most parsimonious, such as encountering an unfilled niche, competitive superiority, or ecological release. Nor do we claim that hybridization is the sole evolutionary pathway to invasiveness. Other evolutionary pathways to invasiveness already have received some attention. For example, weeds have evolved to mimic unrelated crops and have become successful invaders of agroecosystems (Barrett, 1983). Also, Jain and Martins (1979) observed that a single gene mutation apparently is responsible for the appearance of invasiveness of rose clover in California.

At the moment, evolution of invasiveness remains an underappreciated area of research on a topic of great applied and basic importance. We have shown that one way to get a handle on studying such evolution is to use examples that have a genetic signature for reconstructing past events. Any other pathways in which past events can be reconstructed should be equally valuable for study. We anticipate that the study of the evolution of invasiveness should be able to provide answers for why invasiveness occurs in some cases and does not occur in others.

We thank V. Symonds, K. Gallagher, and N. Sherman for discussions early in the development of the manuscript and also thank the following for their comments on draft versions of the manuscript: R. Abbott, D. Crawford, J. Hamrick, I. Parker, L. Rieseberg, A. Snow, P. Soltis, and R. Whitkus. We are grateful for feedback from the 1998 Ecological Genetics Group Meeting and the first workshop of the Collaboratory on the Population Biology of Invasive Species (funded by a grant from the National Science Foundation through a grant to the University of California at Irvine). This work was supported by funding from a University of California Competitive grant (1997–980069) and an Environmental Protection Agency grant (R-826102–01-0) to N.C.E., and by National Science Foundation Grants 9973734 to K.A.S. and 9322795 to K.A.S. and C. D'Antonio.

REFERENCES

Abbott, R. (1992) Plant invasions, interspecific hybridization, and the evolution of new plant taxa. *Trends in Ecology and Evolution* 7, 401–405.

Abbott, R. J., James, J. K., Irwin, J. A. & Comes, H. P. (2000). Allopatric origin of a new diploid hybrid species, the Oxford ragwort, *Senecio squalidus* L (Asteraceae). *Watsonia* 23, 123–138.

Abbott, R. J. & Milne, R. I. (1995) Origins and evolutionary effects of invasive weeds. *BCPC Symp. Proc.* 64, 53–64.

Ainouche, M. L. & Bayer, R. J. (1996) On the origins of two Mediterranean allotetraploid *Bromus* species: *Bromus hordeaceus* L. and *B. lanceolatus* Roth. (Poaceae). *Am. J. Bot.* 83 (SUPPL.), 135.

Anderson, E. (1948) Hybridization of the habitat. *Evolution* 2, 1–9.

Anderson, E. & Stebbins, G. L. (1954) Hybridization as an evolutionary stimulus. *Evolution* 8, 378–388.

Arnold, M. L. (1997) *Natural Hybridization and Evolution.* (Oxford Univ. Press, Oxford).

Ayres, D. R., Garcia-Rossi, D., Davis, H. G. & Strong, D. R. (1999) Extent and degree of hybridization between exotic (*Spartina alterniflora*) and native (*S. foliosa*) cordgrass (Poaceae) in California, USA determined by random amplified polymorphic DNA (RAPDs). *Molec. Ecol.* 8, 1179–1186.

Bailey, J. P., Child, L. E. & Wade, M. (1995) Assessment of the genetic variation and spread of British populations of *Fallopia japonica* and its hybrid *Fallopia x bohemica*. In *Plant Invasions: General Aspects and Special Problems,* eds. Pyšek, P., Prach, K., Rejmanek, M. & Wade, P. M. (SPB Academic Publishing bv, The Hague, Netherlands), pp. 141–150.

Barrett, S. C. H. (1983) Crop mimicry in weeds. *Econ. Bot.* 37, 255–282.

Barrett, S. C. H. & Husband, B. C. (1990) The genetics of plant migration and colonization. In *Plant Population Genetics, Breeding, and Genetic Resources,* eds. Brown, A. H. D., Clegg, M. T., Kahler, A. L. & Weir, B. S. (Sinauer, Sunderland, MA), pp. 254–278.

Bazzaz, F. A. (1986) Life history of colonizing plants: some demographic, genetic and physiological features. In *Ecology of Biological Invasions of North America and Hawaii,* eds. Mooney, H. A. & Drake, J. A. (Springer-Verlag, New York), pp. 96–110.

Binggeli, P., Hall, J. B. & Healey, J. R. (1998) *A Review of Invasive Woody Plant Species in the Tropics.* School of Agriculture and Forestry Sciences Publication Number 13. (University of Wales, Bangor).

Bleeker, W., Hurka, H. & Koch, M.(1997) Presence and morphology of Nasturtium sterile (Airy Shaw) Oef. in southwestern Lower Saxony and neighboring region. *Floristische Rundbriefe* 31, 1–8.

Brashier, C. M., Cooke, D. E. L. & Duncan, J. M. (1999) Origin of a new *Phytophthora* pathogen through interspecific hybridization. *Proc. Natl. Acad. Sci., USA* 96, 5878–5883.

Burdon, J. J. & Brown, A. H. D. (1986) Population genetics of *Echium plantagineum* L.: Target weed for biological control. *Aust. J. Biol. Sci.* 39, 369–378.

Camazine, S. & Morse, R. A. (1988) The Africanized honeybee. *Am. Sci.* 76, 465–471.

Campbell, C. S., Wojciechowski, M. F., Baldwin, B. G., Alice, L. A. & Donoghue, M. J. (1997) Persistent nuclear ribosomal DNA sequence polymorphism in the *Amelanchier* agamic complex (Rosaceae). *Molec. Biol. Evol.* 14, 81–90.

Cleland, R. E. (1972) *Oenothera: Cytogenetics and Evolution* (Academic Press, New York).

Cook, L. M., Soltis, P. S. Brunsfeld, S. J. & Soltis, D. E. (1998) Multiple independent formations of *Tragopogon* tetraploids (Asteraceae): evidence from RAPD markers. *Molec. Ecol.* 7, 1293–1302.

Crawley, M. J. (1986) The population biology of invaders. *Phil. Trans. R. Soc. Lond.* B 314, 711–731.

Daehler, C. C. (1998) The taxonomic distribution of invasive angiosperm plants: Ecological insights and comparison to agricultural weeds. *Biol. Conserv.* 84, 167–180.

Daehler, C. C. & Strong, D. R. (1997) Hybridization between introduced smooth cordgrass (*Spartina alterniflora;* Poacae) and native California (*S. foliosa*) cordgrass in San Francisco Bay, California, USA. *Amer. Jour. Bot.* 84, 607–611.

Darwin, C. (1859) *On the Origin of Species* (Murray, London).

Di Castri, F. (1989) On invading species and invaded ecosystems: the interplay of historical chance and biological necessity. In *Ecology of Biological Invasions: a Global Perspective,* eds. Drake, J. A. & Mooney, H. (John Wiley and Sons, New York), pp. 3–16.

Dodd, A. P. (1959) The biological control of prickly pear in Australia. In *Biogeography and Ecology in Australia,* eds. Keast, A., Crocker, R. L. & Christian, C. S. Monogr. Biol. #8.

Ehrlich, P. R., Dobkin, D. S. & Wheye, D. (1988) *The Birder's Handbook: A Field Guide to the Natural History of North American Birds* (Simon & Schuster, New York).

Elton, C. S. (1958) *The Ecology of Invasion by Animals and Plants* (London, Chapman and Hall).

Ellstrand, N.C., Whitkus, R.W. & Rieseberg, L.H. (1996) Distribution of spontaneous plant hybrids. *Proc. Natl. Acad. Sci. USA* 93, 5090–5093.

Ewel, J. J., O'Dowd, D. J., Bergelson, J., Daehler, C. C., D'Antonio, C. M., Gomez, D., Gordon, D. R., Hobbs, R. J., Holt, A., Hopper, K. R., Hughes, C. E., Lahart, M., Leakey, R. R. B., Lee, W. G., Loope, L. L., Lorence, D. H. Louda, S. M., Lugo, A. E., Mcevoy, P. B., Richardson, D. M. & Vitousek, P. M. (1999) Deliberate introductions of species: Research needs. *BioScience* 49, 619–630.

Gallagher, K. G., Schierenbeck, K. A. & D'Antonio, C. M. (1997) Hybridization and introgression in *Carpobrotus* spp. (Aizoaceae) in California. Allozyme evidence. *Am. J. Bot.* 84, 905–911.

Grant, V. (1975) *Genetics of Flowering Plants* (Columbia Univ. Press, New York)

Grant, V. (1981) *Plant Speciation* (Columbia Univ. Press, New York), 2nd Ed.

Gray, A. J. (1986) Do invading species have definable genetic characteristics? *Phil. Trans. R. Soc. Lond.* B 314, 655–674.

Gray, A. J., Marshall, D. F. & Raybould, A. F. (1991) A century of evolution in *Spartina anglica. Adv. Ecol. Res.* 21, 1–61.

Hamrick, J. L. & Godt, M. J. W. (1990) Allozyme diversity in plant species. In *Plant Population Genetics: Breeding and Genetic Resources,* eds. Brown, A. H. D., Clegg, M. T., Kahler, A. L. & Weir, B. S. (Sinauer, Sunderland), pp. 43–63.

Harper, J. (1977) *Population Biology of Plants* (Academic Press, London).

Horvitz, C. C., Pacarella, J. B., McMann, S., Freedman, A. & Hofstetter, R. H. (1998) Functional roles of invasive non-indigenous plants in hurricane-affected subtropical hardwood forests. *Ecological Applications* 8, 947–974.

Huffaker, C. B. & Kennett, C. E. (1959) A 10 year study of vegetational changes associated with biological control of Klamath weed species. *J. Range Manage.* 12, 69–82.

Jain, S. & Martins, P. (1979) Ecological genetics of the colonizing ability of rose clover (*Trifolium hirtum* All.). *Am. J. Bot.* 66, 361–366.

Kowarik, I. (1995) Time lags in biological invasions with regard to the success and failure of alien species. In *Plant Invasions: General Aspects and Special Problems,* eds. Pyšek, P., Prach, K., Rejmanek, M. & Wade, P. M. (SPB Academic Publishing bv, The Hague, Netherlands), pp. 15–38.

Lande, R. (1995) Mutation and conservation. *Conserv. Biol.* 9, 782–791.

Lewontin, R. C. & Birch, L. C. (1966) Hybridzation as a source of variation for adaptation to new environments. *Evolution* 20, 315–336.

Mack, R. N. (1985) Invading plants: their potential contribution to population biology. In *Studies on Plant Demography: John L. Harper Festschrift,* ed. White, J. (Academic Press, London), pp. 127–142.

Mills, L. S. & Smouse, P. (1994) Demographic consequences of inbreeding in remnant populations. *Am. Natur.* 144, 412–431.

Milne, R. I. & Abbott, R. J. (2000) Origin and evolution of invasive naturalized material of *Rhododendron ponticum* L. in the British Isles. *Molec. Ecol.* 9, 541–556.

Moody, M. E. & Mack, R. N. (1988) Controlling the spread of plant invasions: the importance of nascent foci. *J. Appl. Ecol.* 25, 1009–1021.

Mooney, H. A. & Drake, J. A., eds. (1986) *Ecology of Biological Invasions of North America and Hawaii* (Springer-Verlag, New York).

Neuffer, B., Auge, H., Mesch, H., Amarell, U. & Brandl, R. (1999) Spread of violets in polluted pine forests: morphological and molecular evidence for the ecological importance of interspecific hybridization. *Molec. Ecol.* 8, 365–377.

Novak, S. J. & Mack, R. N. (1993) Genetic variation in *Bromus tectorum* (Poaceae): comparison between native and introduced populations. *Heredity* 71, 167–176.

Novak, S. J., Mack, R. N. & Soltis, P. S. (1993) Genetic variation in *bromus tectorum* (Poaceae) — Introduction dynamics in North America. *Can. J. Bot.* 71, 1441–1448.

Novak, S. J., Soltis, D. E. & Soltis, P. S. (1991) Ownbey's Tragopogons: 40 years later. *Am. J. Bot.* 78, 1586–1600.

O'Hanlon, P. C., Peakall, R. & Briese, D. T. (1999). Amplified fragment length polymorphism (AFLP) reveals introgression in weedy *Onopordum* thistles: hybridization and invasion. *Molec. Ecol.* 8, 1239–1246.

Panetsos, C. A. & Baker, H. G. (1967) The origin of variation in "wild" *Raphanus sativus* (Cruciferae) in California. *Genetica* 38, 243–274.

Parker, I. M. & Bartsch, D. (1996) Recent advances in ecological biosafety research on the risks of transgenic plants: A trans-continental perspective. In *Transgenic Organisms—Biological and Social Implications,* eds., Tomiuk, J., Wöhrmann, K. & Sentker, A. (Birkhauser Verlag, Basel, Switzerland), pp. 147–161.

Paterson, A. H., Schertz, K. F., Lin, Y.-R. & Chang, Y.-L. (1995) The weediness of wild plants: Molecular analysis of genes influencing dispersal and persistence of johnsongrass, *Sorghum halepense* (L.). *Proc. Natl. Acad. Sci. USA* 92, 6127–6131.

Piggin, C. M. & Sheppard, A. W. (1995) *Echium plantagineum* L. In *The Biology of Australian Weeds, Vol. 1,* eds. Groves, R. H. & Shepherd, R. C. H. (R. G. & F. J. Richardson, Melbourne), pp. 87–110.

Pimentel, D., Lach, L., Zuniga, R. & Morrison, D. (2000) Environmental and economic costs of nonindigenous species in the United States. *BioScience* 50, 53–65.

Pyšek, P. (1997) Clonality and plant invasions: Can a trait make a difference? In *The Ecology and Evolution of Clonal Plants,* eds. De Kroon, H. & van Groenendael, J. (Backhuys Publishers, Leiden, The Netherlands), pp. 405–427.

Pyšek, P. (1998) Is there a taxonomic pattern to plant invasions?. *Oikos* 82, 282–294.

Pyšek, P. & Prach, K. (1993) Plant invasions and the role of riparian habitats—a comparison of 4 species alien to Central Europe. *J. Biogeogr.* 204, 413–420.

Rejmanek, M. (1996) A theory of seed plant invasiveness: The first sketch. *Biol. Conserv.* 78, 171–181.

Rieseberg, L. H., Archer, M. A. & Wayne, R. K. (1999) Transgressive segregation, adaptation and speciation. *Heredity* 83, 363–372.

Rieseberg, L. H., Beckstrom-Sternberg, S. & Doan, K. (1990) *Helianthus-annuus* ssp *texanus* has chloroplast DNA and nuclear ribosomal RNA genes of *Helianthus-debilis* ssp *cucumerifolius. Proc. Natl. Acad. Sci. USA* 87, 593–597.

Rieseberg, L. H. & Ellstrand, N. C. (1993) What can molecular and morphological markers tell us about plant hybridization? *Critical Reviews in Plant Sciences* 12, 213–241.

Sauer, J. D. (1988) *Plant Migration: the Dynamics of Geographic Patterning in Seed Plant Species* (Univ. of California Press, Berkeley).

Stace, C. A. (1975) *Hybridization and the Flora of the British Isles* (Academic Press, London).

Stace, C. A. (1991) *New Flora of the British Isles* (Cambridge Univ. Press, Cambridge).

Stebbins, G. L. (1959) The role of hybridization in evolution. *Proc. Amer. Phil. Soc.* 103, 231–251.

Stebbins, G. L. (1969) The significance of hybridization for plant taxonomy and evolution. *Taxon* 18, 26–35.

Stebbins, G. L. (1974) *Flowering Plants: Evolution Above the Species Level* (Belknap Press, Cambridge).

Strefeler, M. S., Darmo, E., Becker, R. L. & Katovich, E. J. (1996) Isozyme characterization of genetic diversity in Minnesota populations of purple loosestrife, *Lythrum salicaria* (Lythraceae). *Am. J. Bot.* 83, 265–273.

Sun, M. & Corke, H. (1992) Population genetics of colonizing success of weedy rye in northern California. *Theor. Appl. Genet.* 83, 321–329.

Thompson, J. D. (1991) The biology of an invasive plant. *BioScience* 41, 393–401.

U.S. Congress, Office of Technology Assessment. (1993) *Harmful Non-indigenous Species in the United States* (U.S. Government Printing Office, Washington, D.C.).

Urbanska, K. M, Hurka, H., Landolt, E., Neuffer, B. & Mummenhoff, K. (1997) Hybridization and evolution in Cardamine (Brassicaceae) at Urnerboden, central Switzerland: Biosystematic and molecular evidence. *Plant Syst. Evol.* 204, 233–256

Vilà, M. & D'Antonio, C. M. (1998) Hybrid vigor for clonal growth in *Carpobrotus* (Aizoaceae) in coastal California. *Ecol. Appl.* 8, 1196–1205.

Waser, N. M. (1993) Population structure, optimal outbreeding, and assortative mating in angiosperms. In *The Natural History of Inbreeding and Outbreeding, Theoretical and Empirical Perspectives,* ed. Thornhill, N. W. (Univ. of Chicago Press, Chicago), pp. 173–199.

Weiner, J. (1994) *The Beak of the Finch* (Vantage Books, New York).

Williamson, M. (1993) Invaders, weeds and the risk from genetically manipulated organisms. *Experientia* 49, 219–224.

Williamson, M. (1999) Invasions. *Ecography* 22, 5–12.

17
The Role of Genetic and Genomic Attributes in the Success of Polyploids

PAMELA S. SOLTIS AND DOUGLAS E. SOLTIS

In 1950, G. Ledyard Stebbins devoted two chapters of his book *Variation and Evolution in Plants* (Columbia Univ. Press, New York) to polyploidy, one on occurrence and nature and one on distribution and significance. Fifty years later, many of the questions Stebbins posed have not been answered, and many new questions have arisen. In this paper, we review some of the genetic attributes of polyploids that have been suggested to account for the tremendous success of polyploid plants. Based on a limited number of studies, we conclude: (i) Polyploids, both individuals and populations, generally maintain higher levels of heterozygosity than do their diploid progenitors. (ii) Polyploids exhibit less inbreeding depression than do their diploid parents and can therefore tolerate higher levels of selfing; polyploid ferns indeed have higher levels of selfing than do their diploid parents, but polyploid angiosperms do not differ in outcrossing rates from their diploid parents. (iii) Most polyploid species are polyphyletic, having formed recurrently from genetically different diploid parents. This mode of formation incorporates genetic diversity from multiple progenitor populations into the polyploid "species"; thus, genetic diversity in polyploid species is much higher than expected

School of Biological Sciences, Washington State University, Pullman, WA 99164-4236

This paper was presented at the National Academy of Sciences colloquium "Variation and Evolution in Plants and Microorganisms: Toward a New Synthesis 50 Years After Stebbins," held January 27–29, 2000, at the Arnold and Mabel Beckman Center in Irvine, CA.

by models of polyploid formation involving a single origin. (iv) Genome rearrangement may be a common attribute of polyploids, based on evidence from genome in situ hybridization (GISH), restriction fragment length polymorphism (RFLP) analysis, and chromosome mapping. (v) Several groups of plants may be ancient polyploids, with large regions of homologous DNA. These duplicated genes and genomes can undergo divergent evolution and evolve new functions. These genetic and genomic attributes of polyploids may have both biochemical and ecological benefits that contribute to the success of polyploids in nature.

Polyploidy, the presence of more than two genomes per cell, is a significant mode of species formation in plants and was one of the topics closest to the heart of Ledyard Stebbins. In *Variation and Evolution in Plants*, Stebbins (1950) devoted two chapters to polyploidy and addressed the following issues: the frequency, taxonomic distribution, and geographic distribution of polyploidy; the origins of polyploidy and factors promoting polyploidy; the direct effects of polyploidy; the polyploid complex; the success of polyploids in extreme habitats (including weeds); ancient polyploidy; and the role of polyploidy in the evolution and improvement of crops. He continued to explore these and other themes in his subsequent work, most notably in *Chromosomal Evolution in Plants* (Stebbins, 1971). In this paper we pay tribute to Ledyard, who was an inspiration and a friend, by exploring some of the questions that he asked about polyploids and by reviewing recent advances in the study of polyploidy.

Estimates of the frequency of polyploid angiosperm species range from ≈30–35% (Stebbins, 1947) to as high as 80% (Masterson, 1994); most estimates are near 50% (Stebbins, 1971; Grant, 1981). Levels of polyploidy may be even higher in pteridophytes, with some estimates of polyploidy in ferns as high as 95% (Grant, 1981). Polyploids often occupy habitats different from those of their diploid parents, and have been proposed to be superior colonizers to diploids. Furthermore, most crop plants are of polyploid origin, as noted by Stebbins (1950). In contrast, although genome doubling has been reported from other major groups of eukaryotes (reviewed in Wendel, 2000), it is not nearly as common in these groups as it is in plants.

The question often has arisen as to *why* polyploids are so common and so successful, and several possible explanations have been proposed. Stebbins (1950) considered vegetative reproduction and the perennial habit to be important factors promoting the establishment of polyploids, along with an outcrossing mating system to allow for hybridization (between species, subspecies, races, populations, etc.) in the formation of the polyploid. Perhaps most important to Stebbins (1950) was the availability of new ecological niches. Additional hypotheses for the success of polyploids include broader ecological amplitude of the polyploid relative to its diploid parents, better colonizing ability, higher selfing rates, and increased heterozygosity.

In fact, many aspects of the genetic systems of polyploids may contribute to the success of polyploid plants. These characteristics range from the molecular level to the population level and include increased heterozygosity, reduced inbreeding depression and an associated increase in selfing rates, increased genetic diversity through multiple formations of a polyploid species, genome rearrangements, and ancient polyploidy and gene silencing. But what role, if any, do these factors really play in the success of polyploids? In this paper, we will explore the evidence for the role of these genetic attributes in the evolutionary success of polyploid plant species.

ALLO- VERSUS AUTOPOLYPLOIDY

We will distinguish among types of polyploids by using Stebbins' (1947, 1950) classification: Allopolyploids are those polyploids that have arisen through the processes of interspecific hybridization and chromosome doubling (not necessarily in that order), autopolyploids are those polyploids that have arisen from conspecific parents, and segmental allopolyploids are those that have arisen from parents with partially divergent chromosome arrangements such that some chromosomal regions are homologous between the parents and others are homoeologous; segmental allopolyploids will not be considered further in this paper. Allopolyploids are characterized by fixed (i.e., nonsegregating) heterozygosity, resulting from the combination of divergent parental genomes; bivalent formation occurs at meiosis, and disomic inheritance operates at each locus. Autopolyploids may exhibit multivalent formation at meiosis and are characterized by polysomic inheritance. Allopolyploids are considered much more prevalent in nature than are autopolyploids, but even a cursory glance at any flora (for example, see Hickman, 1993) or list of plant chromosome numbers (for example, see Federov, 1969) will reveal multiple cytotypes within many species, even though these additional ploidal levels are not typically accorded species status. Thus, autopolyploids in nature likely are much more common than typically is recognized.

INCREASED HETEROZYGOSITY

Roose and Gottlieb (1976) showed that allotetraploids in *Tragopogon* had fixed heterozygosity at isozyme loci, representing the combination of divergent genomes. In the allotetraploids *Tragopogon mirus* and *Tragopogon miscellus*, 33% and 43%, respectively, of the loci examined were duplicated. These values are typical of the levels of duplicated loci observed in many allotetraploid plants (Gottlieb, 1982; Crawford, 1983). Of course, this value varies depending on the extent of allozyme divergence between the diploid progenitors: An allotetraploid derivative of two allozymically similar parents would display lower apparent levels of duplicated loci and fixed heterozygosity than would a derivative of more genetically divergent parents. However, even in those cases where there is no apparent allelic divergence between the parental genomes, the chromosomal segment is still duplicated; the possible fates of duplicated genes are reviewed by Wendel (2000). All allopolyploid individuals are essentially heterozygous through nonsegregating, fixed heterozygosity.

Populations of autopolyploids are expected to maintain higher levels of heterozygosity than do their diploid progenitors (Muller, 1914; Haldane, 1930), and these higher heterozygosities can be attributed simply to polysomic inheritance (Moody *et al.*, 1993). For example, assuming simple tetrasomic segregation, selfing of a heterozygous autotetraploid of genotype *aabb* is expected to produce progeny in the ratio of 1 *aaaa*:34 heterozygotes (of various genotypes):1 *bbbb*, a huge increase over expectations for a diploid with disomic inheritance (i.e., 1*aa*:2*ab*:1*bb*).

Empirical studies have demonstrated that autotetraploids with tetrasomic inheritance do indeed have higher levels of heterozygosity than do their diploid parents (Soltis and Soltis, 1989a). For example, *Tolmiea menziesii*, which occurs along the Pacific Coast of North America from central California to southeastern Alaska, comprises diploid populations, which are distributed in the southern portion of the range, and tetraploid populations, which occupy the northern portion. Various measures of genetic diversity were compared in natural populations of the two cytotypes (Soltis and Soltis, 1989a). At seven polymorphic isozyme loci, a substantially larger number of tetraploid individuals was heterozygous as compared with diploid individuals (Soltis and Soltis, 1989a), and comparisons of diploid and autopolyploid populations in other species show the same pattern (Table 1). Many other polyploids also exhibit polysomic inheritance (Table 2); consequently, these polyploids likely also maintain higher levels of heterozygosity than do their diploid parents, simply because of their mode of inheritance.

OUTCROSSING RATES IN POLYPLOIDS AND THEIR DIPLOID PROGENITORS

Some aspects of polyploid success have been attributed to improved colonizing ability, which may involve higher selfing rates than those of

TABLE 1. Genetic variation (mean values) in diploid (2*n*) and tetraploid (4*n*) populations (Soltis and Soltis, 1993)

	P		H		A	
Species	2*n*	4*n*	2*n*	4*n*	2*n*	4*n*
Tolmiea menziesii	0.240	0.408	0.070	0.237	3.0	3.53
Heuchera grossulariifolia	0.238	0.311	0.058	0.159	1.35	1.55
Heuchera micrantha	0.240	0.383	0.074	0.151	1.41	1.64
Dactylis glomerata	0.70	0.80	0.17	0.43	1.51	2.36
Turnera ulmifolia						
var. *elegans*	0.459	0.653	0.11	0.42	2.20	2.56
var. *intermedia*	0.459	0.201	0.11	0.07	2.20	2.00

P, proportion of loci polymorphic; *H*, observed heterozygosity; *A*, mean number of alleles per locus.

the diploid parents. We will present both theory on why we might expect increased selfing in polyploids and empirical data for selected ferns and angiosperms.

Theoretical models predict reduced inbreeding depression in polyploids relative to their diploid parents, because of the buffering effect of additional genomes: Deleterious alleles are masked by the extra genomes (Stebbins, 1971; Richards, 1986; Barrett and Shore, 1989). Both allopoly-

TABLE 2. Examples of polysomic inheritance (Soltis and Soltis, 1993)

Species	Inheritance	Evidence
Allium nevii	Tetrasomic	Isozymes
Chrysanthemum morifolium	Hexasomic	Morphology
Dactylis glomerata	Tetrasomic	Isozymes
Dahlia variabilis	Tetrasomic	Morphology
Haplopappus spinulosus	Tetrasomic	Isozymes
Heuchera grossulariifolia	Tetrasomic	Isozymes
Heuchera micrantha	Tetrasomic	Isozymes
Lotus corniculatus	Tetrasomic	Cyanogenic markers, isozymes
Lythrum salicaria	Tetrasomic	Morphology
Maclura pomifera	Tetrasomic	Isozymes
Medicago falcata	Tetrasomic	Morphology, isozymes
Medicago sativa	Tetrasomic	Morphology, isozymes
Pachycereus pringlei	Tetrasomic	Isozymes
Phleum pratense	Hexasomic	Morphology
Solanum tuberosum	Tetrasomic	Morphology, isozymes
Tolmiea menziesii	Tetrasomic	Isozymes
Turnera ulmifolia		
var. *elegans*	Tetrasomic	Isozymes
var. *intermedia*	Tetrasomic	Morphology, isozymes
Vaccinium corymbosum	Tetrasomic	Isozymes

ploids and autopolyploids are expected to have reduced inbreeding depression (Charlesworth and Charlesworth, 1987, except under their overdominance model, and Hedrick, 1987), and the magnitude of inbreeding depression is negatively correlated with selfing rates in diploid angiosperms and gymnosperms (Husband and Schemske, 1996). Unfortunately, few studies have addressed levels of inbreeding depression in polyploids empirically; most of the available data come from ferns.

Inbreeding depression in ferns (often referred to as genetic load in these studies) has been estimated by taking advantage of the life cycle that involves a free-living, haploid gametophyte generation that can, in most cases, self-fertilize to produce a completely homozygous diploid sporophyte. These studies have involved culturing gametophytes in isolation, in sib pairs, and in non-sib pairs. The number and survival of sporophytes resulting from these treatments are recorded, and these data can be used to estimate inbreeding depression, outcrossing depression, and the number of lethal equivalents per genome. If a greater number of normal sporophytes is produced by non-sib pairs of gametophytes than by either sib pairs or isolated gametophytes, then the population or species is considered to exhibit inbreeding depression.

Masuyama and Watano (1990) reported two studies of inbreeding depression in diploid and tetraploid pairs of ferns. In *Phegopteris*, 30–60% of selfed gametophytes of the diploid race formed sporophytes, and nearly 100% of all selfed gametophytes of the tetraploid race formed sporophytes. In *Lepisorus*, only 4% of selfed gametophytes of the diploid race produced normal sporophytes, whereas 98–100% of the gametophytes of the tetraploid race formed sporophytes. These data were interpreted as evidence for reduced inbreeding depression in the tetraploid, with the lower inbreeding depression allowing for increased selfing rates.

There are few estimates of selfing rates in polyploid fern species, largely because polyploid fern populations often lack sufficient levels of segregating allozyme markers; however, selfing rates have been estimated in a few diploid-tetraploid pairs. In *Polystichum*, the allotetraploid *Polystichum californicum* has a selfing rate of 0.236, whereas selfing rates in the two diploid progenitors, *Polystichum dudleyi* and *Polystichum imbricans*, are only 2–3% (Soltis and Soltis, 1990). In tetraploid *Pteris dispar*, selfing rates are 0.84, much higher than the rate of 0.01 estimated for the diploid race (Masuyama and Watano, 1990). Limited evidence for ferns suggests reduced inbreeding depression and higher selfing rates in tetraploids than in diploids.

Comparisons of outcrossing rates and levels of inbreeding depression in diploid and polyploid angiosperms also are rare. Both outcrossing rates and inbreeding depression have been estimated for diploid and autotetraploid populations of *Epilobium angustifolium* (Husband and Schemske,

1995, 1997). Outcrossing rates in the two cytotypes were very similar; values (after correcting for inbreeding depression in the diploid) are 0.45 and 0.43 for diploids and tetraploids, respectively. However, the tetraploids have substantially lower inbreeding depression (0.95 for diploids versus 0.67 for tetraploids), as expected from population genetic theory.

Outcrossing rates also have been estimated in diploid and allotetraploid species of *Tragopogon* (Cook and Soltis, 1999, 2000). Outcrossing rates in the allotetraploid *T. mirus* (0.381 and 0.456 for two populations) were higher than those found in the diploid parent *Tragopogon dubius* (0.068 and 0.242), although significantly higher than only one of the two populations; the other parent, *Tragopogon porrifolius* (Ownbey, 1950; Soltis *et al.*, 1995), lacked segregating allozyme variation from which to estimate outcrossing rates. This pattern is exactly the opposite of that predicted by population genetic theory, and one explanation offered to explain it is that rates of outcrossing were underestimated, particularly in *T. dubius*, because of limited polymorphic loci in all populations. To account for this possibility, outcrossing rates were estimated in *T. mirus* and *T. dubius* from artificial arrays constructed to maximize the chances of detecting an outcrossing event if one had occurred. Outcrossing rates ranged from 0 to >1 for diploid and tetraploid families, and the mean values were quite similar (0.696 and 0.633, respectively, for *T. mirus* and *T. dubius*) and higher than those estimated for natural populations, suggesting that some outcrossing events in both species, and especially the diploid *T. dubius*, had gone undetected (Cook and Soltis, 2000). If the outcrossing rates estimated from the artificial arrays are more accurate than are those from natural populations, the discrepancy between predictions and results may be attributable to the recent ancestry of *T. mirus* (most likely post-1928; Ownbey, 1950; Soltis *et al.*, 1995) and to the limited time for the mating systems to have diverged.

THE GENETIC IMPLICATIONS OF RECURRENT POLYPLOID FORMATION

The application of isozyme analysis and DNA techniques to the study of polyploid ancestry dramatically altered our view of polyploid origins. Although morphological or cytological differences among populations of a few polyploid species suggested evidence of repeated polyploid formation (see example in Ownbey, 1950), most polyploid species, until recently, were considered to have had a unique origin. Nearly all polyploid species of plants that have been examined with molecular markers have been shown to be polyphyletic, having arisen multiple times from the same diploid species (reviewed in Soltis and Soltis, 1993, 1999; Soltis *et al.*, 1992). Polyphyletic polyploid species have been reported for mosses

(Wyatt *et al.*, 1988), ferns (Werth *et al.*, 1985a, b; Soltis *et al.*, 1991), and many angiosperms (Soltis *et al.*, 1992; Soltis and Soltis, 1993, 1999), and include both autopolyploids [e.g., *Heuchera grossulariifolia* (Wolf *et al.*, 1989, 1990; Segraves *et al.*, 1999) and *Heuchera micrantha* (Ness *et al.*, 1989; Soltis *et al.*, 1989)] and allopolyploids (Soltis *et al.*, 1992; Soltis and Soltis, 1993, 1999). Recurrent formation of a polyploid species has implications for the taxonomy of polyploids, our understanding of the ease with which and rate at which polyploidization can occur, and, most relevant here, the genetic diversity of polyploid "species." In this section, we will address (*i*) the proportion of polyploid plant species that are known to have formed recurrently, (*ii*) the extent of recurrent formations within a species, and (*iii*) the genetic and evolutionary significance of these multiple origins.

Most polyploid species examined to date have shown evidence of recurrent formation (Soltis *et al.*, 1992; Soltis and Soltis, 1993, 1999). Remarkably, these independent origins have been identified even though sampling strategies typically were not designed to investigate multiple origins but rather to test hypotheses of diploid parentage. In many cases, as few as two or three populations of a polyploid species were sampled; the genetic distinctness of these populations, coupled with additivity of diploid genotypes, strongly supported interpretations of recurrent formation. All available data suggest that nearly all polyploid species analyzed comprise multiple lineages of independent formation.

How many such lineages are present within a given polyploid species? Few studies have explicitly addressed this question. Two allotetraploid species of *Tragopogon*, *T. mirus* and *T. miscellus*, arose within the past century in the Palouse region of eastern Washington and adjacent Idaho from diploid progenitors that had been introduced to the region from Europe in the early 1900s (Ownbey, 1950; Fig. 1). During the past several decades, the ancestries of these two tetraploids have been investigated by using nearly every technique that has become available (Cook *et al.*, 1998), and Ownbey's (1950) interpretations have been confirmed.

Early morphological and cytological data (Ownbey, 1950; Ownbey and McCollum, 1953, 1954) suggested multiple origins of each species, two of *T. miscellus* and three of *T. mirus*, in different locations on the Palouse. Recent isozyme and DNA analyses have supported Ownbey's (1950) original hypotheses of recurrent origin and have identified additional lineages of independent formation (Roose and Gottlieb, 1976; Soltis and Soltis, 1989b; Soltis and Soltis, 1991; Soltis *et al.*, 1995). For example, based on the geographic distribution of isozyme multilocus genotypes, chloroplast DNA haplotypes, and rDNA markers, estimates of the number of lineages in *T. mirus* ranged from 4 to 9 (with an extinct population of independent origin, based on flavonoid markers; Brehm and Ownbey, 1965), and the number in *T. miscellus* ranged from 2 to 21 (Soltis *et al.*,

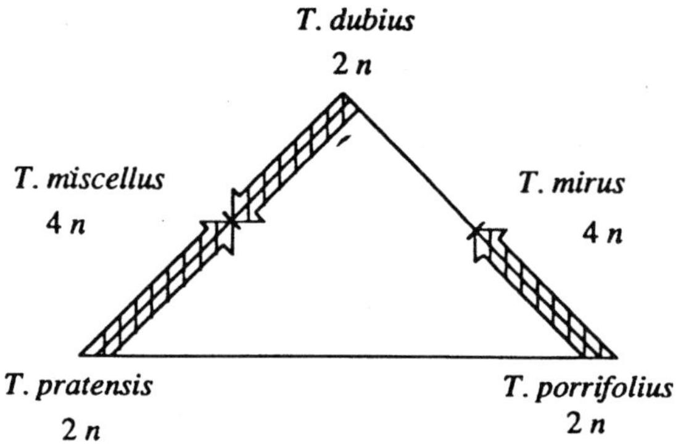

FIGURE 1. Parentage and reciprocal origins of tetraploid species of *Tragopogon* in North America. Hatched lines indicate diploid(s) contributing chloroplast to the tetraploids.

1995). However, several populations of *T. mirus* in different locations had the same isozyme multilocus genotype, chloroplast DNA haplotype, and rDNA repeat, and, in many cases, they co-occurred with the diploid progenitor species, *T. dubius* and *T. porrifolius*; the same was true of *T. miscellus*, which co-occurred in at least some locations with both of its progenitors, *T. dubius* and *Tragopogon pratensis*. It was possible that these separate locations represented independent sites of polyploid formation from genetically identical (based on the markers at hand) diploids. However, this hypothesis could not be tested without the use of more sensitive markers.

Cook *et al.* (1998) used random amplified polymorphic DNA (RAPD) markers to test the hypothesis that isozymically identical populations of *T. mirus* having the same chloroplast DNA haplotype and rDNA repeat were of separate origin and that "identical" populations of *T. miscellus* also were of separate origin. For *T. mirus*, five populations with isozyme multilocus genotype 1 (Soltis *et al.*, 1995) and two populations with isozyme genotype 2 (Soltis *et al.*, 1995) were sampled. Each population had a unique RAPD profile (and, in fact, two populations were polymorphic), suggesting that each population may have had a separate origin. Taken with other data, *T. mirus* may represent a collection of as many as 11 lineages (Cook *et al.*, 1998). RAPD data for three populations of isozyme genotype 1 (Soltis *et al.*, 1995) of *T. miscellus* demonstrated that all three were distinct and possibly of separate origin, raising the number of genetically distinct populations of *T. miscellus* to five (Cook *et al.*, 1998).

The *Tragopogon* tetraploids represent remarkable cases of recurrent formation on a small geographic scale and in a short period, perhaps the last 70 years. Other polyploid species, if examined in sufficient detail, may be similarly grossly polyphyletic. Furthermore, recurrent formation of *T. mirus* and *T. miscellus* also has occurred on a broader geographic scale. Both species have been reported from Flagstaff, AZ (Brown and Schaak, 1972), and *T. miscellus* has been reported from Gardiner, MT, and Sheridan, WY (M. Ownbey, unpublished notes cited in Roose and Gottlieb, 1976; Sheridan site confirmed by P.S.S. and D.E.S. in 1997; *T. miscellus* not observed in Gardiner in 1997).

Although such polyphyly calls into question the meaning of the term "polyploid species," the biological implications of recurrent polyploidizations from the same diploid progenitor species are indeed intriguing. Such multiple formations may play a significant role in shaping the genetic structure of polyploid species, as they are currently recognized. The concept of recurrent formations forces us to consider polyploid species not as genetically uniform, as previous models of polyploid formation imply, but as genetically variable. In fact, multiple formations may represent a significant source of genetic diversity in polyploid species, as a polyploid species may comprise multiple, genetically different lineages. Finally, crossing between individuals of separate origin will break down the distinctions among lineages and may produce novel genotypes through recombination.

The long-term evolutionary significance of recurrent polyploid formations is unclear; however, a host of specific questions can be addressed. For example, do plants of different origins have distinct evolutionary potentials? Does recurrent formation lead to different locally adapted genotypes? How extensive is gene flow between populations of independent origin, and to what extent does gene flow contribute to the genetic diversity of populations? How frequently are new genotypes produced through recombination?

GENOME REARRANGEMENTS IN POLYPLOIDS

Another possible source of genetic novelty in polyploids is genome rearrangements. Evidence for chromosomal changes has been obtained through a number of techniques, including genome *in situ* hybridization (GISH), analysis of restriction fragment length polymorphism (RFLP) loci, and chromosome mapping. Among the earliest studies reporting widespread genomic changes in tetraploids relative to their diploid progenitors is an analysis of tobacco genome structure using GISH (reviewed in Leitch and Bennett, 1997). Tobacco (*Nicotiana tabacum*) is an allotetraploid whose parents are *Nicotiana sylvestris* and a T-genome diploid from section *Tomentosae* (Leitch and Bennett, 1997). GISH clearly revealed numerous chromosomal rearrangements. In fact, nine intergenomic translocations have occurred within the genome of tobacco, that is, translocations between the chromosomes donated by *N. sylvestris* and the T-genome parent. Most of the chromosomes of tobacco are therefore mosaics, composed of regions of both parental chromosome sets.

In *Brassica*, there is evidence that such genome rearrangements may occur very soon after the formation of the tetraploid. Song *et al.* (1995) produced artificial tetraploids resulting from interspecific crosses between *Brassica rapa* and *Brassica nigra* and between *B. rapa* and *Brassica oleracea.* They compared genome structure in the F_5 derivatives of these crosses with their F_2 ancestors and found genetic divergence in these few generations, with distances as high as almost 10%. In addition, Song *et al.* (1995) found evidence of cytoplasmic–nuclear interactions—the maternal genotype had definite control over aspects of the nuclear genome. They concluded that a possible result of polyploid formation is the production of novel genotypes. Furthermore, extensive genetic change can occur in the early generations after polyploid formation and may therefore be important in the formation of a functional polyploid. Chromosome mapping of diploid *Brassica* and comparison with the map of *Arabidopsis thaliana* suggest that the diploid species of *Brassica* (n = 9) may actually be ancient hexaploids (Lagercrantz, 1998, but see Quiros, 1998 for a different interpretation).

Such intergenomic translocations are not limited to tobacco and *Brassica.* Instead, extensive chromosomal changes have been reported in a number of other polyploids, including maize, oats, and soybeans. Such intergenomic translocations may be mediated by transposable elements (Matzke and Matzke, 1998) and may be an important source of genetic novelty in polyploids (see also Wendel, 2000). Furthermore, cytoplasmic–nuclear interactions may be important in the establishment of a fertile polyploid (reviewed in Leitch and Bennett, 1997).

ANCIENT POLYPLOIDY AND GENE SILENCING

Basal Angiosperms

Estimates of ancient polyploidy generally have relied on chromosome number alone; Stebbins (1950), for example, viewed those plants with a base chromosome number of n = 12 or higher to be polyploid, and others (Goldblatt, 1980; Grant, 1981, 1982) used similar criteria. Based on this criterion, a large number of angiosperm families, most of which trace their roots far back into angiosperm phylogeny, are considered to be the products of ancient polyploid events whose diploid ancestors are now extinct. For example, the Illiciales have n = 14, and both the Lauraceae and Calycanthaceae of Laurales have a base number of n = 12. The lowest chromosome number in the Magnoliaceae is $n = 19$, and the family exhibits a range of numbers that are multiples of this base number. Some early eudicots, such as *Trochodendron* and *Tetracentron* (with $n = 19$) and *Platanus* (with $n = 21$), also have high chromosome numbers. Some families of more recent origin [e.g., Salicaceae (willows and poplars), Hippocastanaceae (horse chestnuts and buckeyes), *Fraxinus* (ashes) and other Oleaceae, and *Tilia* (linden and basswood)] also are considered ancient polyploids. Some families of possible ancient polyploid origin, along with their chromosome numbers, are listed in Table 3, and the phylogenetic distribution of these families (on portions of the tree of Soltis *et al.*, 1999a; Soltis *et al.*, 2000) is shown in Fig. 2. Stebbins (1950, 1971) also suggested that the ancestral base chromosome number for angiosperms is $x = 6, 7$, or 8; other, later authors (Ehrendorfer *et al.*, 1968; Raven, 1975; Grant, 1981, 1982) have concurred. Reconstruction of chromosomal evolution across the angiosperms is partially consistent with Stebbins' hypothesis. Although the high chromosome numbers of the basal angiosperm groups make it difficult to infer base chromosome numbers for those groups of angiosperms and therefore for angiosperms as a whole, our reconstructions show an ancestral number of $x = 8$ for the eudicots (D.E.S., unpublished data), that is, the large clade that makes up 75% of angiosperm species. Identifying the ancestral number for all angiosperms will require teasing apart the base numbers of the ancient polyploid groups and will require further work.

Most, if not all, angiosperms may have experienced one or more cycles of genome doubling (Wendel, 2000), and these hypotheses of ancient polyploidy have several implications for the genetics, genomics, and evolutionary biology of these plants. First, if they are indeed polyploids, then these plants should exhibit extra copies of their genes above the level that

TABLE 3. Angiosperm families with high chromosome numbers, suggested to be of ancient polyploid origin (Stebbins, 1950)

Basal angiosperms	Chromosome number, n
Family	
Illiciaceae	14
Schisandraceae	14
Lauraceae	12
Calycanthaceae	12
Magnoliaceae	19
Eudicot families	
Trochodendraceae	19
Platanaceae	21
Cercidiphyllaceae	19
Salicaceae	19
Hippocastanaceae	19

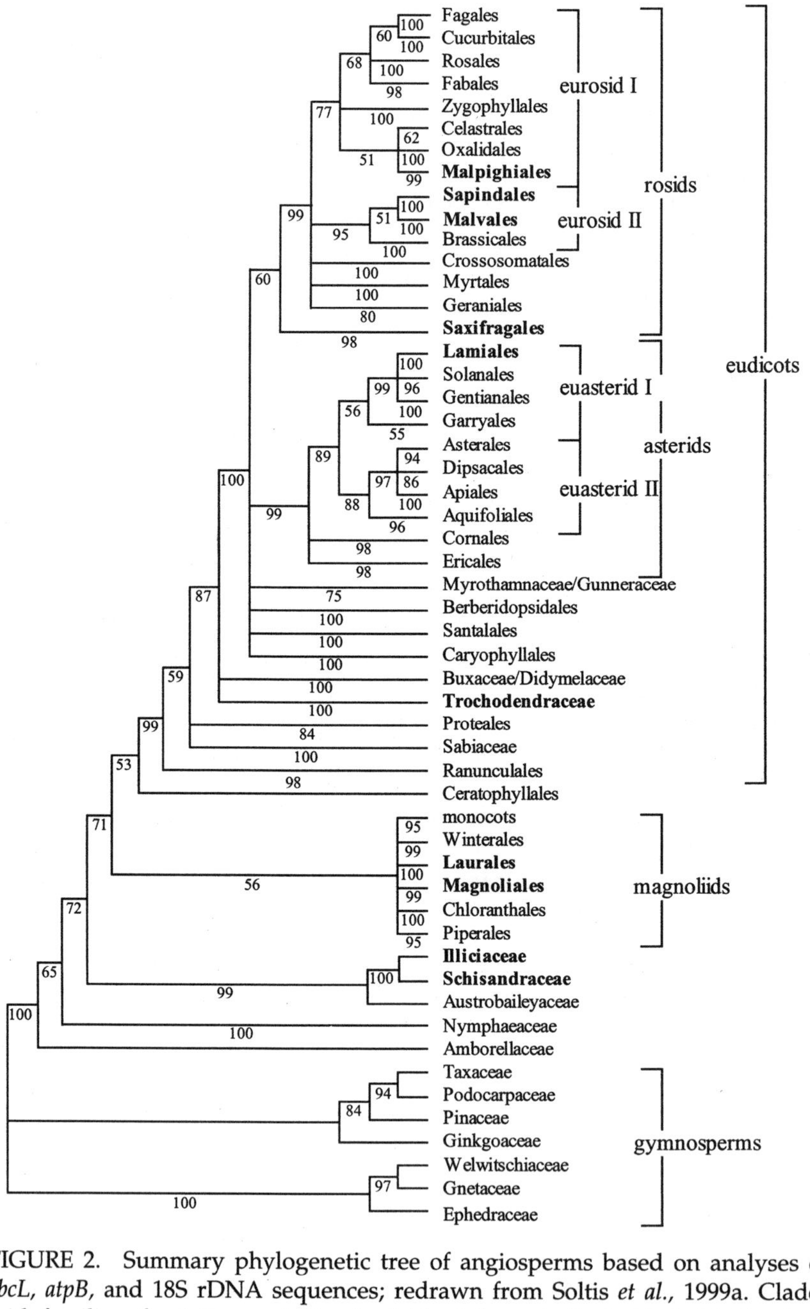

FIGURE 2. Summary phylogenetic tree of angiosperms based on analyses of *rbcL*, *atpB*, and 18S rDNA sequences; redrawn from Soltis *et al.*, 1999a. Clades with families of putative ancient polyploid origin are indicated in bold. Numbers below branches are jackknife support values.

one would expect for diploid plants (Gottlieb, 1982; Crawford, 1983). Analyses of enzyme expression indicate that multiple enzymes are indeed expressed in putatively paleopolyploid angiosperm families, such as those listed in Table 3 (Soltis and Soltis, 1990); issues of the regulation of duplicated genes are discussed by Wendel (1999). Second, some copies of these multiple genes might be expected to be silenced, particularly in the more ancient families (see *Gene Silencing* below). Third, reorganization of the original polyploid genome might have led to a novel genomic arrangement and perhaps to novel phenotypes. Finally, given that all members of a family have chromosome numbers that are multiples of a single lower number, it appears that, after polyploidization, diversification continued at the new polyploid level, with subsequent episodes of polyploidy superimposed on this initial polyploid level. This pattern of divergent speciation at the polyploid level contradicts the view of polyploids as evolutionary dead-ends.

Homosporous Pteridophytes

Homosporous pteridophytes are those ferns (including *Psilotum* and *Tmesipteris*; Manhart, 1994; Wolf, 1997; Soltis *et al.*, 1999b), lycophytes, and *Equisetum* with a homosporous life cycle; all of these groups are the descendants of ancient plant lineages that extend back to the Devonian Period (Kenrick and Crane, 1997). The mean gametic chromosome number for homosporous pteridophytes is $n = 57$; for angiosperms, it is $n = 16$ (Klekowski and Baker, 1966). Despite their high chromosome numbers, however, homosporous pteridophytes exhibit diploid gene expression at isozyme loci (Haufler and Soltis, 1986; Soltis, 1986; D. Soltis and Soltis, 1988; P. Soltis and Soltis, 1988). At least two possible explanations can explain this paradox of high chromosome numbers and genetic diploidy. First, these plants are ancient polyploids that have undergone extensive gene silencing to produce genetic diploids, and second, they may have achieved high chromosome numbers through another mechanism, such as chromosomal fission.

Gene Silencing

Genes duplicated through polyploidy have several possible fates: retention of both copies as functional genes, acquisition of new function by one copy, and gene silencing (Wendel, 2000). Several models of genome evolution, in which a polyploid genome gradually will undergo gene silencing and return to a diploid condition, have been presented (Ohno, 1970; Haufler, 1987). Unfortunately, little empirical evidence is available to support or to refute these models.

Potential examples occur in the homosporous pteridophytes. Data for the ferns *Polystichum munitum* (n = 41) and *Ceratopteris richardii* (n = 39) may address these alternatives. Pichersky *et al.* (1990) studied the genes for the chlorophyll *a*/*b* binding proteins in *P. munitum*. These proteins are important in photosynthesis and are encoded by a small multigene family (Pichersky *et al.*, 1990). *P. munitum* exhibits diploid isozyme expression (Soltis and Soltis, 1987, 1990; Soltis *et al.*, 1991). If this species is of ancient polyploid origin but has since undergone substantial gene silencing, then pseudogenes should be detectable in the genome. Five clones of the *CAB* genes were analyzed by Pichersky *et al.* Three of the five clones were structurally nonfunctional, a fourth clone had a structurally intact sequence but was nonfunctional at the sequence level, and a fifth clone was a functional sequence. Possible explanations for these results (Pichersky *et al.*, 1971) are (*i*) amplification of nonfunctional sequences in the genome of *P. munitum*, regardless of the ploidy of *P. munitum*, (*ii*) *P. munitum* is diploid with a large number of mutant *CAB* genes, and (*iii*) *P. munitum* is polyploid, with silencing of multiple genes that are present because of ancient polyploidy. In *C. richardii*, cDNA clones hybridized to multiple fragments on genomic DNA blots, suggesting that 50% or more of these expressed sequences were present in multiple copies in this fern genome (McGrath *et al.*, 1994). In contrast, a similar experiment with *A. thaliana* detected only 15% duplicated fragments (McGrath *et al.*, 1993). Further characterization of the hybridizing fragments of the genome is necessary to document that they are in fact duplicated sequences. However, this evidence for multiple hybridizing fragments in *C. richardii*, along with the *CAB* gene data for *P. munitum*, suggests that the genomes of homosporous ferns may in fact be anciently polyploid.

Gene silencing remains an underinvestigated area of polyploid research. If it occurs as described in models of wholesale diploidization of the polyploid genome (Haufler, 1987), what are the mechanisms and at what rate does such silencing occur? Or does silencing occur gradually, essentially one locus at a time? Many unanswered questions remain.

CONCLUSIONS

Leitch and Bennett (1997) have suggested that the evolutionary potential of a polyploid depends on a number of factors associated with the formation of the polyploid and with genetic divergence between the parents; unfortunately, the factors involved in the origin and establishment of polyploids in nature are largely unknown (Ramsey and Schemske, 1998). The success of a polyploid may depend, in part, on the parental origin of particular DNA sequences—is the sequence maternal or paternal and does it interact favorably with the organellar genomes? The type of sequence under study also may be important: is it coding or noncoding DNA, is it telomeric or centric in origin, and is it located near heterochromatin? Finally, what is the level of genetic differentiation between the parents?

Although unreduced gamete production and even polyploid formation may be quite common in many groups of plants (Ramsey and Schemske, 1998), there are many obstacles to establishment of a polyploid population. Minority cytotype exclusion (Levin, 1975; Fowler and Levin, 1984; Felber, 1991) may be particularly important in newly formed outcrossing polyploids where there are few potential mates unless there is substantial assortative mating (Husband, 2000); when only one or a few polyploid individuals emerge within a population of diploids, outcrossing polyploid individuals may spend most of their gametes in sterile or partially sterile matings with their diploid parents. The apparent success of polyploids is biased toward those species that have overcome the barrier(s) to establishment, and this success may ultimately derive from a number of the genetic attributes of the polyploids. Polyploids have increased heterozygosity, an attribute that may be beneficial (Mitton and Grant, 1984; Mitton, 1989). Polyploids also harbor higher levels of genetic and genomic diversity than was anticipated, with recurrent formation from genetically divergent diploid parents and possibly genome rearrangements contributing genetic diversity. This genetic diversity results in greater biochemical diversity, which also may be beneficial to the polyploid (Levin, 1983). Finally, these genetic attributes may have ecological consequences. For example, if polyploids have lower inbreeding depression and are more highly selfing, they may be better colonizers, explaining the prevalence of polyploids on the list of the world's worst weeds. Polyploids may have broader ecological amplitudes than do their diploid progenitors because of their increased genetic and biochemical diversity (Levin, 1983). Polyploids may experience new interactions with other species, such as pollinators (Segraves, 1998; Segraves and Thompson, 1999).

What are some of the future directions we see for research on the genetic attributes of polyploids? The general mode of formation of polyploids remains unknown; research into the factors that produce unreduced gametes and bring them together certainly is warranted. Additional studies, both theoretical and empirical, are needed to address expectations of inbreeding depression and outcrossing rates. Furthermore, the levels of gene flow among populations, especially those populations of separate origin, are unknown. Regarding genome rearrangements, how extensive are they within an individual or race? How widespread are they among species? How quickly do such rearrangements occur? Do populations of separate origin exhibit the same or different rearrangements? Finally, are basal angiosperms and homosporous pteridophytes with high chromosome numbers of ancient polyploid origin? If so, what can we learn about gene silencing from these plants? How extensive has gene silencing been, and is there evidence for the cooption of duplicated genes for new function? The study of polyploidy is a dynamic and open area of research, ranging from molecular genetic comparisons to population genetics, with important implications for the biology and evolution of the majority of plant species.

We thank Kent Holsinger and an anonymous reviewer for helpful comments on the manuscript. This research was supported, in part, by the National Science Foundation. This work is dedicated to the memory of G. Ledyard Stebbins.

REFERENCES

Barrett, S. C. H. & Shore, J. S. (1989) Isozyme variation in colonizing plants. In *Isozymes in Plant Biology*, eds. Soltis, D. E. & Soltis, P. S. (Dioscorides Press, Portland), pp. 106–206.

Brehm, B. G. & Ownbey, M. (1965) Variation in the chromatographic patterns in the *Tragopogon dubius-pratensis-porrifolius* complex (Compositae). *American Journal of Botany* 52, 811–818.

Charlesworth, D. & Charlesworth, B. (1987) Inbreeding depression and its evolutionary consequences. *Annual Review of Ecology and Systematics* 18, 237–268.

Cook, L. M. (1998) Mating systems and multiple origins in the North American *Tragopogon* complex. Ph.D. dissertation. Washington State University.

Cook, L. M. & Soltis, P. S. (1999) Mating systems of diploid and allotetraploid populations of *Tragopogon* (Asteraceae) I. Natural populations. *Heredity* 82: 237–244.

Cook, L. M. & Soltis, P. S. (2000) Mating systems of diploid and allotetraploid populations of *Tragopogon* (Asteraceae) II. Artificial populations. *Heredity*: 84, 410–415.

Cook, L. M., Soltis, P. S., Brunsfeld, S. J. & Soltis, D. E. (1998) Multiple independent formations of *Tragopogon* tetraploids (Asteraceae): Evidence from RAPD markers. *Molecular Ecology* 7, 1293–1302.

Crawford, D. J. (1983) Phylogenetic and systematic inference from electrophoretic studies. In *Isozymes in Plant Genetics and Breeding, Part A*, eds. Tanksley, S. D. & Orton, T. G. (Elsevier, Amsterdam), pp. 257–287.

Ehrendorfer, F., Krendl, F., Habeler, E. & Sauer, W. (1968) Chromosome numbers and evolution in primitive angiosperms. *Taxon* 17, 337–468.

Felber, F. (1991) Establishment of a tetraploid cytotype in a diploid poulation: effect of relative fitness of the cytotypes. *Journal of Evolutionary Biology* 4, 195–207.

Federov, A. (ed.). (1969) *Chromosome Numbers in Flowering Plants* (Academy of Sciences of the U.S.S.R., Leningrad).

Fowler, N. L. & Levin, D. A. (1984) Ecological constraints on the establishment of a novel polyploid in competition with its diploid progenitor. *American Naturalist* 124, 703–711.

Goldblatt, P. (1980) Polyploidy in angiosperms: monocotyledons. In *Polyploidy-Biological Relevance*, ed. Lewis, W. H. (Plenum Press, New York), pp. 219–239.

Gottlieb, L. D. (1982) Conservation and duplication of isozymes in plants. *Science* 216, 373–380.

Grant, V. (1981) *Plant Speciation*. 2nd Edition (Columbia University Press, New York).

Grant, V. (1982) Chromosome number patterns in primitive angiosperms. *Botanical Gazette* 143, 390–394.

Haldane, J. B. S. (1930) Theoretical genetics of autopolyploids. *Journal of Genetics* 22, 359–372.

Haufler, C. H. (1987) Electrophoresis is modifying our concepts of evolution in homosporous pteridophytes. *American Journal of Botany* 74, 953–966.

Haufler, C. H. & Soltis, D. E. (1986) Genetic evidence that homosporous ferns with high chromosome numbers are diploid. *Proceedings of the National Academy of Sciences USA* 83, 4389–4393.

Hedrick, P. W. (1987) Genetic load and the mating system in homosporous ferns. *Evolution* 41, 1282–1289.

Hickman, J. C. (ed.) (1993) *The Jepson Manual* (University of California Press, Berkeley).

Husband, B. C. (2000) Constraints on polyploid evolution: a test of the minority cytotype exclusion principle. *Proceedings of the Royal Society of London B* 267, 217–223.

Husband, B. C. & Schemske, D. W. (1995) Magnitude and timing of inbreeding depression in a diploid population of *Epilobium angustifolium* (Onagraceae). *Heredity* 75, 206–215.

Husband, B. C. & Schemske, D. W. (1996) Evolution of the magnitude and timing of inbreeding depression in plants. *Evolution* 50, 54–70.

Husband, B. C. & Schemske, D. W. (1997) The effect of inbreeding in diploid and tetraploid *Epilobium angustifolium* (Onagraceae): Implications for the genetic basis of inbreeding depression. *Evolution* 51, 737–746.

Kenrick, P. & Crane, P. R. (1997) *The Origin and Early Diversification of Land Plants* (Smithsonian Institution Press, Washington, D. C.).

Klekowski, E. J. & Baker, H. G. (1966) Evolutionary significance of polyploidy in the Pteridophyta. *Science* 135, 305–307.

Lagercrantz, U. (1998) Comparative mapping between *Arabidopsis thaliana* and *Brassica nigra* indicates that *Brassica* genomes have evolved through extensive genome replication accompanied by chromosome fusions and frequent rearrangements. *Genetics* 150, 1217–1228.

Leitch, I. J. & Bennett, M. D. (1997) Polyploidy in angiosperms. *Trends in Plant Science* 2, 470–476.

Levin, D. A. (1975) Minority cytotype exclusion in local plant populations. *Taxon* 24, 35–43.

Levin, D. A. (1983) Polyploidy and novelty in flowering plants. *American Naturalist* 122, 1–25.

Manhart, J. R. (1994) Phylogenetic analysis of green plant *rbcL* sequences. *Molecular Phylogenetics and Evolution* 3, 114–127.

Masterson, J. (1994) Stomatal size in fossil plants: Evidence for polyploidy in majority of angiosperms. *Science* 264, 421–423.

Masuyama, S. & Watano, Y. (1990) Trends for inbreeding in polyploid pteridophytes. *Plant Species Biology* 5, 13–17.

Matzke, M. A. & Matzke, A. J. M. (1998) Polyploidy and transposons. *Trends in Ecology and Evolution* 13, 241.

McGrath, J. M., Jancso, M. M. & Pichersky, E. (1993) Duplicate sequences with similarity to expressed genes in the genome of *Arabidopsis thaliana. Theoretical and Applied Genetics* 86, 880–888.

McGrath, J. M., Hickok, L. G. & Pichersky, E. (1994) Assessment of gene copy number in the homosporous ferns *Ceratopteris thalictroides* and *C. richardii* (Parkeriaceae) by restriction fragment length polymorphisms. *Plant Systematics and Evolution* 189, 203–210.

Mitton, J. (1989) Physiological and demographic variation associated with allozyme variation. In *Isozymes in Plant Biology*, eds. Soltis, D. E. & Soltis, P. S. (Dioscorides Press, Portland), pp. 127–145.

Mitton, J. & Grant, M. C. (1984) Relationships among protein heterozygosity, growth rate, and developmental stability. *Annual Review of Ecology and Systematics* 15, 479–499.

Moody, M. E., Mueller, L. D., & Soltis, D. E. (1993) Genetic variation and random drift in autotetraploid populations. *Genetics* 134, 649–657.

Muller, H. J. (1914) A new mode of segregation in Gregory's tetraploid primulas. *American Naturalist* 48, 508–512.

Ness, B. D., Soltis, D. E. & Soltis, P. S. (1989) Autopolyploidy in *Heuchera micrantha* Dougl. (Saxifragaceae). *American Journal of Botany* 76, 614–626.

Ohno, S. (1970) *Evolution by Gene Duplication* (Springer, New York).

Ownbey, M. (1950) Natural hybridization and amphiploidy in the genus *Tragopogon*. *American Journal of Botany* 37, 487–499.

Ownbey, M. & McCollum, G. D. (1953) Cytoplasmic inheritance and reciprocal amphiploidy in *Tragopogon*. *American Journal of Botany* 40, 788–796.

Ownbey, M. & McCollum, G. D. (1954) The chromosomes of *Tragopogon*. *Rhodora* 56, 7–21.

Pichersky, E., Soltis, D. E., & Soltis, P. S. (1990) Defective CAB genes in the genome of a homosporous fern. *Proceedings of the National Academy of Sciences USA* 87, 195–199.

Quiros, C. F. (1998) Molecular markers and their applications to genetics, breeding and the evolution of *Brassica*. *Journal of the Japanese Society of Horticultural Science* 67, 1180–1185.

Ramsey, J. & Schemske, D. W. (1998) Pathways, mechanisms and rates of polyploid formation in flowering plants. *Annual Review of Ecology and Systematics* 29, 467–501.

Raven, P. H. (1975) The basis of angiosperm phylogeny: Cytology. *Annals of the Missouri Botanical Garden* 62, 724–764.

Richards, A. J. (1986) *Plant Breeding Systems* (George Allen and Unwin, London).

Roose, M. L. & Gottlieb, L. D. (1976) Genetic and biochemical consequences of polyploidy in *Tragopogon*. *Evolution* 30, 818–830.

Segraves, K. A. (1998) Plant polyploidy and the divergence of floral traits and pollinator–plant interactions. M. S. Thesis, Washington State University, Pullman.

Segraves, K. A. & Thompson, J. N. (1999) Plant polyploidy and pollination: floral traits and insect visits to diploid and tetraploid *Heuchera grossulariifolia*. *Evolution* 53, 1114–1127.

Segraves, K. A., Thompson, J. N., Soltis, P. S. & Soltis, D. E. (1999) Multiple origins of polyploidy and the geographic structure of *Heuchera grossulariifolia*. *Molecular Ecology* 8, 253–262.

Soltis, D. E. (1986) Genetic diploidy in *Equisetum*. *American Journal of Botany* 73, 908–913.

Soltis, D. E. & Soltis, P. S. (1988) Are lycopods with high chromosome numbers ancient polyploids? *American Journal of Botany* 75, 238–247.

Soltis, D. E. & Soltis, P. S. (1989a) Genetic consequences of autopolyploidy in *Tolmiea* (Saxifragaceae). *Evolution* 43, 586–594.

Soltis, D. E. & Soltis, P. S. (1989b) Allopolyploid speciation in *Tragopogon*: insights from chloroplast DNA. *American Journal of Botany* 76, 1119–1124.

Soltis, D. E. & Soltis, P. S. (1990) Genetic evidence for ancient polyploidy in primitive angiosperms. *Systematic Botany* 15, 328–337.

Soltis, D. E. & Soltis, P. S. (1993) Molecular data and the dynamic nature of polyploidy. *Critical Reviews in Plant Sciences* 12, 243–273.

Soltis, D. E. & Soltis, P. S. (1999) Polyploidy: Origins of species and genome evolution. *Trends in Ecology and Evolution* 14, 348–352.

Soltis, D. E., Soltis, P. S., Chase, M. W., Mort, M. E., Albach, D. C., Zanis, M., Savolainen, V., Hahn, W. H., Hoot, S. B., Axtell, M., Swensen, S. M., Nixon, K. C. & Farris, J. S. (2000) Angiosperm phylogeny inferred from a combined data set of 18S rDNA, *rbcL*, and *atpB* sequences. *Botanical Journal of the Linnean Society*: In press.

Soltis, D. E., Soltis, P. S. & Ness, B. D. (1989) Chloroplast DNA variation and multiple origins of autopolyploidy in *Heuchera micrantha* (Saxifragaceae). *Evolution* 43, 650–656.

Soltis, P. S., Doyle, J. J. & Soltis, D. E. (1992) Molecular data and polyploidy in plants. In *Molecular Systematics of Plants*, eds. Soltis, P. S., Soltis, D. E. & Doyle, J. J. (Chapman and Hall, New York), pp. 177–201.

Soltis, P. S., Plunkett, G. M., Novak, S. J. & Soltis, D. E. (1995) Genetic variation in *Tragopogon* species: Additional origins of the allotetraploids *T. mirus* and *T. miscellus* (Compositae). *American Journal of Botany* 82, 1329–1341.

Soltis, P. S. & Soltis, D. E. (1987) Population structure and estimates of gene flow in the homosporous fern *Polystichum munitum*. *Evolution* 41, 620–629.

Soltis, P. S. & Soltis, D. E. (1988) Electrophoretic evidence for genetic diploidy in *Psilotum nudum*. *American Journal of Botany* 75, 1667–1671.

Soltis, P. S. & Soltis, D. E. (1990) Evolution of inbreeding and outcrossing in ferns and fern–allies. *Plant Species Biology* 5, 1–11.

Soltis, P. S. & Soltis, D. E. (1991) Multiple origins of the allotetraploid *Tragopogon mirus* (Compositae): rDNA evidence. *Systematic Botany* 16, 407–413.

Soltis, P. S., Soltis, D. E. & Chase, M. W. (1999a) Angiosperm phylogeny inferred from multiple genes: A research tool for comparative biology. *Nature* 402, 402–404.

Soltis, P. S., Soltis, D. E. & Wolf, P. G. (1991) Allozymic and chloroplast DNA analyses of polyploidy in *Polystichum* (Dryopteridaceae). I. The origin of *P. californicum* and *P. scopulinum*. *Systematic Botany* 16, 245–256.

Soltis, P. S., Soltis, D. E., Wolf, P. G., Nickrent, D. L., Chaw, S-M. & Chapman, R. L. (1999b) Land plant phylogeny inferred from 18S rDNA sequences: Pushing the limits of rDNA sequences? *Molecular Biology and Evolution* 16, 1774–1784.

Song, K., Lu, P., Tang, K. & Osborn, T. C. (1995) Rapid genome change in synthetic polyploids of *Brassica* and its implications for polyploid evolution. *Proceedings of the National Academy of Sciences USA* 92, 7719–7723.

Stebbins, G. L. (1947) Types of polyploidy: their classification and significance. *Advances in Genetics* 1, 403–429.

Stebbins, G. L. (1950) *Variation and Evolution in Plants* (Columbia University Press, New York).

Stebbins, G. L. (1971) *Chromosomal Evolution in Higher Plants* (Edward Arnold, London).

Wendel, J. F. (2000) Genome evolution in polyploids. *Plant Molecular Biology* 42, 225–249.

Werth, C. R., Guttman, S. I. & Eshbaugh, W. H. (1985a) Recurring origins of allopolyploid species of *Asplenium*. *Science* 228, 731–733.

Werth, C. R., Guttman, S. I. & Eshbaugh, W. H. (1985b) Electrophoretic evidence of reticulate evolution in the Appalachian *Asplenium* complex. *Systematic Botany* 10, 184–192.

Wolf, P. G. (1997) Evaluation of atpB nucleotide sequences for phylogenetic studies of ferns and other pteridophytes. *American Journal of Botany* 84, 1429–1440.

Wolf, P. G., Soltis, D. E. & Soltis, P. S. (1990) Chloroplast-DNA and allozymic variation in diploid and autotetraploid *Heuchera grossulariifolia* (Saxifraceae). *American Journal of Botany* 77, 232–244.

Wolf, P. G., Soltis, P. S. & Soltis, D. E. (1989) Tetrasomic inheritance and chromosome pairing behaviour in the naturally occurring autotetraploid *Heuchera grossulariifolia* (Saxifragaceae). *Genome* 32, 655–659

Wyatt, R., Odrzykoski, I. J., Stoneburner, A., Bass, W. H. & Galau, G. A. (1988) Allopolyploidy in bryophytes: multiple origins of *Plagiomnium medium*. *Proceedings of the National Academy of Sciences USA* 85, 5601–5604.

Index

E

F

G

H

I

S

T

V